SpringerWienNewYork

CISM COURSES AND LECTURES

The series presents lecture notes, monographs, edited works and proceedings in the field of Mechanics, Engineering, Computer Science and Applied Mathematics.
Purpose of the series is to make known in the international scientific and technical community results obtained in some of the activities organized by CISM, the International Centre for Mechanical Sciences.

INTERNATIONAL CENTRE FOR MECHANICAL SCIENCES

COURSES AND LECTURES - No. 525

DAMAGE MECHANICS AND MICROMECHANICS OF LOCALIZED FRACTURE PHENOMENA IN INELASTIC SOLIDS

EDITED BY

GEORGE Z. VOYIADJIS
LOUISIANA STATE UNIVERSITY, BATON ROUGE, LA, USA

SpringerWienNewYork

This volume contains 234 illustrations

Printed in Italy
SPIN 80025726

All contributions have been typeset by the authors.

ISBN 978-3-7091-0426-2 SpringerWienNewYork

PREFACE

This book resulted from a series of lecture notes presented in CISM, Udine. It has two major contributions: Damage Mechanics and Localization in Inelastic Deformation. The course is intended to provide researchers and graduate students with a clear and thorough presentation of the recent advances in continuum damage mechanics for both metals and metal matrix composites as well as the micromechanics of localization in inelastic solids.

Damage Mechanics and Applications: *The major goal of this part of the course is to present many of the different constitutive damage models that have recently appeared in the literature. Another goal is to clearly present the different approaches to this topic in a single complete course that is easily accessible to researchers and graduate students in civil engineering, mechanical engineering, engineering mechanics, aerospace engineering, and material science. The course material was delivered in well-organized lectures that started with the preliminaries and proceeded to advanced topics.*

Micromechanics of Localization in Inelastic Solids and Applications: *The main objective of this contribution is to discuss very efficient procedures of the numerical investigation of localized fracture in inelastic solids generated by impact-loaded adiabatic processes. Particular attention is focused on the proper description of a ductile mode of fracture propagating along the shear band for high impact velocities. This procedure of investigation is based on utilization the finite element or finite difference methods for regularized thermo-elasto-viscoplastic constitutive model of damage material using both rate dependency and non-local approaches.*

It is noteworthy to stress that all considered numerical examples are motivated by recent experimental observations. Qualitative comparison of numerical results with experimental observation data is presented. The numerical results obtained have proven the usefulness of the thermo-elasto-viscoplastic theory in the numerical investigation of dynamic shear band propagation and localized fracture using rate dependency and nonlocal formulations.

Invited Lecturers:
Tomasz Lodygowski (Poznan University of Technology, Poland), Piotr Perzyna (Institute of Fundamental Technological Research, Poland), Antonio Rinaldi (University of Rome "Tor Vergata", Italy), and George Z. Voyiadjis (Louisiana State University, USA).

Coordinator: *George Z. Voyiadjis*

*The authors present in **Part I** the micro-mechanical damage model that accounts for the nonlocal microscopic interactions in the simulation of metal/composite impact and severe contact stress problems. This is achieved by introducing the contributions of damage and its corresponding gradients in the virtual power relations as measures of micro motion of damage within the bulk. By using these internal state variables together with displacement and temperature, the constitutive model is formulated with state laws based on the free energies and the complimentary laws based on the dissipation potentials.*

In this work also the mechanics of small damage is presented using a consistent mathematical and mechanical framework based on the equations of damage mechanics. In this regard, the new scalar damage variable is investigated in detail. The investigation in this work has been carried out for seeking a physical basis is sought for the damage tensor [M] that is used to link the damage state of the material with effective undamaged configuration. The approach presented here provides for a strong physical basis for this missing link. In particular, the authors have made an important link between the damage tensor and fabric tensors.

Computational aspects of the presented theory are also discussed. Numerical integration algorithms, verification and validation process of the theory are discussed. The finite element simulations are also performed by implementing the presented model in the commercial finite element code ABAQUS [6.8.3] as a user defined subroutine (VUMAT).

*The outline of the material presented in **Part I** is as follows: in section 2, general mechanisms of the perforation and penetration mechanisms are discussed. A coupled rate-dependent (viscoplasticity) continuum damage theory is presented. In section 3, mechanics of small damage in fiber reinforced composite material is presented. In section 4, a comparative study has been made on the damage variables of the continuum. In section 5, computational aspects of the theory are discussed. The elastic predictor and coupled viscoplastic-viscodamage corrector algorithm that allows for total uncoupling of geometrical and material nonlinearities are presented. The nonlinear algebraic system of equations is solved by consistnt linearization and the Newton–Raphson iteration. A derivation of a new definition for the consistent tangent stiffness matrix that is essential to maintaining the asymptotic quadratic rate of convergence is also presented in section 5. In section 6, numerical simulations of material instability are introduced in order to validate and test the proposed finite strain approach along with the proposed algorithm and its implementation in the ABAQUS finite element code. In section 7, various numerical examples are presented in order to validate the reliability and capability of the theory in simulating various impact and contact stress problems. Experimental comparisons of the adiabatic shear localizations between the proposed model simulations and other independent results are presented. Effect of initial temperatures and loading rates on the development of shear localizations is also* *investigated in this section. Model capabilities are preliminarily illustrated for the dynamic localization of inelastic flow in adiabatic shear bands and compared with the experimental results of steel plates by deformable blunt projectiles at various impact speeds.*

A handful of methodologies can be pursued in alternative to (and in combination with) continuum mechanics to develop advanced damage models that are intrinsically suited to address complex issues, such as strain localization phenomena and sample-size dependence of structural failure, which are intimately related to microscale phenomena. Statistical Damage Mechanics (SDM) is one of them.

In ***Part II*** *a brief discourse about advances in SDM introduces this methodology and shows its potential. In general, SDM is a multidisciplinary field that seeks to devise non-conventional approaches to fracture and damage by means of discrete damage models accounting for the essential microscale properties of a material or structural system. The ultimate goal is the development of multiscale methodologies that can reliably predict the materials macroscale response in consideration of the microstructural evolution caused by the damage process. Starting from the lower length scale characteristic of the microstructure, the SDM multiscale approach delivers a surprising amount of new insight and makes for a precious companion to the micromechanics methods discussed previously in the book. The bottom-up approach proves to be very effective in understanding the physics of the damage localization process. Although SDM is not yet a mature discipline, by now it has a clear footprint, after nearly two decades of research. This part of the book describes the state of the art and exposes recent trends in SDM for consideration and discussion by the larger solid mechanics community. Some of the concepts discussed herein – yet at their infancy – possess the potential to develop into full blown, innovative, and successful engineering tools. Designing the materials from the microstructure is already a current industry practice to achieve superior damage tolerance, partly driven by widespread usage of composites. SDM may help to do it even better. Another brand new application field for SDM is represented by nanotechnology, especially as far as the design of small (micro- and nano-sized) mechanical systems, such as MEMS and NEMS, where sample-size effects and discreteness are pervasive aspects of the design. After the introductory section 1, Part II unfolds a logical overview of concepts and solutions that begins with simpler one-dimensional models (i.e. the parallel bar system or PBS) in section 2 and ends with multidimensional systems (i.e. spring networks) in section 3. Great emphasis is placed on the derivation of physically-based definitions of the damage parameter for quasi-brittle systems that stem from a pervasive statistical rationale. In section 2, though, a coupled damage-plasticity model is also covered for ductile systems, limited to the one dimensional case. Finally, the last portion of section 3 focuses on sample-size effects in quasi-brittle materials and* *shows how to obtain and harness scaling laws into fractal-based constitutive relations for engineering applications.*

The main objective of ***Part III*** *is to show the broad application of the thermodynamic theory of elasto–viscoplasicity for the description of important problems in modern manufacturing processes, and particularly for meso-, micro-, and nano-mechanical issues. This description is particularly needed for the investigation by using the numerical methods how to avoid unexpected plastic strain localization and localized fracture phenomena in new manufacturing technology.*

In the first part the development of thermo–elasto–viscoplastic constitutive model of a material which takes into consideration the induced anisotropy effects as well as observed contribution to strain rate effects generated by microshear banding is presented. Analysis of recent experimental observations concerning investigations of fracture phenomena under dynamic loading processes suggests that there are two kind of induced anisotropy: (i) the first caused by the residual type stresses produced by the heterogeneous nature of the finite plastic deformation in polycrystalline solids and (ii) the second called the fracture induced anisotropy generated by the evolution of the microdamage mechanisms. It is noteworthy to stress that both these induced anisotropy effects are coupled and this property has to be taken into account in the proposed constitutive description. On the other hand we very well know from recent experimental observations concerning investigation of dynamic loading processes that formation of microshear bands influences the evolution of microstructure of a material. The basic assumption is that microshear banding contributes to viscoplastic strain rate effects.The model is developed within the thermodynamic framework of a unique, covariance constitutive structure with a finite set of the internal state variables. A set of internal state variables consists of one scalar and two tensors, namely the equivalent inelastic deformation, the second order microdamage tensor with the physical interpretation that defines the volume fraction porosity and the residual stress tensor (the back stress). To describe suitably the influence of both induced anisotropy effects and the stress triaxiality observed experimentally the new kinetic equations for the microdamage tensor and for the back stress tensor are proposed. To describe the contribution to strain rate effects generated by microshear banding we propose to introduce certain scalar function which affects the relaxation time in the viscoplastic flow rule. The relaxation time is used as a regularization parameter. Fracture criterion based on the evolution of the anisotropic intrinsic microdamage is formulated. The fundamental features of the proposed constitutive theory have been carefully discussed.

The objective of the second part is to discuss very efficient procedure of the numerical investigation of localized fracture in inelastic solids generated by impact–loaded adiabatic processes. Particular attention is focused on the proper description of a ductile mode of fracture propagating along the shear band for high impact velocities. This procedure of investigation is based on utilization the finite difference and finite element methods for regularized thermo-elasto-viscoplastic model of damaged material. The viscoplastic regularization procedure assures the stable integration algorithm by using the finite difference or finite element methods. Particular attention is focused on the well-posedness of the evolution problem (the initial–boundary value problem) as well as on its numerical solutions.

*In **Part IV** the behavior of selected brittle materials and structures (concrete and masonry) subjected to explosive loadings is discussed. For concrete the accepted Cumulative Fracture*

Criterion (CFC) is exposed. It describes the degradation of the material under fast dynamic processes accompanied by the strong waves propagation phenomenon and large strain rates of deformation. To overcome the computational difficulty in the analyses of such complex problems, the sub-modeling technique as well as splitting of the calculations into two separate parts: analysis of acoustic wave in the air and the propagation of stresses in structures, are used. Some instructive numerical examples of concrete and masonry walls are in focus of the presentation. The numerical tools and computer simulations allow for proper estimation of the structures safety and for taking the design decisions on how to ensure their expected strength.

George Z. Voyiadjis , Baton Rouge, 2010

~

The authors dedicate this volume to the Memory of Dusan Krajcinovic

a Friend, Mentor and Inspirator

CONTENTS

Part I: Consistent Non Local Coupled Damage Model and Its Application in Impact Response of Composite Materials **1**

1 Prologue **3**

2 Nonlocal formulation of the highvelocity impact induced damage **5**

2.1 Phyical foundation of penetration and perforation mechanisms 5

2.2 A coupled rate dependent (Viscoplastic) continuum damage theory 7

3 Mechanics of small damage in fiber reinforced composite materials **13**

3.1 New scalar damage variable 15

3.2 Damage evolution 19

3.3 Mechanics of small damage 24

3.4 Overall and local approaches to damage in composite materials 27

3.4.1 Overall approach 28

3.4.2 Local approach 30

3.4.3 Equivalence of two approaches 33

4 A Comparative Study of Damage Variables in Continuum Damage Mechanics **34**

4.1 Scalar damage variables 34

4.1.1 Hypothesis of elastic strain equivalence . . 36

4.1.2 Hypothesis of elastic strain energy equivalence 38

4.2 Other damage variable 40

4.3 Stress and strains 44

4.4 Tensorial damage variables 44

4.4.1 Hypothesis of elastic strain equivalence . . 46

4.4.2 Hypothesis of elastic strain energy equivalence 47
4.4.3 Isotropic elasticity 49
4.4.4 Plane stress 52
4.4.5 Plane strain 54

5 Computational aspects of the proposed theory 58
5.1 Numerical integration 58
5.2 Return mapping algorithm 59
5.3 Coupled viscoplastic damage corrector algorithms 61

6 Verification and validation of the theoretical formulation 65
6.1 Calibration of the metallic material constants and model parameters 66
6.2 Calibration of the composite material constants and model parameters 69

7 Application of the theory in studying high velocity contact problems 74
7.1 Simulation of experimental work on the study of friction between metallic surfaces 74
7.2 Shear localization in strip tension 81
7.3 Shear localizations in cylindrical hat-shaped samples and comparisons with experimental results . 84
7.4 A blunt projectile impact in a target 86
7.5 A blunt projectile impact in a composite plate . . 90

8 Conclusions 91

Part II: Advances in Statistical Damage Mechanics (SDM): New Modeling Strategies **103**

1 Introduction: from Continuum to Discrete Modeling **105**

2 One-Dimensional Models: Parallel Bundles Systems (PBS) **112**
2.1 Pure Damage: Quasi-Brittle PBS 113
2.2 Damage and Plasticity: Generalized PBS 118
2.2.1 Correlation between p_p and p_f 124
2.3 Final Thoughts about PBS Applications 125

3 Higher-Dimensional Models: Spring Networks **126**
3.1 Lattice Representation and Model Definition . . . 128
3.2 Simulations Results from Tensile and Multiaxial Tests 132
3.3 Physical Foundations of $\bar{D}$ 151
3.3.1 A Perspective from Thermodynamics . . . 151
3.3.2 Energetics of the Lattice in Uniaxial Tests 156
3.3.3 Rational Damage Model for Uniaxial Tests 166
3.3.4 Simplified Closed-Form Models of $\bar{D}$ . . . 179
3.4 Sample-Size Effects 200
3.4.1 Hardening Phase 201
3.4.2 Softening Phase 204
3.4.3 Model Validation and Final Remarks . . . 209

4 More Related Work **212**

5 Concluding Remarks **213**
Appendix 1: Notes About Numerical Solution Schemes 214
Appendix 2: The One-way Correlation Hypothesis 218

References **221**

Part III: Application of the Thermodynamical Theory of ElastoViscoplasticity in Modern Manufacturing Processes **225**

1 Prologue **229**

2 The Thermodynamical Theory of Elasto-Viscoplasticity **236**
2.1 Physical Foundations 236
2.2 Experimental Motivation for Microshear Banding Effects 238
2.3 Experimental Motivation for Induced Anisotropy Effects 248
2.4 Kinematics of finite deformation and fundamental definitions 255
2.4.1 Fundamental measures of total deformation 255
2.5 Finite elasto-viscoplastic deformation 256
2.5.1 Rates of the deformation tensor 259
2.5.2 Rates of the stress tensors 260
2.5.3 Fundamental properties of the Lie derivatives 261
2.6 Thermo-Elasto-Viscoplasticity Constitutive Model 262
2.6.1 Constitutive postulate 262
2.6.2 Fundamental assumptions 263
2.6.3 Microshear bending effects 264
2.6.4 Constitutive assumption for the plastic spin 265
2.6.5 Anisotropic intrinsic microdamage mechanisms 265
2.6.6 Kinematic hardening 266
2.6.7 Thermodynamic restrictions and rate type constitutive relations 267
2.6.8 Fracture criterion based on the evolution of microdamage 268
2.7 Analysis of the Fundamental Features of the Model 268
2.7.1 Invariance with respect to diffeomorphism 268

2.7.2 Finite plastic deformation and plastic spin effects 268
2.7.3 Plastic non-normality 269
2.7.4 Softening effects generated by microdamage mechanisms 269
2.7.5 Anisotropic effects in microdamage mechanisms 269
2.7.6 Plastic deformation induced anisotropic effects 269
2.7.7 Thermomechanical couplings (thermal plastic softening and thermal expansion) . . . 270
2.7.8 Influence of stress triaxiality on the evolution of microdamage 270
2.7.9 Rate sensitivity 270
2.7.10 Length-scale sensitivity of the constitutive model 271
2.7.11 Dissipation effects 272
2.7.12 Dispersion effects 273
2.7.13 Regularization of the evolution problem . . 274
2.7.14 Synergetic effects generated by cooperative phenomena 274

3 Constitutive Modelling for Dynamic Cyclic Loadings **275**
3.2 Experimental and physical motivations 275
3.2 Fundamental assumptions 280
3.3 Intrinsic micro-damage process 280
3.4 Kinematic hardening 282
3.5 Thermodynamic restrictions and rate type constitutive relations 284
3.6 Fracture criterion based on the evolution of porosity 285
3.7 Rate independent plastic response 285
3.7.1 Rate independent plastic response as a limit case 285

3.7.2 Rate independent intrinsic micro-damage process 286
3.7.3 Fundamental rate type constitutive equations 287

4 Localization and Localized Fracture Phenomena in Cyclic Dynamic Loading Processes 289
4.1 Formulation of the evolution problem 289
4.2 Application of finite difference method 290
4.3 Stability criterion and the Lax-Richtmyer equivalence theorem 293
4.4 Identification procedure 293
4.4.1 Assumptions of the material functions . . 293
4.4.2 Identification of the material constants . . 295
4.4.3 Investigation of stability and convergence . 296
4.4.4 Investigation of localization and fracture phenomena 297
4.5 Particular example 297
4.5.1 Dynamic, adiabatic, cyclic loading processes for a thin plate with sharp notch 297
4.5.2 Discussion of localization and localized fracture phenomena 298

5 Numerical Analysis of Macrocrack Propagation along a Bimaterial Interface under Dynamic Loading Processes 310
5.1 Experimental motivation 310
5.2 Thermo-elasto-viscoplastic model of a material . . 312
5.3 Identification procedure 314
5.3.1 Assumption of the material functions for an adiabatic process 314
5.3.2 Assumption of the material functions for an isothermal process 315
5.3.3 Determination of the material constants . 316

5.4 Numerical solution of the initial boundary-value problems . . . 318
5.4.1 Formulation of the initial boundary-value problem . . . 318
5.4.2 Investigation of the macro crack propagation along a bimaterial interface . . . 320
5.4.3 Influence of temperature effects for an adiabatic process . . . 329

6 Numerical Investigation of Dynamic Shear Bands in Inelastic Solids as a Problem of Mesomechanics 332
6.1 Physical and experimental motivation . . . 332
6.1.1 Analysis of mesomechanical problems . . . 332
6.1.2 Experimental investigation of the initiation and propagation of shear bands . . . 333
6.1.3 Fracture phenomena along localized shear bands . . . 335
6.1.4 Temperature measuremenet . . . 337
6.1.5 Possible shear band branching . . . 338
6.2 Formulation of the evolution problem . . . 338
6.2.1 Thermodynamic theory of elasto-viscoplasticity . . . 338
6.2.2 Initial-boundary value problem (evolution problem) . . . 340
6.2.3 Numerical solution of the evolution problem 340
6.3 Identification procedure . . . 341
6.3.1 Assumption of the material functions for an adiabatic process . . . 341
6.3.2 Determination of the material constants . 341
6.4 Numerical examples . . . 347
6.4.1 Propagation of shear bands . . . 347
6.4.2 Discussion of the numerical results . . . 347

7 Numerical Investigation of Localized Fracture Phenomena in Inelastic Solids 352
7.1 Thermo-elasto-viscoplastic model of a material . . 352
7.2 Formulation of the evolution problem 352
7.2.1 Initial-boundary value problem (evolution problem) 352
7.2.2 Numerical solution of the evolution problem 353
7.3 Identification procedure 353
7.3.1 Assumption of the material functions for an adiabatic process 353
7.3.2 Determination of the material constatnts . 355
7.4 Numerical example. Investigation of localized fracture phenomena 356
7.4.1 Application of a finite difference method . 356
7.4.2 Numerical solution of the initial boundary-value problem 357
7.4.3 Discussion of the numerical results 363

8 Epilogue 365

References 368

Part IV: Safety of Concrete and Masonry Structures under Unusual Loadings 377

1 Some remarks on damage and localized fracture 379

2 Introduction 381

3 Cumulative Fracture Criterion for brittle materials 382
3.1 Dynamic concrete behavior in codes 382
3.2 Cumulative Fracture Criterion (CFC) 383
3.3 Identification and the influence of used parameters 385

4 Numerical simulation 388
4.1 One-dimensional verification 388
4.2 Explosive loading 393
4.3 Three-dimensional verification for circular concrete plate 398

5 Engineering structure subjected to explosion 399
5.1 Pressure distribution in the air 399
5.2 Pressure distribution on the wall front surface . . 401

6 Conclusions 405

Bibliography 410

PART I

Consistent Non Local Coupled Damage Model and Its Application in Impact Response of Composite Materials

by

George Z. Voyiadjis , Babur Deliktas and Peter I. Kattan

Consistent Non Local Coupled Damage Model and Its Application in Impact Response of Composite Materials

George Z. Voyiadjis[1*], Babur Deliktas[2] and Peter I. Kattan[1]

[1] Department of Civil and Environmental Engineering, Louisiana State University, Baton Rouge, LA 70803, USA

[2] Department of Civil Engineering, Mustafa Kemal University, Hatay, Antakya 31034, Turkey

[*]Voyiadjis@eng.lsu.edu

Abstract: In this work also the mechanics of small damage are also presented using a consistent mathematical and mechanical framework based on the equations of damage mechanics. In this regard, the new scalar damage variable is investigated in detail. The investigation in this work has been carried out for seeking a physical basis is sought for the damage tensor $[M]$ that is used to link the damage state of the material with effective undamaged configuration. The approach presented here provides for a strong physical basis for this missing link. In particular, the authors have made an important link between the damage tensor and fabric tensors.Computational aspects of the presented theory are also discussed. Numerical integration algorithms, verification and validation process of the theory are discussed. The finite element simulations are also performed by implementing the presented model in the commercial finite element code ABAQUS [6.8.3] as a user defined subroutine (VUMAT).

1. PROLOGUE

The recent advances in aerospace and war capabilities have made necessary the modification of the design of structures that are proned to hypervelocity impact in order to increase their resistancy againts the penetration and perforation by projectiles with much higher impact energies. In this respect, high performance materials need to be developed so that they can offer significant advantages over the currently used materials. Metals and composites are among those materials that are oftenly used in various part of the structural components of the engineering structure in aerospace, automible and defense industries. In general, these materials are subjeced to tremendeous microstructural changes due to perforation and penetration phenomena and have complex material response due to a strong rate and temperature dependence when deformed non-uniformly into the inelastic range. Therefore, the high velocity impacting mechanism needs to be understood properly in order to be able to design materials of high ballistic resistant response. However, the exact mechanism by which the impacting target materials undergo fracture and ablation is a relatively complex process (Zukas, (1990). Generally, strong shock wave-material interactions are generated and propagated along both the projectile and the target, which can lead to fracture at low global inelastic strains.

The key role in the numerical simulation of the impact damage related problems is the accurate modeling of the material behavior at high strain rates and temperatures. Many researchers, therefore, have investigated the material failure mechanism during high velocity impact conditions with the ultimate goal of developing a micromechanical constitutive model that can effectively simulate the impact damage problem (Armstrong and Zerilli, 1994, Bammann, et al., 1990, Borvik, et al., 2004, Borvik, et al., 2006, Borvik, et al., 2002, Camacho and Ortiz, 1997, Curran, et al., 1987, Eftis, et al., 2003, Johnson and Cook, 1985, Steinberg and Lund, 1989). It is noted that none of these constitutive models address the problem of describing high shock compression and subsequent material degradation and failure in which the latter is expressed as an evolving micro-flaw having a damage rate determined from micro-mechanical analysis. Moreover, these models cannot consider the actual sizes, shapes, and orientations of the individual micro-voids and micro-cracks, which may have a predominant influence on the mechanical response of the material. The mechanical behavior of these applications cannot be characterized by classical (rate-independent) continuum theories because they incorporate no 'material length scales.' It is therefore necessary to develop a coupled rate-dependent (viscoplasticity) continuum damage theory bridging the gap between the classical continuum theories and the microstructure simulations.

The authors of this work have recently recognized the need for a micro-mechanical damage model (Abu Al-Rub and Voyiadjis, 2006, Voyiadjis, et al., 2003, Voyiadjis, et al., 2004) that accounts for the nonlocal microscopic interactions between material points (i.e. to take into account the influence of an internal state variable at a point on its neighborhood) in the simulation of metal impact problems This nonlocal microdamage model is formulated based on the enhanced gradient-dependent theory which is successful in explaining the size effects encountered at the micron scale and in preserving the well-posed nature of the initial boundary value problem that governs the solution of material instability triggering strain localization. Moreover, the viscoplasticity theory (rate-dependency) allows the spatial difference operator in the governing equations to retain its ellipticity and consequently the initial boundary value problem is hence well-posed (Batra, 2006, Batra and Chen, 1999, Batra and Kim, 1988, Batra and Kim, 1990, Loret and Prevost, 1990, Molinari, 1997, Needleman, 1988, Sluys, 1992). However, the gradient dependent theories enhance a stronger regularization of the localization problem than the rate-dependent theory. Moreover, the rate-dependent theory cannot explain the size effect of the microdamage zone (i.e. the void/crack size and spacing) on the material failure while the gradient theory can address that.

Therefore, the objective of this work is to present for high speed impact damage problems a novel microdamage constitutive model that possesses several material length scales. This model can be used to produce physically meaningful and numerically converging results within strain localization computations by finite element codes. Moreover, the algorithmic aspects and numerical implementation of this model in finite element codes are presented in this chapter.

The outline of the presented materials is as follows: in section 2, general mechanisms of the perforation and penetration mechanisms are discussed. A coupled rate-dependent (viscoplasticity) continuum damage theory is presented. In section 3, mechanics of small damage in fiber reinforced composite material is presented. In section 4, a comaparative study has been made on the damage variables of the continuum. In section 5, computational aspects of the theory are discussed. The elastic predictor and coupled viscoplastic-viscodamage corrector algorithm that allows for total uncoupling of geometrical and material nonlinearities are presented. The nonlinear algebraic system of equations is solved by consistnt linearization and the Newton–Raphson iteration. A derivation of a new definition for the consistent tangent stiffness matrix that is essential to maintaining the asymptotic quadratic rate of convergence is also presented in section 5. In section 6, numerical simulations of material instability are introduced in order to validate and test the proposed finite strain approach along with the proposed algorithm and its implementation in the ABAQUS finite element code. In section 7, various numerical example are presented in order to validate the reliability and capability of the the theory in simulating various impact and contact stress problems. Experimental comparisons of the adiabatic shear localizations between the proposed model simulations and other independent results are presented. Effect of initial temperatures and loading rates on the development of shear localizations is also investigated in this section. Model capabilities are preliminarily illustrated for the dynamic localization of inelastic flow in adiabatic shear bands and compared with the experimental results of Borvik et al. (2002) for the perforation of 12mm thick Weldox 460E steel plates by deformable blunt projectiles at various impact speeds.

2. Nonlocal Formulation of High Velocity Impact Induced Damage

2.1. Penetration and perforation mechanisms

The penetration and perforation mechanisms are interdisciplinary subjects and are based on the laws of conservation and compatibility. Penetration implies movement of a projectile into a target. Phenomenologically, as illustrated in

Figure 1, the penetration can be viewed as a process to generate a coneshaped macro crack in the material, in which, the kinetic energy of the penetrator is dissipated.

Perforation implies penetration all the way through a target. Projectile exit through a finite target is often accompanied by delamination and plugging. Delamination refers to a tensile failure parallel to the target rear surface and is often initiated by spall. However, it can also occur in quite thick targets, where it seems to be caused by *shear bands* around the projectile head that can propagate as cracks near the exit surface (Chelluru, 2007). In general, a penetration equation is a set of equations that are used to predict the outcome of an important event, such as the residual velocity or mass of the projectile after impact. Empirical penetration equations are essentially curve fits and take the general form $f(x_1, x_2, \ldots x_n)$ where $x_1, x_2, \ldots x_n$ are parameters such as projectile size and target thickness.

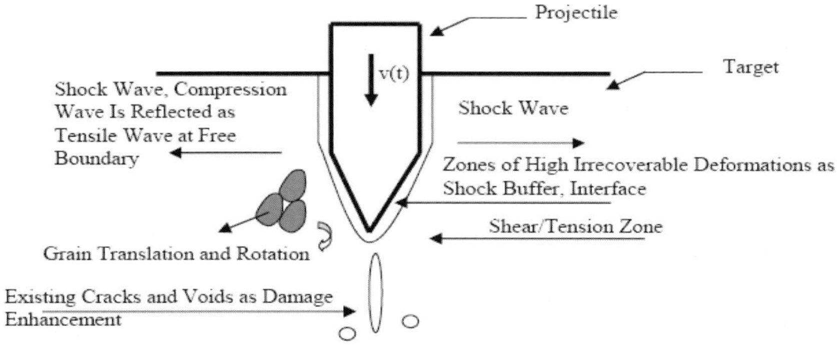

Figure 1. Schematics Illustration of the Penetration Process

It is important to note that results of these equations are only accurate if the case for which it is being used is close to that of the experimental data to which the equations are fit (Christiansen and Friesen, 1997). Prediction of the Ballistic limit is a very difficult task for which complete success may not be possible. Therefore, extensive experimental work is being done to date to understand which parameters affect the impact in composite structures. The analysis is still complex because the events that occur at the projectile/target interface are somewhat unknown. Although many studies have been performed, only highly controlled velocities, shape, sizes and trajectories have been examined. As a result, numerous approximations and assumptions must be made in order to apply to these analyses. Impact is a much localized phenomenon. Stress and strain effects are usually limited to within 3-6 projectile diameters of the impacted zone (Chelluru, 2007). Impacted target materials may fail by a combination of several modes including spalling, plugging, petaling, ductile or brittle fracture, and adiabatic shearing.

Spalling is the tensile failure of the target material due to reflection of the initial compressive waves from the far side of the target. Failure by spalling can occur on either the front or back of a target and is characterized by the formation of petals or ejects. In the event of impact there is an exchange of energy that takes place:

$$E_{tans} = E_{plate}^{internal} + E_{plate}^{kinetic} + E_{eroded}^{internal} + E_{eroded}^{kinetics} \tag{1}$$

Law of conservation is observed in any physical phenomena. Kinetic energy of the projectile is spent in raising the internal energy and kinetic energy of the plate and some part of the energy is lost in the form of eroded material. The amount of energy dissipated also differs with geometry. Blunt projectiles like cylinders are found to cause plugging because of pure shear failure, while the conical projectiles are found to cause petaling effects.

High velocity impact will localize compression of the composite and subsequently shearing the fiber and spalling of the resin during impact. Once the projectile has slowed, the composite deforms causing fiber stretching, pullout and delamination of the composite layers and thus lower the load carrying capacity.

2.2 A coupled rate-dependent (viscoplasticity) continuum damage theory

The theoretical model presented in this section is considered within a thermodynamic framework, where it is assured that the principles of thermodynamics are satisfied. Therfore, the virtual power relations are first defined and the principle of virtual power along with varitational formulations are used to develop the governing differential equations and their corresponding boundary conditions of the proposed theory. The principle of virtual power used by Voyiadjis and Deliktas (2009, 2009) is different than those used by Voyiadjis and Co-workers (Abu Al-Rub and Voyiadjis, 2004, Abu Al-Rub and Voyiadjis, 2006, Abu Al-Rub, et al., 2007, Dorgan and Voyiadjis, 2003, Dorgan and Voyiadjis, 2006, Voyiadjis and Abu Al-Rub, 2005, Voyiadjis and Abu Al-Rub, 2007, Voyiadjis, et al., 2003, Voyiadjis and Almasri, 2008, Voyiadjis and Deliktas, 2000, Voyiadjis and Deliktas, 2000), Gudmundson and co-workers (Fredriksson and Gudmundson, 2005, Fredriksson and Gudmundson, 2007, Gudmundson, 2004, Nygards and Gudmundson, 2004, Tjernlund, et al., 2006), Willis and coworkers (Aifantis and Willis, 2005, Aifantis and Willis, 2006, Fleck and Willis, 2009, Fleck and Willis, 2009) and, Gurtin and coworkers (Anand, et al., 2005, Bittencourt, et al., 2003, Cermelli, et al., 2004, Gudmundson, 2004, Gurtin, 2000, Gurtin, 2004, Gurtin and Anand, 2009, Gurtin and Needleman, 2005) where the principle of virtual power is modified by adding the contributions from damage and its corresponding gradients as a measure of micro motion of damage within the bulk. In addition two internal state variables are introduced on the contact interface, one measuring the tangential slip and another measuring the wear. By using these internal state variables together with displacements and temperature, the constitutive model is formulated with state laws based on the free energies and the complimentary laws based on the dissipation potentials. This model provides a potential feature for enabling one to relate the non-local continuum plasticity and damage of the bulk material to friction and wear at the contact interfaces.

One now defines a region $V \subset \Re^d$ (d=2, 3) with a piecewise smooth boundary Ω that occupies a continuously deformable body (Figure 2). The boundary Ω is divided into three disjoint parts; Ω_t is the part of the boundary where tractions are prescribed whereas the displacements are prescribed at the boundary Ω_u, and the unilateral contact interface is defined by the boundary Ω_c. The contact interfaces surface energy is considered as material boundary following the concept of Voyiadjis and Deliktas (2009) and enhanced by the following pioneering work of Fremond (1996) and its corresponding modification by Iremena et al.(2003).

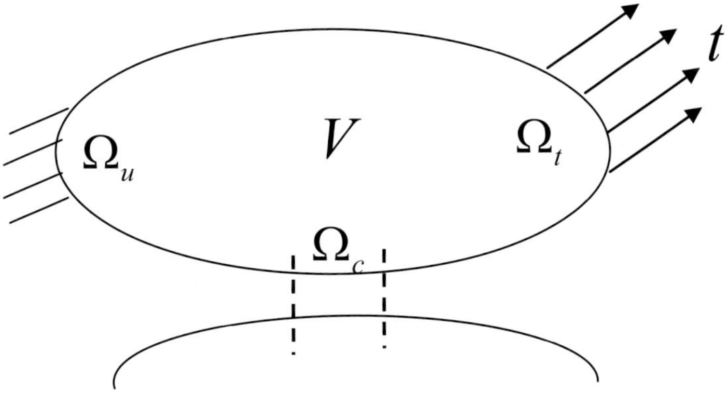

Figure 2. A Deformable Body with Unilateral Contact

The modified form of the internal virtual power is expressed here as follows:

$$P_{\text{int}} = \int_V \left(\sigma_{ij}\dot{\varepsilon}^e_{ij} + X_{ij}\dot{\varepsilon}^p_{ij} + S_{ijk}\dot{\varepsilon}^p_{ij,k} + Y_{ij}\dot{\phi}_{ij} + \Gamma_{ijk}\dot{\phi}_{ij,k}\right)dV - \int_{\Omega_c} p_i v_i \, d\Omega + \int_{\Omega_c} M^c_{ij}\dot{\varepsilon}^c_{ij} d\Omega_c + \int_{\Omega_c} \aleph^c_{ij}\dot{\phi}^c_{ij} d\Omega_c + \int_{\Omega_c} Q^c \varphi^c d\Omega_{cc} + \int_{\Omega_c} q_i v^t_i d\Omega_c \tag{2}$$

The superscripts are used to describe c for contact, n for normal and t for traction. The tensor X_{ij} is the driving stress work conjugate to the plastic strain, $\dot{\varepsilon}^p_{ij}$, and S_{ijk} is the micro-stress work conjugate to the plastic strain gradient, $\dot{\varepsilon}^p_{ij,k}$. The tensors Y_{ij} and Γ_{ijk} are the damage related internal forces work conjugate to ϕ_{ij} and $\phi_{ij,k}$, respectively. The external power is expressed as

$$P_{ext} = \int_{\Omega_t} t_i v_i \, d\Omega_t + \int_{\Omega_t} m^t_{ij}\dot{\varepsilon}^p_{ij} d\Omega_t + \int_{\Omega_c} \eta^t_{ij}\dot{\phi}_{ij} d\Omega_t \tag{3}$$

Furthermore, forces associated with friction and wear are also introduced. The non local differential equation of the flow rule and its relevant nonstandard boundary conditions are derived by using the principle of virtual power $P_{ext} - P_{\text{int}} = 0$ and imposing integration by parts, and the divergence theorem

$$\begin{aligned}
&\int_{\Omega_t} \left(t_i - \sigma_{ij}n_j\right)v_i d\Omega_t + \int_{\Omega_t} \left(-p_i - \sigma_{ij}n_j\right)v_i d\Omega_t + \\
&\int_V \left(\sigma_{ij,j}\right)v_i dV + \int_V \left(\tau_{ij} - X_{ij} + S_{ijk,k}\right)\dot{\varepsilon}^p_{ij} dV + \int_V \left(\Gamma_{ijk,k} - Y_{ij}\right)\dot{\phi}_{ij} dV + \\
&\int_{\Omega_c} \left(-M_{ik} - S_{ijk}n_k\right)\dot{\varepsilon}^p_{ij} d\Omega_c + \int_{\Omega_c} \left(-\aleph_{ij} - \Gamma_{ijk}n_k\right)\dot{\phi}_{ij} d\Omega_c + \\
&\int_{\Omega_c} Q^c \dot{\varphi} d\Omega_c + \int_{\Omega_c} q_i v^t_i d\Omega_c + \int_{\Omega_t} \left(m_{ik} - S_{ijk}n_k\right)\dot{\varepsilon}^p_{ij} d\Omega_c + \int_{\Omega t} \left(\eta_{ij} - \Gamma_{ijk}n_k\right)\dot{\phi}_{ij} d\Omega_t = 0
\end{aligned} \tag{4}$$

From Eq. (3) one can obtain the balance laws as follows

$$\left.\begin{aligned}
\sigma_{ij,j} &= 0 \\
\tau_{ij} &= X_{ij} + S_{ijk,k} \\
\Gamma_{ijk,k} - Y_{ij} &= 0
\end{aligned}\right\} \text{ in } V \tag{5}$$

Standard and non standard boundary conditions on the boundaries, Ω_t and Ω_c are obtained respectively

$$\left.\begin{aligned} t_i &= \sigma_{ij} n_j \\ m_{ij} - S_{ijk} n_k &= 0 \\ \eta_{ij} - \Gamma_{ijk} n_k &= 0 \end{aligned}\right\} \quad \text{on } \Omega_t \tag{6}$$

$$\left.\begin{aligned} M_{ik} + S_{ijk} n_k &= 0 \\ \aleph_{ij} + \Gamma_{ijk} n_k &= 0 \\ -p_i &= \sigma_{ij} n_j \\ Q^c &= 0 \\ q_i &= 0 \end{aligned}\right\} \quad \text{on } \Omega_c \tag{7}$$

Consequently forces Q^c and q_i are simply equal to zero. In addition to balance laws one needs constitutive assumptions that couple internal forces to the state variables. A general framework for the thermodynamical forces and state variables are obtained by using free energies and the dissipation potential in the style of the standard material. Such an approach is consistent with satisfying the second law of thermodynamics which states that the rate change of the total free energy must be less than the external power i.e. $\dot{\Upsilon} \leq P_{ext}$, where the total free energy can be defined as

$$\Upsilon = \int_V \rho \psi_v dV + \int_{\Omega_c} \psi_c d\Omega_c \tag{8}$$

where ρ is the mass density, $\psi_v = \psi_v\left(\varepsilon^e, \varepsilon^p, \phi, \nabla\varepsilon^p, \nabla\phi\right)$ is the free energy in the volume, V and $\psi_c = \psi_c\left(u, \varpi, \varphi, \varepsilon^p, \phi\right)$ is the free energy on the contact surface Ω_c. Thermodynamical Clausius-Duhem inequalities for both the bulk and the contact interface are given as follows

$$\begin{gathered} \Phi_v = \sigma : \dot{\varepsilon}^e + X : \dot{\varepsilon}^p + Y : \dot{\phi} + S : \nabla\dot{\varepsilon}^p + \Gamma : \nabla\dot{\phi} - \rho\dot{\psi}_v \geq 0 \\ \Phi_c = p : \dot{u} + \beta\dot{\varepsilon}^p + \varsigma\dot{\phi} + q_t\dot{\varpi} + \chi\dot{\varphi} - \dot{\psi}_c \geq 0 \end{gathered} \tag{9}$$

These contact surfaces interpret the grain interactions with reference to separation and sliding as an additional source of form of the inelastic deformation. The vector $\boldsymbol{p}$ is the interface traction force vector. In addition χ and ς are two new thermodynamic forces work conjugate to the measure of grain boundary separation defined by the flux, φ, and the measure of the grain boundary sliding defined by the flux, ϖ , respectively. The term β represents the interfacial strength parameter associated with the plastic strain jump at the sub interface surface.

The number of unknown internal state variables of the model can be determined by solving the constraint minimization problem of the energy dissipation by using the Lagrange multiplier method and assuming the existence of the constraint surfaces for plasticity and damage respectively such that

$$
\begin{aligned}
L_p(X,S) &= \Phi_v - \lambda^p f(X,S) \\
L_d(Y,\Gamma) &= \Phi_v - \lambda^d g(Y,\Gamma) \\
L_c(p,\beta,\varsigma,q_t,\chi) &= \Phi_c - \lambda^c f(p,\beta,\varsigma,q_t,\chi)
\end{aligned}
\tag{10}
$$

The mathematical aspects of the thermodynamical consideration of the theory for such form of the free and dissipation potential, dissipation inequalities, minimization problem and existence of the uniqueness of the solutions are discussed in detail in recent studies by Voyiadjis and Deliktas (2009). Here, the governing equations for the coupled viscoplastic damage behavior are defined by the following constitutive relations :

$$
\begin{aligned}
\sigma_{ij} &= E_{ijkl}(\phi)\varepsilon^e_{kl} \\
\tau_{ij} &= X_{ij} + S_{ijk,k} \\
f &= \left(X_{ij}X_{ij} + \ell^{-2}S_{ijk}S_{ijk}\right)^{1/2} - \sigma_f\left(\dot{E}^p\right) \le 0 \\
&\dot{\lambda}^{cp} \ge 0 \quad \dot{\lambda}^{cp} f^c \ge 0 \quad \dot{\lambda}^{cp}\dot{f}^c = 0 \\
g &= \left(Y_{ij}Y_{ij} + \ell_d^{-2}\Gamma_{ijk}\Gamma_{ijk}\right)^{1/2} - \sigma\ (\dot{\kappa}) \le 0 \\
&\Gamma_{ijk,k} - Y_{ij} = 0 \\
&\dot{\lambda}^d \ge 0 \quad \dot{\lambda}^d g \ge 0 \quad \dot{\lambda}^d \dot{g} = 0
\end{aligned}
\tag{11}
$$

and for the contact interface

$$\begin{aligned} & M_{ik} + S_{ijk} n_k = 0 \\ & f^c = \left(M_{ij} M_{ij} \right)^{1/2} - \sigma_f \left(\dot{\varepsilon}^p \right) \leq 0 \\ & \dot{\lambda}^p \geq 0 \;\; \dot{\lambda}^p f^c \geq 0 \;\; \dot{\lambda}^p \dot{f}^c = 0 \\ & \aleph_{ij} + \Gamma_{ijk} n_k = 0 \\ & g^c = \left(\aleph_{ij} \aleph_{uj} \right)^{1/2} - \sigma \left(\dot{\phi} \right) \leq 0 \\ & \dot{\lambda}^d \geq 0 \;\; \dot{\lambda}^d g^c \geq 0 \;\; \dot{\lambda}^d \dot{g}^c = 0 \end{aligned} \tag{12}$$

where $\dot{E}^p$ is the effective nonlocal flow rate such that $\dot{E}^p = \sqrt{\dot{\varepsilon}^p_{ij}\dot{\varepsilon}^p_{ij} + \ell^2_p \dot{\varepsilon}^p_{ij,k}\dot{\varepsilon}^p_{ij,k}}$. The nonlocal effective damage flow rate is defined by $\dot{\kappa} = \sqrt{\dot{\phi}_{ij}\dot{\phi}_{ij} + \ell^2_d \dot{\phi}_{ij,k}\dot{\phi}_{ij,k}}$. In this constitutive modeling two characteristic material length scales are introduced. They are the plastic length scale, ℓ_p and the damage length scale, ℓ_d respectively. The term α is the thermal expansion coefficient; $\dot{T}$ is the rate of absolute temperature; $\boldsymbol{I}$ is the second order identity tensor; and $\boldsymbol{A}$ is the fourth order tensor defined as

$$\boldsymbol{A} = \frac{\partial \boldsymbol{M}^{-1}}{\partial \emptyset} : \boldsymbol{M} : \boldsymbol{\sigma} + \boldsymbol{E} : \boldsymbol{M} : \frac{\partial \boldsymbol{M}^{-1}}{\partial \emptyset} : \boldsymbol{E}^{-1} : [\boldsymbol{\sigma} + \alpha(T - T_r)\boldsymbol{I}] \tag{13}$$

The fourth order damage tensor $\boldsymbol{M}$ is the function of the second order damage tensor, $\emptyset$. Its explicit form can be found in the works of Voyiadjis and co-workers (Abu Al-Rub and Voyiadjis, 2003, Abu Al-Rub and Voyiadjis, 2005, Abu Al-Rub and Voyiadjis, 2006, Abu Al-Rub, et al., 2007, Dorgan and Voyiadjis, 2003, Dorgan and Voyiadjis, 2006, Dorgan and Voyiadjis, 2007, Dorgan and Voyiadjis, 2007, Kattan and Voyiadjis, 1993, Kattan and Voyiadjis, 1993, Kattan and Voyiadjis, 1996, Park and Voyiadjis, 1997, Park and Voyiadjis, 1998, Voyiadjis and Kattan, 1992, Voyiadjis and Park, 1997). The functional form of the flow rule of plasticity is given as follows :

$$\sigma(\dot{E}^p) = \sigma_y + R(\dot{E}^p)\left[1 + \left(c_p(T - T_r)\dot{E}^p\right)^{1/m}\right] \tag{14}$$

where σ_y is the initial yield strength; and c_p is the specific heat. The evolution equation for isotropic and hardening laws are given as follows

$$\dot{R} = \mu \dot{E}^p q e^{-\mu E^p} \tag{15}$$

and

$$\dot{\boldsymbol{X}} = \frac{2}{3} C \dot{\boldsymbol{\varepsilon}}^p - \gamma \boldsymbol{X} \dot{E}^p + \beta \dot{\boldsymbol{\sigma}} \tag{16}$$

Voyiadjis and Abu Al-Rub (2003) defined, q as

$$q = q_m + (q_m + q_o) e^{-2\mu q} \tag{17}$$

and $m, q_m, q_o,$, C, γ and β are the material constant to be calibrated from available experimental data. Similary one can define the flow rules of damage as follows

$$\sigma(\dot{\kappa}) = Y_o + K\big((\dot{\kappa})\big)\left[1 + \big(c_p(T - T_r)\dot{\kappa}\big)^{1/m}\right] \tag{18}$$

where Y_o is the initial damage threshold; $\left(\dot{\kappa} = \sqrt{\dot{\tilde{\boldsymbol{\phi}}} : \dot{\tilde{\boldsymbol{\phi}}}}\right)$; and the nonlocal damage tensor is defned as $\tilde{\boldsymbol{\phi}} = \boldsymbol{\phi} + {}^1/_2\, \ell^2 \nabla^2 \boldsymbol{\phi}$. The evolution of the damage anisotropic hardening equation is given by Voyiadjis and Deliktas (2000) as follows

$$K(\kappa) = \lambda \zeta \left(\frac{\kappa}{\lambda}\right)^{\xi} \delta_{ij} \tilde{\phi}_{ij} + \delta_{ij} \lambda Y_o^2 \tag{19}$$

where λ is the Lame constant and is defined by Voyiadjis and Park (1997, 1995, Voyiadjis and Park, 1995, Voyiadjis and Park, 1997). The exponent, ζ represents the damage hardening parameter, and ξ denotes the damage growth rate. The influences of these parameters on the response of the material was studied by Voyiadjis and Deliktas (Voyiadjis and Deliktas, 1997)

3. Mechanics of Small Damage in Fiber-Reinforced Composite Materials

In this section the new concept of small damage is examined within the framework of continuum damage mechanics. In particular, special emphasis is given to a new damage variable that is defined in terms of the elastic stiffness of the material. Only the scalar case is studied here. The investigation of the new scalar damage variable and the new concept of small damage is carried out on fiber-reinforced composite materials. Furthermore, the two approaches to damage analysis in composite materials are re-examined in this work using the new damage variable. These are the well known overall and local approaches established in the literature of damage mechanics. It is noted that the examination of these two approaches that is presented here applies to both small and large damage mechanics. Finally, the two approaches are compared mathematically and are shown to be equivalent.

The damage variable (or tensor), based on the effective stress concept, represents average material degradation which reflects the various types of damage at the micro-scale level like nucleation and growth of voids, cracks, cavities, micro-cracks, and other microscopic defects (Budiansky and O'Connell, 1976, Krajcinovic, 1996, Lubarda and Krajcinovic, 1993). For the case of isotropic damage mechanics, the damage variable is scalar and the evolution equations are easy to handle. However, it has been shown by Ju (1990) and Cauvin and Testa (1999) that two independent damage variables must be used in order to describe accurately and consistently the special case of isotropic damage. It has been argued (Lemaitre, 1984) that the assumption of isotropic damage is sufficient to give good predictions of the load carrying capacity, the number of cycles or the time to local failure in structural components. However, the development of anisotropic damage has been confirmed experimentally (Chow and Wang, 1987, Lee, et al., 1985) even if the virgin material is isotropic. This has prompted several researchers to investigate the general case of anisotropic damage (Kattan and Voyiadjis, 1996, Kattan and Voyiadjis, 2001, Murakami, 1983, Voyiadjis and Kattan, 2006) In continuum damage mechanics, usually a phenomenological approach is adopted. In this approach, the most important concept is that of the Representative Volume Element (RVE). The discontinuous and discrete elements of damage are not considered within the RVE; rather their combined effects are lumped together through the use of a macroscopic internal variable. In this way, the formulation may be derived consistently using sound mechanical and thermodynamic principles (Doghri, 2000, Hansen and Schreyer, 1994, Luccioni and Oller, 2003).

Kachanov(1958) and Rabotnov (1969) introduced the concept of effective stress for the case of uniaxial tension. This concept was later generalized to three-dimensional states of stress by Lemaitre (1970) and Chaboche (1981). Let $\boldsymbol{\sigma}$ be the second-rank Cauchy stress tensor and $\overline{\boldsymbol{\sigma}}$ be the corresponding effective stress tensor. The effective stress $\overline{\boldsymbol{\sigma}}$ is the stress applied to a fictitious state of the material which is totally undamaged, i.e. all damage in this state has been removed. This fictitious state is assumed to be mechanically equivalent to the actual damaged state of the material. In this regard, one of two hypotheses (elastic strain equivalence or elastic energy equivalence) is usually used (Lemaitre, 1984, Sidoroff, 1981).

Ju (1990) pointed out that even for isotropic damage one should employ a damage tensor (not a scalar damage variable) to characterize the state of damage in materials. However, the damage generally is anisotropic due to the external agency condition or the material nature itself. Although the fourth-rank damage tensor (Murakami, 1983) can be used directly as a linear transformation tensor to define the effective stress tensor, it is not easy to characterize physically the fourth-rank damage tensor compared to the second-rank damage tensor.

In this work the new concept of small damage is examined within the framework of continuum damage mechanics. In particular, special emphasis is given to a new damage variable that is defined in terms of the elastic stiffness of the material. Only the scalar case is studied in this work. The scalar definition of the new damage variable was used recently by many researchers. The investigation of the new scalar damage variable and the new concept of small damage is carried out on fiber-reinforced composite materials.

Furthermore, the two approaches to damage analysis in composite materials are re-examined in this work using the new damage variable. These are the well known overall and local approaches established in the literature of damage mechanics. It is noted that the examination of these two approaches that is presented here applies to both small and large damage mechanics. Finally, the two approaches are compared mathematically and are shown to be equivalent.

3.1. New Scalar Damage Variable

This section addresses a new scalar damage variable that was used by researchers recently. For other damage variables, see the work of Lemaitre (1971, 1984) and Lemaitre and Chaboche (1990). This scalar damage variable ℓ may be defined in terms of the reduction in the elastic modulus as follows:

$$\ell = \frac{\overline{E} - E}{E} \tag{20}$$

where E is the elastic modulus in the damaged state while $\overline{E}$ is the effective elastic modulus (in the fictitious state) with $\overline{E} > E$ (see Figure 3). This damage variable was used recently by Celentano et. al (2004) and Nichols and Abell (2003) and Nichols and Totoev (1999). It was also used by Voyiadjis and Kattan (2009) in a comparative study of damage variables within the context of continuum damage mechanics. It should also be mentioned that Voyiadjis (1988) used a similar relation but in the context of elasto-plastic deformation.

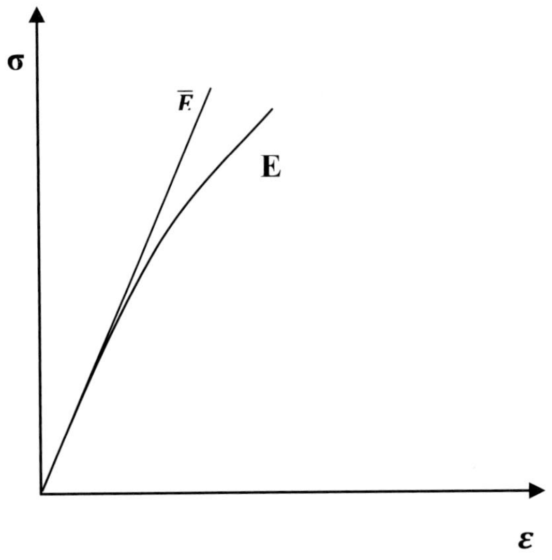

Figure 3. Damaged and Effective Moduli of Elasticity

The definition of the new damage variable of equation (20) may be re-written in the following more appropriate form:

$$\overline{E} = E(1+\ell) \tag{21}$$

It is clear from the definition in equation (21) that $\ell = 0$ when the body is undamaged, i.e. when $\overline{E} = E$. For a composite material consisting of a matrix and fibers, two local scalar damage variables may be similarly defined as follows:

$$\ell^m = \frac{\overline{E}^m - E^m}{E^m} \tag{22}$$

$$\ell^f = \frac{\overline{E}^f - E^f}{E^f} \tag{23}$$

where E^m and E^f are the elastic moduli for the matrix and fibers, respectively, while $\overline{E}^m$ and $\overline{E}^f$ are their effective counterparts. The two scalar local damage variables are denoted above by ℓ^m and ℓ^f for the matrix and fibers, respectively.

Using the hypothesis of elastic energy equivalence (see Voyiadjis and Kattan, 2009), one now can write the following transformation relation between the stress σ and the effective stress $\overline{\sigma}$:

$$\overline{\sigma} = \sigma\sqrt{1+\ell} \tag{24}$$

In the same way, another transformation relation can be obtained between the strain ε and the effective strain $\overline{\varepsilon}$ as follows:

$$\overline{\varepsilon} = \frac{\varepsilon}{\sqrt{1+\ell}} \tag{25}$$

For a composite material consisting of a matrix and fibers, equivalent local relations to equations (24) and (25) may be written as follows:

$$\overline{\sigma}^m = \sigma^m \sqrt{1+\ell^m} \tag{26}$$

$$\overline{\sigma}^f = \sigma^f \sqrt{1+\ell^f} \tag{27}$$

$$\overline{\varepsilon}^m = \frac{\varepsilon^m}{\sqrt{1+\ell^m}} \tag{28}$$

$$\overline{\varepsilon}^f = \frac{\varepsilon^f}{\sqrt{1+\ell^f}} \tag{29}$$

For small values of the scalar damage variable ℓ, the square root appearing in equations (24) – (29) can be approximated as follows using the Taylor series expansion:

$$\sqrt{1+\ell} \approx 1+\frac{1}{2}\ell \tag{30}$$

Next, one uses the rule of mixtures for the total effective stress as follows:

$$\bar{\sigma} = \bar{c}^m \bar{\sigma}^m + \bar{c}^f \bar{\sigma}^f \tag{31}$$

where $\bar{c}^m$ and $\bar{c}^f$ are the effective volume fractions for the matrix and fibers, respectively. Substituting equations (24) and (27) into equation (31) and simplifying, one obtains the relation between the total stress and the local stresses as follows:

$$\sigma = \bar{c}^m \sigma^m \sqrt{\frac{1+\ell^m}{1+\ell}} + \bar{c}^f \sigma^f \sqrt{\frac{1+\ell^f}{1+\ell}} \tag{32}$$

For the case of small damage and using the approximation in equation (30), the relation in equation (32) becomes:

$$\sigma + \frac{1}{2}\sigma \ell = \bar{c}^m \sigma^m + \bar{c}^f \sigma^f + \frac{1}{2}\bar{c}^m \sigma^m \ell^m + \frac{1}{2}\bar{c}^f \sigma^f \ell^f \tag{33}$$

It is clear that the above special relation for the case of small damage is a linear relation in the respective variables. It should be noted that the above relation cannot be simplified further without making certain limiting assumptions. For example, for the special case where the values of the three damage variables appearing in the equation are equal, i.e. when $\ell^m = \ell^f = \ell$, one may obtain the following simplified relations only for this very special case:

$$\sigma = \bar{c}^m \sigma^m + \bar{c}^f \sigma^f \tag{34}$$

$$\sigma \ell = \bar{c}^m \sigma^m \ell^m + \bar{c}^f \sigma^f \ell^f \tag{35}$$

Going back now to the general case of damage, one uses the following two relations between the local stresses σ^m, σ^f and the total stress σ:

$$\sigma^m = B^m \sigma \tag{36}$$

$$\sigma^f = B^f \sigma \tag{37}$$

where B^m and B^f are the matrix and fiber stress concentration factors, respectively. Substituting equations (36) and (37) into equation (32), one obtains the following relation :

$$\bar{c}^m B^m \sqrt{1+\ell^m} + \bar{c}^f B^f \sqrt{1+\ell^f} = \sqrt{1+\ell} \tag{38}$$

The above relation is an exact relation between the local damage variables and the overall damage variable. For the special case of small damage and equal values of the three damage variables given in equations (34) and (35), the relation in equation (38) becomes:

$$\bar{c}^m B^m + \bar{c}^f B^f = 1 \tag{39}$$

$$\bar{c}^m B^m \ell^m + \bar{c}^f B^f \ell^f = \ell \tag{40}$$

It should be noted that the above simplified relations apply only to this special case.

3.2. Damage Evolution

In this section one derives the general relation for the evolution of the new damage variable. The evolution equation will be written in terms of the increments of stress or strain based on sound thermodynamic principles for a uniaxial state of stress. In general, the elastic strain energy for the effective configuration is given as follows:

$$U = \frac{1}{2} \bar{\sigma} \bar{\varepsilon} \tag{41}$$

where $\bar{\sigma}$ and $\bar{\varepsilon}$ are the effective stress and effective strain, respectively. In addition, one can also write the elastic constitutive relation in the effective configuration as follows;

$$\bar{\sigma} = \bar{E} \bar{\varepsilon} \tag{42}$$

where $\overline{E}$ is the effective elastic modulus. Substituting equation (42) into equation (41), one obtains:

$$U = \frac{1}{2}\overline{E}\,\overline{\varepsilon}^{2} \tag{43}$$

Next, one substitutes equation (25) into equation (43) to obtain:

$$U = \frac{\overline{E}\,\varepsilon^{2}}{2(1+\ell)} \tag{44}$$

The above relation provides an expression for the elastic strain energy in terms of the new damage variable. Next, one takes the derivative of U in the above relation to obtain:

$$dU = \frac{\overline{E}\,\varepsilon\, d\varepsilon}{1+\ell} - \frac{\overline{E}\,\varepsilon^{2}\, d\ell}{2(1+\ell)^{2}} \tag{45}$$

Next, one defines the variable y to be the general thermodynamic force associated with the scalar damage variable ℓ as follows:

$$y = \frac{\partial U}{\partial \ell} \tag{46}$$

Substituting equation (44) into equation (46), one obtains:

$$y = -\frac{\overline{E}\,\varepsilon^{2}}{2(1+\ell)^{2}} \tag{47}$$

Next, one introduces $g(y,B)$ as the damage function or criterion given as follows:

$$g(y,B) \equiv \frac{1}{2}y^{2} - B(\beta) = 0 \tag{48}$$

where $B(\beta)$ is the overall threshold of damage and β is an overall damage parameter. The power of dissipation Π can now be written as follows:

$$\Pi = -y\, d\ell - B\, d\beta \tag{49}$$

Next, one constructs an objective function Ψ as follows:

$$\Psi = \Pi - d\lambda \cdot g \tag{50}$$

where $d\lambda$ is a Lagrange multiplier. In order to extremize the function Ψ, the following two conditions must be satisfied:

$$\frac{\partial \Psi}{\partial y} = 0 \tag{51}$$

$$\frac{\partial \Psi}{\partial B} = 0 \tag{52}$$

Substituting equation (50) into equation (51) and using equation (49), one obtains;

$$d\ell = -d\lambda \frac{\partial g}{\partial y} \tag{53}$$

Substituting equation (50) into equation (52) and using equation (49), one obtains:

$$d\beta = d\ell \tag{54}$$

Next, one substitutes equation (54) into equation (53) to obtain:

$$d\ell = -d\beta \frac{\partial g}{\partial y} \tag{55}$$

Finally, one invokes the consistency condition $dg = 0$ and applies it to equation (48) to obtain:

$$d\ell = -\frac{y\,dy\dfrac{\partial g}{\partial y}}{\left(\dfrac{\partial B}{\partial \beta}\right)} \tag{56}$$

Substituting the equation for the function g from equation (48) into equation (56) and simplifying, one obtains the evolution equation for the new damage variable;

$$d\ell = -\frac{y^2 dy}{\left(\dfrac{\partial B}{\partial \beta}\right)} \tag{57}$$

It is clear from the above equation that the evolution of the damage variable is an ordinary differential equation between the variables ℓ and y. It should be noted that the term $\frac{\partial B}{\partial \beta}$ appearing in the denominator can be taken to be a constant (Voyiadjis and Kattan, 2006). In order to solve the above governing differential equation, one substitutes equation (47) into equation (57) to obtain:

$$\left(\frac{\partial B}{\partial \beta}\right) d\ell = \frac{\overline{E}^3 \varepsilon^5 d\varepsilon}{4(1+\ell)^6} - \frac{\overline{E}^3 \varepsilon^6 d\ell}{4(1+\ell)^7} \tag{58}$$

The above relation is an explicit differential equation between the new damage variable and the strain. In order to solve the above differential equation, one uses the substitution $x = \left(\frac{\varepsilon}{1+\ell}\right)^6$ and one obtains the final solution as follows:

$$\left(\frac{\partial B}{\partial \beta}\right) \ell = \frac{1}{24} \overline{E}^3 \left(\frac{\varepsilon}{1+\ell}\right)^6 \tag{59}$$

Note that the above equation is the sought explicit relation between the new damage variable ℓ and the strain ε. It is clear that this relation is nonlinear.

Using the elastic constitutive relation of equation (42) but written in the damaged configuration as $\sigma = E\,\varepsilon$, and substituting it into equation (59), one obtains the following relation between the stress and the new damage variable:

$$\sigma = \left[24 \left(\frac{\partial B}{\partial \beta} \right) \overline{E}^{\,3} \ell \right]^{1/6} \tag{60}$$

The above relation is an explicit relation between the stress and the new damage variable. It is clearly a nonlinear relation.

Next, one plots the two relations given in equations (59) and (60). Figure 4 shows the relation between the stress and the damage variable based on equation (60) while Figure 5 shows the relation between the strain and the damage variable based on equation (59). In plotting these graphs, one uses a SiC-Ti-Al composite lamina with the following material properties: $\overline{E} = 199$ GPa and $\left(\frac{\partial B}{\partial \beta} \right) = 1000$ $(\text{kPa})^3$.

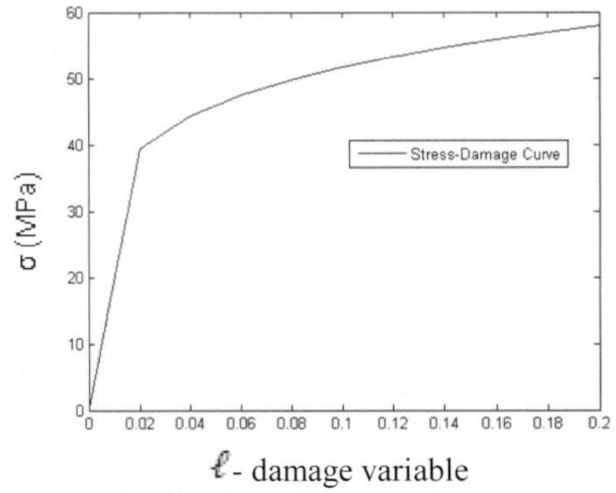

Figure 4. Stress-Damage Relation

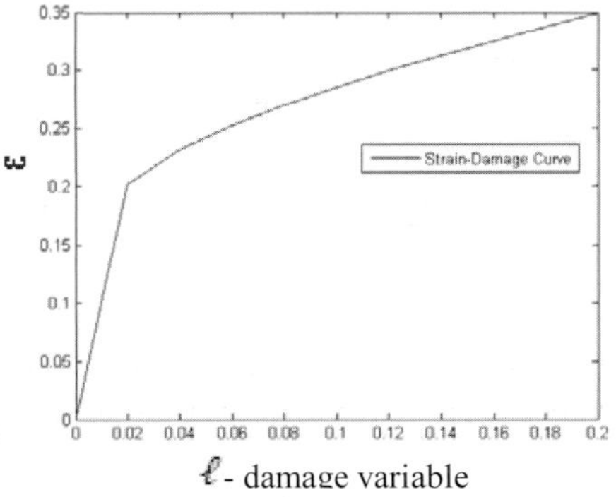

Figure 5. Strain-Damage Relation

It should be noted from Figures 4 and 5 that the stress-damage and strain-damage relations are linear for very small values of the damage variable, i.e for $\ell < 0.02$. For this small range of damage values, a linear and simplistic theory of damage mechanics may be formulated.

3.3. Mechanics of Small Damage

In this section, one provides certain equations for the mechanics of small damage, i.e. when the value of the damage variable is small. One will start by re-writing equation (60) as follows:

$$\sigma = (\alpha \ell)^{1/6} \tag{61}$$

where the coefficient α is given by:

$$\alpha = 24\left(\frac{\partial B}{\partial \beta}\right)\bar{E}^{3} \tag{62}$$

Similarly, one can write the following two relations for the matrix and fiber stresses:

$$\begin{aligned} \sigma^{m} &= \left(\alpha^{m} \ell^{m}\right)^{1/6} \\ \sigma^{f} &= \left(\alpha^{f} \ell^{f}\right)^{1/6} \end{aligned} \tag{63}$$

where the coefficients α^{m} and α^{f} are given by (see equation (62)):

$$\begin{aligned} \alpha^{m} &= 24\left(\frac{\partial B^{m}}{\partial \beta^{m}}\right)\bar{E}^{m3} \\ \alpha^{f} &= 24\left(\frac{\partial B^{f}}{\partial \beta^{f}}\right)\bar{E}^{f3} \end{aligned} \tag{64}$$

Next, one uses the law of mixtures for the damaged configuration as follows:

$$\sigma = c^{m}\sigma^{m} + c^{f}\sigma^{f} \tag{65}$$

where c^{m} and c^{f} are the matrix and fiber volume fractions in the damaged configuration, respectively. Substituting equations (61) and (63) into equation (65) and simplifying, one obtains;

$$\ell^{1/6} = c^{m}\left(\frac{\alpha^{m}\ell^{m}}{\alpha}\right)^{1/6} + c^{f}\left(\frac{\alpha^{f}\ell^{f}}{\alpha}\right)^{1/6} \tag{66}$$

For small values of the damage variables ℓ, ℓ^{m}, and ℓ^{f}, and combining equations (66), (33), and (38), one obtains the following relation after some tedious mathematical operations involving some approximations for small damage:

$$\frac{\ell^{f}}{\ell^{m}} = \frac{\left[\left(\bar{c}^{m}B^{m}\right)^{1/6} - c^{m}\left(\frac{\alpha^{m}}{\alpha}\right)^{1/6}\right]^{6}}{c^{f}\left(\frac{\alpha^{f}}{\alpha}\right)^{1/6} - \left(\bar{c}^{m}B^{m}\right)^{-5/6}\bar{c}^{f}B^{f}\left[\left(\bar{c}^{m}B^{m}\right)^{1/6} - c^{m}\left(\frac{\alpha^{m}}{\alpha}\right)^{1/6}\right]^{5}} \tag{67}$$

The above relation is valid for small values of the parameter $\frac{\bar{c}^{m}B^{m}\ell^{m}}{\bar{c}^{f}B^{f}\ell^{f}}$.

Similarly, another relation can be obtained as follows:

$$\frac{\ell^{m}}{\ell^{f}} = \frac{\left[\left(\bar{c}^{f}B^{f}\right)^{1/6} - c^{f}\left(\frac{\alpha^{f}}{\alpha}\right)^{1/6}\right]^{6}}{c^{m}\left(\frac{\alpha^{m}}{\alpha}\right)^{1/6} - \left(\bar{c}^{f}B^{f}\right)^{-5/6}\bar{c}^{m}B^{m}\left[\left(\bar{c}^{f}B^{f}\right)^{1/6} - c^{f}\left(\frac{\alpha^{f}}{\alpha}\right)^{1/6}\right]^{5}} \tag{68}$$

The above relation is valid for small values of the parameter $\frac{\bar{c}^{f}B^{f}\ell^{f}}{\bar{c}^{m}B^{m}\ell^{m}}$.

In deriving equations (67) and (68), the following approximations have been used based on the Taylor series expansion of each term. For any variables x and y, and constants A, B, C, D, the following equation

$$Ax^{1/6} \cdot Bx^{-5/6}y = Cx^{1/6} + Dy^{1/6} \tag{69}$$

can be approximated by the following simpler relation:

$$(A-C)^{1/6}x \approx \left\lfloor D - 6B(A-C)^{5}\right\rfloor y \tag{70}$$

for small values of the parameter $\frac{By}{(A-C)x}$. Similarly, the equation

$$Ay^{1/6} \cdot By^{-5/6}x = Cy^{1/6} + Dx^{1/6} \tag{71}$$

can be approximated by the following simpler relation:

$$(A-C)^{1/6} y \approx \left[D - 6B(A-C)^5\right]x \tag{72}$$

for small values of the parameter $\frac{Bx}{(A-C)y}$.

Next, one writes the expression for the elastic strain energy U as defined in equation (20) in the damaged configuration while using equations (59) and (60):

$$U = \left[24\left(\frac{\partial B}{\partial \beta}\right)\right]^{2/6} (1+\ell)\ell^{2/6} \tag{73}$$

Similar relations to the above can be written for the matrix and fiber strain energies. The relation between the strain energy and the new damage variable is plotted in Figure 6 based on equation (73). Clearly, the relation is nonlinear.

It should be noted from Figure 6 that for very small values of the damage variable, the strain energy-damage curve is linear. This effect is again noted for values of the damage variable $\ell < 0.02$. For this small range of values, a linear and simplistic theory of damage mechanics may be formulated.

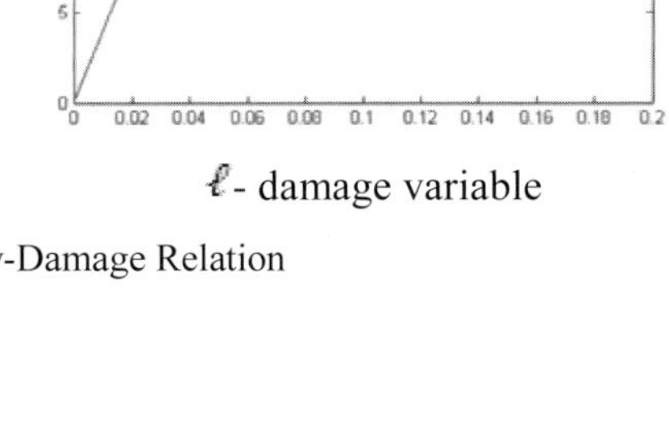

Figure 6. Strain Energy-Damage Relation

3.4. Overall and Local Approaches to Damage in Composite Materials

In this section one formulates the overall and local approaches to damage characterization in fiber-reinforced composite materials. The final outcome will be to show that under similar assumptions both the local and overall approaches yield similar results when applied to fiber-reinforced metal matrix composites. This demonstration will be shown for the elastic case in this work. The fibers are assumed to be continuous and perfectly aligned. In addition, a perfect bond is assumed to exist at the matrix-fiber interface. A consistent mathematical formulation is used to show the equivalence of the two approaches in this case. The elastic stiffness matrices are derived using both approaches based on the new damage variable and are shown to be identical in both cases.

The equivalence of the overall and local damage approaches based on the new damage variable will be shown for elastic composites. For simplicity, the composite system is assumed to consist of a matrix reinforced with continuous fibers. Both the matrix and fibers are linearly elastic with different material constants. Let $\overline{C}$ denote the configuration of the undamaged composite system and let $\overline{C}^m$ and $\overline{C}^f$ denote the configurations of the undamaged matrix and fibers, respectively. Since the composite system assumes a perfect bond at the matrix-fiber interface, it is clear that $\overline{C}^m \cap \overline{C}^f = \phi$ and $\overline{C}^m \cup \overline{C}^f = \overline{C}$. In the overall approach, the problem reduces to transforming the undamaged configuration $\overline{C}$ into the damaged configuration C. In contrast, two intermediate configurations C^m and C^f are considered in the local approach for the matrix and fibers, respectively. In the latter approach, the problem is reduced to transforming each of the undamaged configurations $\overline{C}^m$ and $\overline{C}^f$ into the damaged configurations C^m and C^f, respectively.

In the case of elastic fiber-reinforced composites, the following linear relations are used for each constituent in its respective undamaged configuration:

$$
\begin{aligned}
\overline{\sigma}^m &= \overline{E}^m \overline{\varepsilon}^m \\
\overline{\sigma}^f &= \overline{E}^f \overline{\varepsilon}^f
\end{aligned}
\tag{74}
$$

where $\overline{\sigma}^m$ and $\overline{\sigma}^f$ are the effective constituent stresses, $\overline{\varepsilon}^m$ and $\overline{\varepsilon}^f$ are the effective constituent strains, and $\overline{E}^m$ and $\overline{E}^f$ are the effective constituent elasticity moduli.

Within the framework of the micromechanical analysis of composite materials, the effective constituent stresses $\overline{\sigma}^m$ and $\overline{\sigma}^f$ are related to the effective composite stress $\overline{\sigma}$ by:

$$\overline{\sigma}^m = \overline{B}^m \overline{\sigma}$$
$$\overline{\sigma}^f = \overline{B}^f \overline{\sigma} \qquad (75)$$

where $\overline{B}^m$ and $\overline{B}^f$ are the constituent stress concentration factors. These concentration factors can be determined through several available models such as the Voigt and Mori-Tanaka models (Mori and Tanaka, 1973; Lagoudas et. al., 1991). The effective constituent strains $\overline{\varepsilon}^m$ and $\overline{\varepsilon}^f$ are determined in a similar way by the equations:

$$\overline{\varepsilon}^m = \overline{A}^m \overline{\varepsilon}$$
$$\overline{\varepsilon}^f = \overline{A}^f \overline{\varepsilon} \qquad (76)$$

where $\overline{\varepsilon}$ is the effective composite strain and $\overline{A}^m$ and $\overline{A}^f$ are the strain concentration factors.

Next, the overall and local approaches to damage in elastic composites are examined and compared based on the new damage variable.

3.4.1 Overall Approach

In this approach, damage is incorporated in the composite system as a whole through one damage variable called the overall damage variable. The two steps needed in this approach are shown schematically in Figure 7 for a two-phase composite system consisting of a matrix and fibers. In the first step, the elastic equations are formulated in an undamaged composite system. This is performed here using the law of mixtures as follows:

$$\overline{\sigma} = \overline{c}^m \overline{\sigma}^m + \overline{c}^f \overline{\sigma}^f \qquad (77)$$

where $\overline{c}^m$ and $\overline{c}^f$ are the effective matrix and fiber volume fractions, respectively.

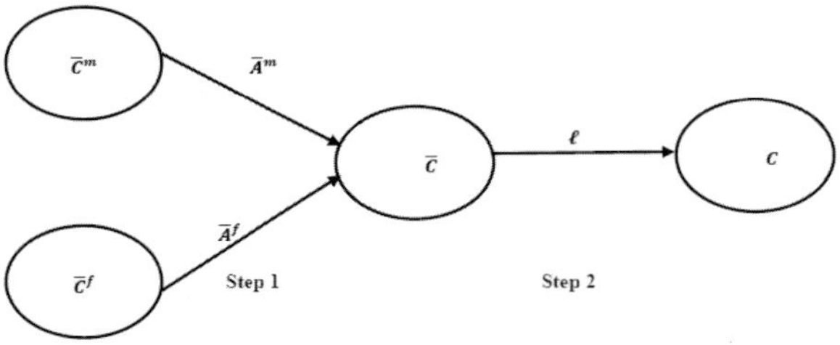

Figure 7. Schematic Diagram of the Overall Approach

In the effective composite configuration $\overline{C}$, the following linear elastic relation holds:

$$\overline{\sigma} = \overline{E}\overline{\varepsilon} \tag{78}$$

where $\overline{E}$ is the constant elasticity modulus. Substituting equations (74), (76) and (78) into equation (77) and simplifying, one obtains the following expression for $\overline{E}$:

$$\overline{E} = \overline{c}^m \overline{E}^m \overline{A}^m + \overline{c}^f \overline{E}^f \overline{A}^f \tag{79}$$

In the second step of the formulation, damage is induced through the new damage variable ℓ as follows:

$$\overline{\sigma} = \sigma\sqrt{1+\ell} \tag{80}$$

where σ is the composite stress. Equation (80) represents the damage transformation for the stress. In order to derive similar relation for the strain, one needs to use the hypothesis of elastic energy equivalence. In this hypothesis, the elastic energy of the damaged system is equal to the elastic energy of the effective system. Applying this hypothesis to the composite system by equating the two elastic energies, one obtains:

$$\frac{1}{2}\sigma\,\varepsilon = \frac{1}{2}\overline{\sigma}\,\overline{\varepsilon} \tag{81}$$

where ε is the composite strain. Substituting equation (80) into equation (81) and simplifying, one obtains the damage transformation equation for the strain as follows:

$$\overline{\varepsilon} = \frac{\varepsilon}{\sqrt{1+\ell}} \tag{82}$$

In order to derive the final elastic relation in the damage composite system, one substitutes equations (80) and (82) into equation (78) to obtain:

$$\sigma = E\varepsilon \tag{83}$$

where the damaged elastic modulus E is given by:

$$E = \frac{\overline{E}}{1+\ell} \tag{84}$$

Substituting for $\overline{E}$ from equation (79) into equation (85), one obtains:

$$E = \frac{\overline{c}^m \overline{E}^m \overline{A}^m + \overline{c}^f \overline{E}^f \overline{A}^f}{1+\ell} \tag{85}$$

The above expression represents the elastic modulus in the damaged composite system according to the overall approach.

3.4.2 Local Approach

In this approach, damage is introduced in the first step of the formulation using two independent damage variables for the matrix and fibers. However, more damage variables may be introduced to account for other types of damage such as debonding and delamination (Voyiadjis and Park, 1997) The two steps involved in this approach are shown schematically in Figure 8. One first introduces the new matrix and fiber damage variables ℓ^m and ℓ^f, respectively, as follows:

$$\begin{aligned} \overline{\sigma}^m &= \sigma^m \sqrt{1+\ell^m} \\ \overline{\sigma}^f &= \sigma^f \sqrt{1+\ell^f} \end{aligned} \tag{86}$$

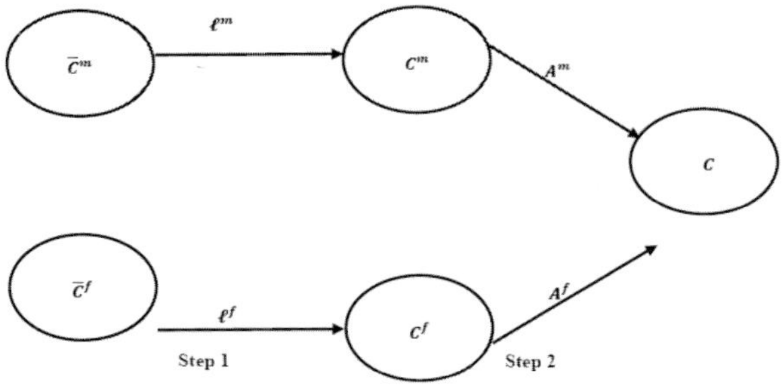

Figure 8. Schematic Diagram of the Local Approach

The above equations can be interpreted in a similar way to equation (80) except that they apply at the constituent level. They also represent the damage transformation equations for each constituent stress. In order to derive a similar transformation equation for the constituent strain, one applies the hypothesis of elastic energy equivalence to each constituent separately as follows:

$$\frac{1}{2}\sigma^m \varepsilon^m = \frac{1}{2}\bar{\sigma}^m \bar{\varepsilon}^m$$
$$\frac{1}{2}\sigma^f \varepsilon^f = \frac{1}{2}\bar{\sigma}^f \bar{\varepsilon}^f \tag{87}$$

In using equations (87), one assumes that there are no micromechanical or constituent elastic interactions between the matrix and fibers. This assumption is not valid in general. From micromechanical considerations, there should be interactions between the elastic energies of the matrix and fibers. However, such interactions are beyond the scope of this work as the resulting equations will be complicated and the sought equivalence relations will consequently be unattainable. It should be clear to the reader that equations (87) are the single most important assumption that is needed to prove the equivalence between the overall and local (constituent) approaches. In fact, the proposed proof hinges entirely on the assumption given by equations (87). Therefore, the sought equivalence between overall and local (constituent) approaches is a very special case when equations (87) are valid.

Substituting equations (86) into equations (87) and simplifying, one obtains the required transformation equation for the constituent strains as follows:

$$\bar{\varepsilon}^m = \frac{\varepsilon^m}{\sqrt{1+\ell^m}}$$
$$\bar{\varepsilon}^f = \frac{\varepsilon^f}{\sqrt{1+\ell^f}} \tag{88}$$

The above equations imply a decoupling between the elastic energy in the matrix and fibers. Other methods may be used that include some form of coupling but they will lead to complicated transformation equations that are beyond the scope of this work.

Substituting equations (86) and (88) into equations (74) and simplifying, one obtains:

$$\sigma^m = E^m \varepsilon^m$$
$$\sigma^f = E^f \varepsilon^f \tag{89}$$

where the constituent elasticity moduli E^m and E^f are given by:

$$E^m = \frac{\overline{E}^m}{1+\ell^m} \tag{90}$$
$$E^f = \frac{\overline{E}^f}{1+\ell^f}$$

Equations (89) and (90) represent the elasticity equations for the damaged constituents. The second step of the formulation involves transforming equations (90) into the whole composite system using the law of mixtures as follows:

$$\sigma = c^m \sigma^m + c^f \sigma^f \tag{91}$$

where c^m and c^f are the matrix and fiber volume fractions, respectively, in the damaged composite system. Before proceeding with equation (91), one needs to derive strain constituent equations similar to equations (76). Substituting equations (82) and (88) into equations (76) and simplifying, one obtains:

$$\varepsilon^m = A^m \varepsilon \tag{92}$$
$$\varepsilon^f = A^f \varepsilon$$

where the constituent strain concentration factors A^m and A^f in the damaged state are given by:

$$A^m = \overline{A}^m \sqrt{\frac{1+\ell^m}{1+\ell}} \tag{93}$$
$$A^f = \overline{A}^f \sqrt{\frac{1+\ell^f}{1+\ell}}$$

The above equations represent the damage transformation equations for the strain concentration factors. Finally, one substitutes equations (83), (89), and (92) into equation (91) to obtain:

$$E = \frac{c^m \overline{E}^m A^m}{1+\ell^m} + \frac{c^f \overline{E}^f A^f}{1+\ell^f} \tag{94}$$

Equation (94) represents the elastic modulus in the damaged composite system according to the overall approach.

3.4.3. Equivalence of the Two Approaches

In this section, it is shown that both the overall and local approaches are equivalent for elastic composites. This is performed by showing that both the elasticity moduli given in equations (85) and (94) are the same. In fact, it is shown that equation (94) reduces to equation (85) after making the appropriate substitutions.

First, one needs to find a damage transformation equation for the volume fractions. This is performed by substituting equations (80) and (86) into equation (77), simplifying and comparing the result with equation (91). One therefore obtains:

$$c^m = \overline{c}^m \sqrt{\frac{1+\ell^m}{1+\ell}} \qquad (95)$$

$$c^f = \overline{c}^f \sqrt{\frac{1+\ell^f}{1+\ell}}$$

Substituting equations (90) and (93) into equation (94) and simplifying, one obtains:

$$E = \frac{c^m \overline{E}^m \overline{A}^m}{\sqrt{(1+\ell^m)(1+\ell)}} + \frac{c^f \overline{E}^f \overline{A}^f}{\sqrt{(1+\ell^f)(1+\ell)}} \qquad (96)$$

Finally, one substitutes equations (95) into equation (96) and simplifies to obtain:

$$E = \frac{\overline{c}^m \overline{E}^m \overline{A}^m + \overline{c}^f \overline{E}^f \overline{A}^f}{1+\ell} \qquad (97)$$

It is clear now that the above equation is exactly the same as equation (85). Therefore, both the overall and local approaches yield the same elasticity modulus in the damaged composite system.

The mechanics of small damage are presented using a consistent mathematical and mechanical framework based on the equations of damage mechanics. In this regard, the new scalar damage variable is investigated in detail. In the second half of this work, two different approaches to damage characterization in composite materials are illustrated. It is shown that both the overall and local approaches to damage yield exactly the same results for elastic composite materials. The investigation in this work has been carried out using scalars and it is hoped that the generalization of this theory to general states of deformation and damage involving tensors will be carried out in a forthcoming work.

4. A Comparative Study of Damage Variables in Continuum Damage Mechanics

In this section various definitions of the damage variables are examined and compared. In particular, special emphasis is given to a new damage variable that is defined in terms of the elastic stiffness of the material. Both the scalar and tensorial cases are investigated. The scalar definition of the new damage variable was used recently by many researchers. However, the generalization to tensors and general states of deformation and damage is new (Voyiadjis and Kattan, 2009). In addition, transformation laws for various elastic constants are derived. Finally, the cases of plane stress, plane strain, and isotropic elasticity are examined in detail. In these cases it is shown that two independent damage parameters are needed to describe the complete state of damage in isotropic materials.

In this work a physical basis is sought for the damage tensor $[M]$ that is used to link the damage state of the material with effective undamaged configuration. The authors and numerous other researchers have used different paths including fabric tensors (Voyiadjis and Kattan, 2006a; Voyiadjis et. al. 2007) to connect the two configurations. However, the approach presented here provides for a strong physical basis for this missing link.

4.1. Scalar Damage Variables

In this section one discusses two major scalar damage variables used by researchers today. For other damage variables, see the works of Lemaitre (1971, 1984) and Lemaitre and Chaboche (1990). The first scalar damage variable is defined in terms of cross-sectional area reduction while the second scalar damage variable is defined in terms of the reduction in the elastic modulus or elastic stiffness.

Consider a body (in the form of a cylinder) in the initial undeformed and undamaged configuration. Consider also the configuration of the body that is both deformed and damaged after a set of external agencies act on it (see Figure 9). Next, consider a fictitious configuration of the body obtained from the damaged configuration by removing all the damage that the body has undergone, i.e. this is the state of the body after it had only deformed without damage (see Figure 8). Therefore, in defining a damage variable φ, its value must vanish in the fictitious configuration.

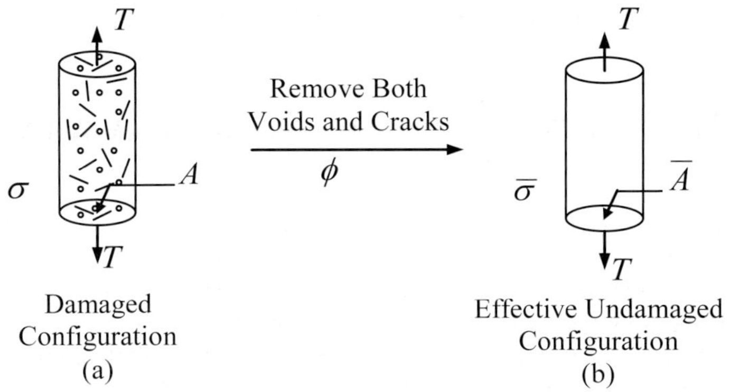

Figure 9. Damaged and Effective Undamaged Configurations

The fundamental damage variable is usually defined as follows:

$$\phi = \frac{A - \overline{A}}{A} \tag{98}$$

where A is the cross-sectional area in the damaged configuration while $\overline{A}$ is the cross-sectional area in the fictitious configuration with $A > \overline{A}$. It is clear that when a body is undamaged, i.e. when $A = \overline{A}$, then $\phi = 0$. The stress in the fictitious configuration is called the effective stress and is denoted by $\overline{\sigma}$. The value of the effective stress $\overline{\sigma}$ may be obtained using the relation $\overline{\sigma}\,\overline{A} = \sigma A$ where σ is the stress in the damaged configuration. Therefore, using this relation along with the definition in equation (98), one obtains:

$$\overline{\sigma} = \frac{\sigma}{1-\phi} \tag{99}$$

It should be mentioned that the equilibrium condition in the paragraph above reflects a mean-field type of assumption on the stress redistribution (uniform over the resistive section) and therefore appears to be appropriate only in the dilute damage regime, away from the stress-strain peak where cooperate effects dominate and damage localization takes place. This holds true especially for the tensorial case.

In order to compute the effective elastic modulus $\overline{E}$ in this case, one needs to make a certain assumption regarding the strains in the two configurations. Usually, one of the following two hypotheses is assumed:

(1) Hypothesis of elastic strain equivalence: in this case, the strain in the damaged state ε is assumed to be equal to the effective strain $\overline{\varepsilon}$ (in the fictitious state).

(2) Hypothesis of elastic energy equivalence: in this case the elastic strain energy is assumed to be equal in both configurations.

While both types of hypotheses are usually used by researchers in the field of damage mechanics, it is believed that the hypothesis of elastic energy equivalence is more valid than the hypothesis of elastic strain equivalence, primarily because it involves some form of energy formulation. Therefore, it is recommended to use the second hypothesis listed above.

The second scalar damage variable ℓ may be defined in terms of the reduction in the elastic modulus as follows:

$$\ell = \frac{\overline{E} - E}{E} \tag{100}$$

where E is the elastic modulus in the damaged state while $\overline{E}$ is the effective elastic modulus (in the fictitious state) with $\overline{E} > E$. The definition of the alternative damage variable of equation (100) may be re-written in the following more appropriate form:

$$\overline{E} = E(1 + \ell) \tag{101}$$

It is clear from the definition in equation (100) that $\ell = 0$ when the body is undamaged, i.e. when $\overline{E} = E$.

4.1.1. Hypothesis of Elastic Strain Equivalence

Using the hypothesis of elastic strain equivalence one assumes:

$$\overline{\varepsilon} = \varepsilon \tag{102}$$

Using the elastic constitutive relations in both configurations as follows:

$$\sigma = E\varepsilon \tag{103}$$

and

$$\overline{\sigma} = \overline{E}\,\overline{\varepsilon} \tag{104}$$

and substituting for $\overline{\sigma}$ and $\overline{\varepsilon}$ using equations (99) and (102), respectively, into equation (104), one compares the result with equation (103) to obtain:

$$\overline{E} = \frac{E}{1-\phi} \tag{105}$$

In order to find an appropriate relation between the two damage variables ϕ and ℓ, one equates equations (101) and (105) to obtain:

$$\ell = \frac{\phi}{1-\phi} \tag{106}$$

The above expression defines the exact relation between the two damage variables ℓ and ϕ in the case of the hypothesis of elastic strain equivalence. The relation in equation (106) may be re-written in the following form:

$$\phi = \frac{\ell}{1+\ell} \tag{107}$$

One now plots the relation of equation (106) between ϕ and ℓ as shown in Figure 10.

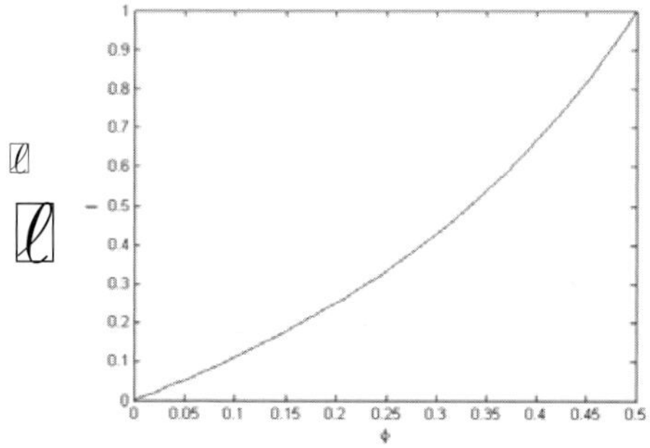

Figure 10. Relation between the Two Damage Variables in the Case of Elastic Strain Equivalence

In this case, it is clear that $\ell = 0$ when $\phi = 0$. However, when one sets ℓ to its maximum value of 1, one obtains $\phi = 0.5$ so one has the valid ranges $0 \le \ell \le 1$ and $0 \le \phi \le 0.5$. This is obviously a limiting maximum value for ϕ based on the definition of stiffness reduction of ℓ. Again, one should stress that this holds for the hypothesis of elastic strain equivalence. It should be noted that equation (100) does not impose any limitations on the value of ℓ. However, it is assumed that the damage variable ℓ has a fractional numerical value between 0 and 1. Thus, the statement that ℓ has a maximum value of 1 arises based on this assumption.

4.1.2. Hypothesis of Elastic Energy Equivalence

Using the hypothesis of elastic energy equivalence one assumes the complementary elastic strain energy ($\frac{\sigma^2}{2E}$) to be equal in both configurations, i.e.

$$\frac{\sigma^2}{2E} = \frac{\overline{\sigma}^2}{2\overline{E}} \tag{108}$$

Substituting for $\overline{\sigma}$ from equations (99) into equation (108) and simplifying, one obtains:

$$\overline{E} = \frac{E}{(1-\phi)^2} \tag{109}$$

Next, in order to derive the relation between the two damage variables ϕ and ℓ, one equates equations (101) and (109) to obtain:

$$\ell = \frac{\phi(2-\phi)}{(1-\phi)^2} \tag{110}$$

The above expression defines the exact relation between the two damage variables ℓ and ϕ in the case of the hypothesis of elastic energy equivalence. The relation in equation (110) may be re-written in the following form:

$$\phi = \frac{\sqrt{1+\ell}-1}{\sqrt{1+\ell}} \tag{111}$$

One now plots the relation of equation (110) between ϕ and ℓ as shown in Figure 11. In this case, it is clear that $\ell = 0$ when $\phi = 0$.

However, when one sets ℓ to its maximum value of 1, one obtains $\phi = 0.293 \approx 0.3$, so one has the valid ranges $0 \le \ell \le 1$ and $0 \le \phi \le 0.293$. This is obviously a limiting maximum value for ϕ based on the definition of stiffness reduction of ℓ. Again, one stresses that this holds for the hypothesis of elastic energy equivalence. It should be noted that equation (100) does not impose any limitations on the value of ℓ. However, it is assumed that the damage variable ℓ has a fractional numerical value between 0 and 1. Thus, the statement that ℓ has a maximum value of 1 arises based on this assumption.

Table 1 summarizes the results for the two hypotheses.

Table 1. Limiting Values of ϕ and ℓ

		Hypothesis of Elastic Strain Equivalence	Hypothesis of Elastic Energy Equivalence
ℓ	Min. Value	0.0	0.0
	Max. Value	1.0	1.0
ϕ	Min. Value	0.0	0.0
	Max. Value	0.5	0.293

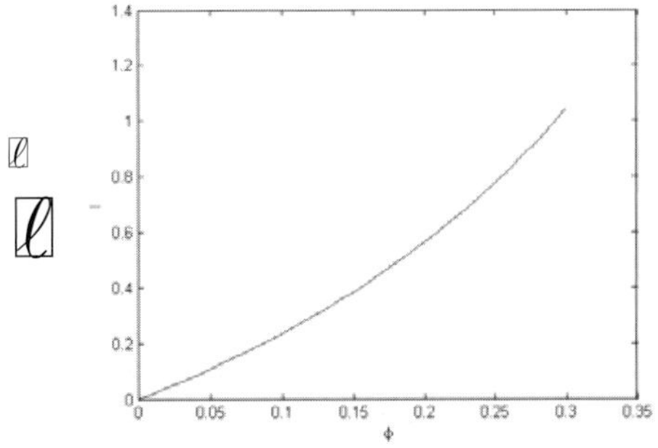

Figure 11. Relation between the Two Damage Variables for the Case of Elastic Energy Equivalence

4.2. Other Scalar Damage Variables

In this section one defines three other scalar damage variables in terms of the three elastic constants: shear modulus G, Poisson's ratio ν, and the bulk modulus K. One also derives the relations between these three variables and ℓ.

One now defines the scalar damage variable m in terms of the reduction in the shear modulus G as follows:

$$m = \frac{\overline{G} - G}{G} \tag{112}$$

where $\overline{G}$ is the effective shear modulus (in the fictitious configuration). The definition of equation (112) may be re-written in the following more appropriate form:

$$\overline{G} = G(1 + m) \tag{113}$$

with $0 \leq m \leq 1$.

One now defines the scalar damage variable n in terms of the reduction in Poisson's ratio ν as follows:

$$n = \frac{\overline{\nu} - \nu}{\nu} \tag{114}$$

where $\overline{\nu}$ is the effective Poisson's ratio (in the fictitious configuration). The definition of equation (114) may be re-written in the following more appropriate form:

$$\overline{\nu} = \nu(1 + n) \tag{115}$$

with $0 \leq n \leq 1$.

Next, onel derives the exact relation between the three damage variables $\ell, m,$ and n. In order to accomplish this, one will use the relation between the three elastic constants in both configurations as follows:

$$G = \frac{E}{2(1 + \nu)} \tag{116}$$

and

$$\overline{G} = \frac{\overline{E}}{2(1+\overline{\nu})} \tag{117}$$

It should be noted that equations (116) and (117) apply only in the case of isotropic solids and isotropic damage. Dividing equation (117) by equation (116), using the definitions in equations (101) and (113), and simplifying, one obtains the following relation:

$$\frac{1+\overline{\nu}}{1+\nu} = \frac{1+\ell}{1+m} \tag{118}$$

Next, one substitutes the definition of equation (115) into equation (118). Simplifying the resulting equation, one obtains:

$$n = \left(\frac{1+\nu}{\nu}\right)\left(\frac{\ell - m}{1+m}\right) \tag{119}$$

The above equation provides the exact relation between the three damage variables $\ell, m,$ and n.

Considering equation (118), one notes that when $\ell > m$ (i.e. the damage due to E is greater than the damage due to G), then $\overline{\nu} > \nu$. This means that in this case there is a reduction in Poisson's ratio ν due to damage effects of E and G. On the other hand, when $\ell < m$ (i.e. the damage due to E is less than the damage due to G), then $\overline{\nu} < \nu$. This means that in this case there is an increase in Poisson's ratio ν due to damage effects of E and G. The effect of damage on the elastic constants including Poisson's ratio has been investigated by Ju (1989, 1990), Ladeveze and Lemaitre(1984) , Ladeveze et al. (1982) and Allix et al.(1989) .

Next, one considers the special case when there is no damage in the shear modulus, i.e. when $m = 0$ (i.e. $\overline{G} = G$). In this special case, one assumes that G is unaffected by the damage process. Therefore, equations (118) and (119) reduce to:

$$\frac{1+\overline{\nu}}{1+\nu} = 1 + \ell \tag{120}$$

$$n = \ell\left(\frac{1+\nu}{\nu}\right) \tag{121}$$

In this special case and since $\ell > 0$, it is clear from equation (120) that there will be a reduction in the value of ν due to the damage effects of E, i.e. $\overline{\nu} > \nu$, even though G is unaffected.

Another special case is when $\ell = m$ (i.e. when the damage due to E is equal to the damage due to G), then equation (120) becomes

$$\overline{\nu} = \nu \tag{122}$$

This means that in this case there is no reduction or change in Poisson's ratio, i.e. $n = 0$. This is also evident from equation (119). Next, one defines the scalar damage variable p due to the reduction in the bulk modulus K. One now defines p as follows:

$$p = \frac{\overline{K} - K}{K} \tag{123}$$

where $\overline{K}$ is the effective bulk modulus (in the fictitious configuration). The definition of equation (123) may be re-written in the following more appropriate form:

$$\overline{K} = K(1+p) \tag{124}$$

where $0 \le p \le 1$.

Using the following relations between the elastic constants involving K in both configurations

$$K = \frac{GE}{9G - 3E} \tag{125}$$

$$\overline{K} = \frac{\overline{G}\overline{E}}{9\overline{G} - 3\overline{E}} \tag{126}$$

and dividing equation (126) by equation (125) while using the definition in equations (101), (113), and (124), along with equations (116) and (117), one obtains the following relation:

$$\frac{1-2\overline{\nu}}{1-2\nu} = \frac{1+\ell}{1+p} \tag{127}$$

Substituting for $\overline{\nu}$ using equation (115), one obtains:

$$p = \frac{(1-2\nu)\ell + 2\nu n}{1-2\nu(1+n)} \tag{128}$$

Equation (128) represents the exact relation between the three damage variables $\ell, n,$ and p Considering equation (31), one obtains the following three special cases:

(1) When $\ell = n$ (i.e. damage effect due to E is equal to the damage effect due to ν), then

$$p = \frac{1}{(1-2\nu)(1+n)} \tag{129}$$

This means that in this case, the damage effect due to K depends only on the damage effect due to ν .

(2) When $n = 0$, (i.e. damage effect due to ν is nonexistent), then one obtains $p = \ell$. This means that in this case, the damage effect due to K is equal to the damage effect due to E .

(3) When $\ell = 0$, (i.e. when there is no damage due to E), then one obtains:

$$p = \left(\frac{2\nu}{1-2\nu}\right)\left(\frac{n}{1+n}\right) \tag{130}$$

This means that in this case, the damage effect due to K depends only on the damage effect due to ν.

4.3. Stresses and Strains

In this section, one will derive the relations between the stresses and strains in the two configurations. One now has two cases as follows:

(1) For the hypothesis of elastic strain equivalence and using equations (102), one obtains:

$$\bar{\sigma} = \sigma(1+\ell) \tag{131}$$

(2) For the hypothesis of elastic energy equivalences one may obtain the following two relations:

$$\bar{\sigma} = \sigma\sqrt{1+\ell} \tag{132}$$

and

$$\bar{\varepsilon} = \frac{\varepsilon}{\sqrt{1+\ell}} \tag{133}$$

The various scalar damage variables used in this section are summarized as shown in Table 2.

Table 2: Elastic Constants and Corresponding Damage Variables

Elastic Constant	Corresponding Damage Variable
E	ℓ
ν	n
G	m
K	p

4.4. Tensorial Damage Variables

In this section, one generalizes the scalar damage variables appearing in equations (99) and (100). One now defines the corresponding tensorial damage variables that can be used in general states of deformation and damage. Tensors will now be used instead of scalars. The tensors are represented by vectors and matrices. In this work, a vector will be denoted by braces and a tensor will be denoted by brackets.

Equation (99) is usually generalized in the damage mechanics literature as follows (Lemaitre and Chaboche, 1990; Voyiadjis and Kattan, 1990, 1992, 1999, 2005, 2006a; Kattan and Voyiadjis, 1990, 1993.):

$$\{\overline{\sigma}\} = [M]\{\sigma\} \tag{134}$$

where $\{\sigma\}$ and $\{\overline{\sigma}\}$ are the vectorial representations of the stress tensor and effective stress tensor, respectively, and $[M]$ is the matrix representation of the damage effect tensor $[M]$. The damage effect tensor $[M]$ is a fourth-rank tensor that is defined in the literature (Voyiadjis and Kattan, 1999, 2005, 2006a).

The scalar damage variable ℓ defined in equation (100) can be easily generalized to three dimensions. The generalization of equation (100) is proposed using two different ways as follows:

$$[L]^{(1)} = \left([\overline{E}] - [E]\right)[E]^{-1} \tag{135}$$

$$[L]^{(2)} = [E]^{-1}\left([\overline{E}] - [E]\right) \tag{136}$$

All the matrices appearing in equations (135) and (136) are representations of fourth-rank tensors. The matrices $[E]$ and $[\overline{E}]$ are representations of the fourth-rank elasticity tensor and its effective counterpart. The matrices $[L]^{(1)}$ and $[L]^{(2)}$ are two different representations of the fourth-rank generalized damage tensor corresponding to the scalar damage variable ℓ. It is seen from equations (135) and (136) that both $[L]^{(1)}$ and $[L]^{(2)}$ are defined in terms of the reduction of the elastic stiffness as was the case with ℓ.

One can also define a fourth-rank damage tensor by taking the average of $[L]^{(1)}$ and $[L]^{(2)}$ as follows:

$$[L] = \frac{1}{2}\left([L]^{(1)} + [L]^{(2)}\right) \tag{137}$$

where $[L]$ is the matrix representation of the new generalized fourth-rank damage tensor. Substituting equations (135) and (136) into equation (137) and simplifying, one obtains the following explicit expression for $[L]$:

$$[L] = \frac{1}{2}\left([\overline{E}][E]^{-1} + [E]^{-1}[\overline{E}]\right) - [I] \tag{138}$$

where $[I]$ is the matrix representation of the fourth-rank identity tensor. Next, one studies the relationship between the fourth-rank damage tensors $[L]$ and $[M]$

. Actually, one will study the relationship between $[L]^{(1)}$ and $[M]$ on the one hand, and between $[L]^{(2)}$ and $[M]$ on the other hand. In this investigation, one will use the generalizations of the elastic constitutive relations of equations (103) and (104) as follows:

$$\{\sigma\} = [E]\{\varepsilon\} \tag{139}$$

$$\{\bar{\sigma}\} = [\bar{E}]\{\bar{\varepsilon}\} \tag{140}$$

where $\{\varepsilon\}$ and $\{\bar{\varepsilon}\}$ are the vectorial representations of the strain tensor and the effective strain tensor, respectively. The relations will be obtained for the two cases of the hypothesis of elastic strain equivalence and the hypothesis of elastic energy equivalence. One can re-write the definitions of equations (135) and (136) in the following more appropriate form:

$$[\bar{E}] = ([I] + [L]^{(1)})[E] \tag{141}$$

$$[\bar{E}] = [E]([I] + [L]^{(2)}) \tag{142}$$

4.4.1. Hypothesis of Elastic Strain Equivalence

In this case, the generalization of equation (102) becomes:

$$\{\bar{\varepsilon}\} = \{\varepsilon\} \tag{143}$$

Substituting equations (141) and (143) into equation (140) and simplifying along with using equation (139) one obtains:

$$\{\bar{\sigma}\} = ([I] + [L]^{(1)})\{\sigma\} \tag{144}$$

Comparing the above equations with equation (134), one obtain:

$$[M]^{(1)} = [I] + [L]^{(1)} \tag{145}$$

where $[M]^{(1)}$ is a fourth-rank damage tensor corresponding to $[L]^{(1)}$. Performing the same operations above again but using equation (142) instead of equation (141), one obtains:

$$\{\overline{\sigma}\} = \left([E]\left([I] + [L]^{(2)}\right)[E]^{-1}\right)\{\sigma\} \tag{146}$$

Comparing the above equation with equation (134), one concudes:

$$[M]^{(2)} = [E]\left([I] + [L]^{(2)}\right)[E]^{-1} \tag{147}$$

where $[M]^{(2)}$ is a fourth-rank damage tensor corresponding to $[L]^{(2)}$. Substituting equations (135) and (136) into equations (146) and (147), respectively, and simplifying, one obtains the following explicit expressions for $[M]^{(1)}$ and $[M]^{(2)}$:

$$[M]^{(1)} = [\overline{E}][E]^{-1} \tag{148}$$

$$[M]^{(2)} = [\overline{E}][E]^{-1} \tag{149}$$

Thus, it is clear that one obtains the same expression for both $[M]^{(1)}$ and $[M]^{(2)}$. This may mean that the definitions in equations (135) and (136) are equivalent. However, the authors are not aware of a mathematical method to show this equivalence except as shown above. Therefore, based on equations (148) and (149), one can define the fourth-rank damage tensor $[M]$ of equation (134) as follows:

$$[M] = [\overline{E}][E]^{-1} \tag{150}$$

In the special case where there is no damage in the material, $[\overline{E}] = [E]$. The damage tensor $[M]$ reduces to the fourth-rank identity tensor $[I]$. This is also consistent with the definition of equation (134).

4.4.2. Hypothesis of Elastic Energy Equivalence

In this case, one generalizes equation (108) as follows:

$$\frac{1}{2}\{\sigma\}^T[E]^{-1}\{\sigma\} = \frac{1}{2}\{\overline{\sigma}\}^T[\overline{E}]^{-1}\{\overline{\sigma}\} \tag{151}$$

Substituting for $\{\overline{\sigma}\}$ and $[\overline{E}]$ using equations (134) and (141), respectively, and simplifying, one obtains:

$$[E] = [M]^{(1)^{-1}}\left([I] + [L]^{(1)}\right)[E][M]^{(1)^{-T}} \tag{152}$$

where one substitutes $[M]^{(1)}$ for $[M]$ to correspond to $[L]^{(1)}$. Performing the same operations again on equation (151) but this time using $[M]^{(2)}$ along with equation (142), one obtains:

$$[E]=[M]^{(2)^{-1}}[E]([I]+[L]^{(2)})[M]^{(2)^{-T}} \tag{153}$$

It is clear from equations (152) and (153) that one cannot solve for an explicit expression for $[M]^{(1)}$ or $[M]^{(2)}$ in terms of $[E]$ and $[\overline{E}]$. Finally, substituting equations (135) and (136) into equations (152) and (153), respectively, one obtains implicit expressions for $[M]^{(1)}$ and $[M]^{(2)}$ in terms of $[E]$ and $[\overline{E}]$:

$$[E]=[M]^{(1)^{-1}}[\overline{E}][M]^{(1)^{-T}} \tag{154}$$

$$[E]=[M]^{(2)^{-1}}[\overline{E}][M]^{(2)^{-1}} \tag{155}$$

Thus, it is clear that the above two expressions are identical for $[M]^{(1)}$ and $[M]^{(2)}$. Therefore, based on equations (154) and (155), one can define the fourth-rank damage tensor $[M]$ of equations (134) as follows:

$$[E]=[M]^{-1}[\overline{E}][M]^{-T} \tag{156}$$

The above relation is a well known expression in the damage mechanics literature. In the special case where there is no damage in the material, $[\overline{E}]=[E]$, the damage tensor $[M]$ reduces to the fourth-rank identity tensor $[I]$. This is also consistent with the definition of equation (134).

Finally, one will write equations (135) and (136) in the following simplified form:

$$[L]^{(1)}=[\overline{E}][E]^{-1}-[I] \tag{157}$$

$$[L]^{(2)}=[E]^{-1}[\overline{E}]-[I] \tag{158}$$

Next, one illustrates the previous equations using three different examples. The special cases of isotropic elasticity, plane stress, and plane strain are now studied.

4.4.3. Isotropic Elasticity

In the case of isotropic elasticity, one has the following matrix representations for the elasticity tensors $[E]$ and $[\overline{E}]$:

$$[\mathrm{E}]^{-1} = \begin{bmatrix} \frac{1}{E} & \frac{-\nu}{E} & \frac{-\nu}{E} & 0 & 0 & 0 \\ \frac{-\nu}{E} & \frac{1}{E} & \frac{-\nu}{E} & 0 & 0 & 0 \\ \frac{-\nu}{E} & \frac{-\nu}{E} & \frac{1}{E} & 0 & 0 & 0 \\ 0 & 0 & 0 & \frac{1}{2G} & 0 & 0 \\ 0 & 0 & 0 & 0 & \frac{1}{2G} & 0 \\ 0 & 0 & 0 & 0 & 0 & \frac{1}{2G} \end{bmatrix} \tag{159}$$

$$[\overline{\mathrm{E}}] = \begin{bmatrix} \overline{\lambda}+2\overline{\mu} & \overline{\lambda} & \overline{\lambda} & 0 & 0 & 0 \\ \overline{\lambda} & \overline{\lambda}+2\overline{\mu} & \overline{\lambda} & 0 & 0 & 0 \\ \overline{\lambda} & \overline{\lambda} & \overline{\lambda}+2\overline{\mu} & 0 & 0 & 0 \\ 0 & 0 & 0 & 2\overline{\mu} & 0 & 0 \\ 0 & 0 & 0 & 0 & 2\overline{\mu} & 0 \\ 0 & 0 & 0 & 0 & 0 & 2\overline{\mu} \end{bmatrix} \tag{160}$$

where $\overline{\lambda}$ and $\overline{\mu}$ are the effective Lame's constants given by:

$$\overline{\lambda} = \frac{\overline{\nu}\overline{\mathrm{E}}}{(1+\overline{\nu})(1-2\overline{\nu})} \tag{161}$$

$$\overline{\mu} = \overline{G} = \frac{\overline{\mathrm{E}}}{2(1+\overline{\nu})} \tag{162}$$

Substituting the matrices of equations (159) and (160) into equation (157), along with using equations (161), (162), (101), (115), (118), and (127), and simplifying, one obtains explicit expressions for the components of the fourth-rank damage tensor $[L]^{(1)}$ as follows:

$$[L]^{(1)} = \begin{bmatrix} \ell_{11} & \ell_{12} & \ell_{13} & \ell_{14} & \ell_{15} & \ell_{16} \\ \ell_{21} & \ell_{22} & \ell_{23} & \ell_{24} & \ell_{25} & \ell_{26} \\ \ell_{31} & \ell_{32} & \ell_{33} & \ell_{34} & \ell_{35} & \ell_{36} \\ \ell_{41} & \ell_{42} & \ell_{43} & \ell_{44} & \ell_{45} & \ell_{46} \\ \ell_{51} & \ell_{52} & \ell_{53} & \ell_{54} & \ell_{55} & \ell_{56} \\ \ell_{61} & \ell_{62} & \ell_{63} & \ell_{64} & \ell_{65} & \ell_{66} \end{bmatrix} \tag{163}$$

where

$$\ell_{11} = \frac{(1+p)(1+n)v + (1+\ell)}{1+(1+n)v} - 1 \tag{164}$$

$$\ell_{12} = \frac{(p+n-pn-\ell)v}{1+(1+n)v} \tag{165}$$

$$\ell_{13} = \ell_{12} = \ell_{21} = \ell_{23} = \ell_{32} = \ell_{31} \tag{166}$$

$$\ell_{22} = \ell_{11} = \ell_{33} \tag{167}$$

$$\ell_{14} = \ell_{41} = \ell_{15} = \ell_{51} = \ell_{16} = \ell_{61} = 0 \tag{168}$$

$$\ell_{24} = \ell_{42} = \ell_{25} = \ell_{52} = \ell_{26} = \ell_{62} = 0 \tag{169}$$

$$\ell_{34} = \ell_{43} = \ell_{35} = \ell_{53} = \ell_{36} = \ell_{63} = 0 \tag{170}$$

$$\ell_{44} = m \tag{171}$$

$$\ell_{55} = \ell_{66} = \ell_{44} = m \tag{172}$$

$$\ell_{45} = \ell_{54} = \ell_{46} = \ell_{64} = \ell_{56} = \ell_{65} = 0 \tag{173}$$

Therefore, the matrix representation of $[L]^{(1)}$ of equation (163) reduces to the following:

$$[L]^{(1)}=\begin{bmatrix} \ell_{11} & \ell_{12} & \ell_{12} & 0 & 0 & 0 \\ \ell_{12} & \ell_{11} & \ell_{12} & 0 & 0 & 0 \\ \ell_{12} & \ell_{12} & \ell_{11} & 0 & 0 & 0 \\ 0 & 0 & 0 & \ell_{44} & 0 & 0 \\ 0 & 0 & 0 & 0 & \ell_{44} & 0 \\ 0 & 0 & 0 & 0 & 0 & \ell_{44} \end{bmatrix} \tag{174}$$

where ℓ_{11}, ℓ_{12} and ℓ_{44} are given by:

$$\ell_{11} = \frac{(1+p)(1+n)v + (1+\ell)}{1+(1+n)v} - 1 \tag{175}$$

$$\ell_{12} = \frac{(p+n-pn-\ell)v}{1+(1+n)v} \tag{176}$$

$$\ell_{44} = m \tag{177}$$

Notice that $\ell_{11} = \ell$ when $v = 0$. This is consistent with the scalar formulation described previously in Section 4.1. It may appear from equation (174) that one needs three independent damage parameters $(\ell_{11}, \ell_{12}, \ell_{44})$ to describe the complete state of damage for the case of isotropic elasticity. But one will show now that only ℓ_{11} and ℓ_{12} are independent. That is, one will show that ℓ_{44} depends on ℓ_{11} and ℓ_{12} and one only needs two independent damage parameters (ℓ_{11}, ℓ_{12}) to describe the complete state of damage for this case of isotropic elasticity.

One now assumes ℓ_{44} to be written in the following form:

$$\ell_{44} = \alpha\ell_{11} + \beta\ell_{12} \tag{178}$$

where α and β are functions of the scalar damage variables m, n, ℓ, and p. Substituting equations (175), (176), and (177) into equations (178) and manipulating the result, one can solve for α and β as follows:

$$\alpha = \frac{m}{\ell} \tag{179}$$

$$\beta = \frac{m(1+n)(\ell - p)}{\ell(p + n - pn - \ell)} \tag{180}$$

Thus ℓ_{44} depends on ℓ_{11} and ℓ_{12}. Therefore, one needs only two independent damage parameters to describe the complete state of damage in the case of isotropic elasticity.

4.4.4 Plane Stress

For the special case of plane stress, one has the following matrix representation for the elasticity tensors $[\mathrm{E}]$ and $[\overline{\mathrm{E}}]$:

$$[\mathrm{E}]^{-1} = \begin{bmatrix} \frac{1}{E} & \frac{-\nu}{E} & 0 \\ \frac{-\nu}{E} & \frac{1}{E} & 0 \\ 0 & 0 & \frac{1}{G} \end{bmatrix} \tag{181}$$

$$\left[\overline{E}\right] = \frac{\overline{E}}{1-\overline{\nu}^2} \begin{bmatrix} 1 & \overline{\nu} & 0 \\ \overline{\nu} & 1 & 0 \\ 0 & 0 & \frac{1-\overline{\nu}}{2} \end{bmatrix} \tag{182}$$

Substituting the matrices of equations (181) and (182) into equation (157) along with using equations (101) and (115), and simplifying, one obtains explicit expressions for the components of the fourth-rank damage tensor $[L]^{(1)}$ as follows:

$$[L]^{(1)} = \begin{bmatrix} \ell_{11} & \ell_{12} & \ell_{13} \\ \ell_{21} & \ell_{22} & \ell_{23} \\ \ell_{31} & \ell_{32} & \ell_{33} \end{bmatrix} \tag{183}$$

where

$$\ell_{11} = \frac{(1+\ell)\left[1-(1+n)v^2\right]}{1-(1+n)^2 v^2} - 1 \tag{184}$$

$$\ell_{22} = \ell_{11} \tag{185}$$

$$\ell_{12} = \frac{(1+\ell)\,v\,n}{1-(1+n)^2 v^2} \tag{186}$$

$$\ell_{21} = \ell_{12} \tag{187}$$

$$\ell_{33} = m \tag{188}$$

$$\ell_{13} = \ell_{23} = \ell_{31} = \ell_{32} = 0 \tag{189}$$

Thus, the matrix representation of $[L]^{(1)}$ of equation (183) reduces to:

$$[L]^{(1)} = \begin{bmatrix} \ell_{11} & \ell_{12} & 0 \\ \ell_{12} & \ell_{11} & 0 \\ 0 & 0 & \ell_{33} \end{bmatrix} \tag{190}$$

where

$$\ell_{11} = \frac{(1+\ell)\left[1-(1+n)v^2\right]}{1-(1+n)^2 v^2} - 1 \tag{191}$$

$$\ell_{12} = \frac{(1+\ell)\nu n}{1-(1+n)^2 \nu^2} \tag{192}$$

$$\ell_{33} = m \tag{193}$$

Notice that $\ell_{11} = \ell$ when $\nu = 0$. This is consistent with the scalar formulation described previously in Section 4.1. It may appear from equation (190) that one needs three independent damage parameters $(\ell_{11}, \ell_{12}, \ell_{33})$ to describe the complete state of damage for the case of plane stress. But one can now show that only ℓ_{11} and ℓ_{12} are independent. Thus, one can show that ℓ_{33} depends on ℓ_{11} and ℓ_{12} and one only needs two independent damage parameters (ℓ_{11}, ℓ_{12}) to describe the complete state of damage for this case of plane stress.

The component ℓ_{33} is assumed to be written in the following form:

$$\ell_{33} = \alpha \ell_{11} + \beta \ell_{12} \tag{194}$$

where α and β are functions of the scalar damage variables m, n, ℓ, and p. Substituting equations (191), (192), and (193) into equations (194) and manipulating the result, one can solve for α and β as follows:

$$\alpha = \frac{m}{\ell} \tag{195}$$

$$\beta = 0 \tag{196}$$

Thus ℓ_{33} depends on ℓ_{11} and ℓ_{12}. Therefore, one needs only two independent damage parameters to describe the complete state of damage in the case of plane stress.

4.4.5. Plane Strain

For the case of plane strain, one has the following matrix representations for the elasticity tensors $[E]$ and $[\overline{\mathrm{E}}]$:

$$[E]^{-1} = \begin{bmatrix} \frac{1}{E} & \frac{-v}{E} & 0 \\ \frac{-v}{E} & \frac{1}{E} & 0 \\ 0 & 0 & \frac{1}{G} \end{bmatrix} \tag{197}$$

$$[\overline{E}] = \frac{\overline{E}}{(1+\overline{v})(1-2\overline{v})} \begin{bmatrix} 1-\overline{v} & \overline{v} & 0 \\ \overline{v} & 1-\overline{v} & 0 \\ 0 & 0 & \frac{1-2\overline{v}}{2} \end{bmatrix} \tag{198}$$

Substituting the matrices of equations (197) and (198) into equation (157) along with using equations (101) and (115), and simplifying, one obtains explicit expressions for thc components of the fourth-rank damage tensor $[L]^{(1)}$ as follows:

$$[L]^{(1)} = \begin{bmatrix} \ell_{11} & \ell_{12} & \ell_{13} \\ \ell_{21} & \ell_{22} & \ell_{23} \\ \ell_{31} & \ell_{32} & \ell_{33} \end{bmatrix} \tag{199}$$

where

$$\ell_{11} = \frac{(1+\ell)\left[1-(1+n)v-(1+n)v^2\right]}{[1+(1+n)v][1-2(1+n)v]} - 1 \tag{200}$$

$$\ell_{22} = \ell_{11} \tag{201}$$

$$\ell_{12} = \frac{(1+\ell)\left[-v+(1+n)v+(1+n)v^2\right]}{[1+(1+n)v][1-2(1+n)v]} \tag{202}$$

$$\ell_{21} = \ell_{12} \tag{203}$$

$$\ell_{13} = \ell_{31} = \ell_{23} = \ell_{32} = 0 \tag{204}$$

$$\ell_{33} = m \tag{205}$$

Thus, the matrix representation of $[L]^{(1)}$ of equation (199) reduces to:

$$[L]^{(1)} = \begin{bmatrix} \ell_{11} & \ell_{12} & 0 \\ \ell_{12} & \ell_{11} & 0 \\ 0 & 0 & \ell_{33} \end{bmatrix} \tag{206}$$

where

$$\ell_{11} = \frac{(1+\ell)\left[1-(1+n)v-(1+n)v^2\right]}{[1+(1+n)v][1-2(1+n)v]} - 1 \tag{207}$$

$$\ell_{12} = \frac{(1+\ell)\left[-v+(1+n)v+(1+n)v^2\right]}{[1+(1+n)v][1-2(1+n)v]} \tag{208}$$

$$\ell_{33} = m \tag{209}$$

Notice that $\ell_{11} = \ell$ when $v = 0$. This is consistent with the scalar formulation described previously in Section 4.1. It may appear from equation (206) that one needs three independent damage parameters $(\ell_{11}, \ell_{12}, \ell_{33})$ to describe the complete state of damage for the case of plane strain. But one canl now show that only ℓ_{11} and ℓ_{12} are independent. Thus, one can show that ℓ_{33} depends on ℓ_{11} and ℓ_{12} and one only needs two independent damage parameters (ℓ_{11}, ℓ_{12}) to describe the complete state of damage for this case of plane strain.

One now assumes ℓ_{33} to be written in the following form:

$$\ell_{33} = \alpha \ell_{11} + \beta \ell_{12} \tag{210}$$

where α and β are functions of the scalar damage variables m, n, ℓ, and p. Substituting equations (191), (192), and (193) into equations (194) and manipulating the result, one can solve for α and β as follows:

$$\alpha = 1 \tag{211}$$

$$\beta = -1 \tag{212}$$

Thus ℓ_{33} depends on ℓ_{11} and ℓ_{12}. Therefore, one needs only two independent damage parameters to describe the complete state of damage in the case of plane strain. Finally it should be noted that one has solved all three cases and derived explicit expressions for the components of the fourth-rank damage tensor $[L]^{(1)}$ using equation (157). One can also use equation (158) to derive explicit expressions for the components of the fourth-rank damage tensor $[L]^{(2)}$. However, in this case, since all the matrices involved are symmetric, one will obtain the same results as $[L]^{(1)}$. Thus $[L]^{(2)}=[L]^{(1)}$ for the three cases of isotropic elasticity, plane stress, and plane strain.

Alternatively, one can use equations (150) and (156) to derive explicit expressions for the components of the fourth-rank damage tensor $[M]$. For the case of the hypothesis of elastic strain equivalence of equation (150), it is very simple and simple explicit expressions for $[M]$ may be obtained. However, for the case of the hypothesis of elastic energy equivalence of equation (156), it is not simple at all. In this case, the equation for $[M]$ is implicit and one will get a set of simultaneous linear algebraic equations that needs to solved in order to obtain the components of $[M]$. It is recommended to use one of the computer algebra systems in this case to obtain the results.

Finally, based on equation (135) and (136), one can write the following relation between the damage tensors $[L]^{(1)}$ and $[L]^{(2)}$:

$$[L]^{(2)}=[E]^{-1}[L]^{(1)}[E] \tag{213}$$

Several definitions of the damage variables have been investigated using both scalar and tensorial formulations. In the scalar case, one derives the exact relationships between several scalar damage variables. Also, one derives an upper bound for the classical scalar damage variable ϕ.

In the tensorial case, two possible generalizations of the scalar damage variable ℓ are proposed that are defined in terms of the reduction of the elastic modulus. The exact relationship between the fourth-rank damage tensors $[L]^{(1)}$, $[L]^{(2)}$ and $[M]$ are also derived. Finally, three examples are solved in detail obtaining explicit expressions for the components of the damage tensor for the special cases of isotropic elasticity, plane stress, and plane strain. In these cases it is shown that only two independent damage parameters are needed to describe the complete state of damage in the material.

In this work (Voyiadjis and Kattan, 2009) a physical basis is sought for the damage tensor $[M]$ that is used to link the damage state of the material with effective undamaged configuration. The authors and numerous other researchers have used different paths including fabric tensors (Voyiadjis and Kattan, 2006) to connect the two configurations. However, the approach presented here provides for a strong physical basis for this missing link. In particular, the authors have made an important link between the damage tensor and fabric tensors (Voyiadjis and Kattan, 2006, 2007, 2007, 2007, Voyiadjis, et al., 2007, 2007).

5. Computational aspects of the proposed theory

The development of computational algorithms that are consistent with the proposed theoretical formulation is given in detail in this section. The problem of numerically integrating the constitutive equations in the context of the finite element method is also addressed. The standard return mapping algorithms of rate-independent problems are extended to rate-dependent problems. Moreover, since the numerical implementation of the gradient-dependent constitutive equations is not a direct task, because of the higher order of the governing equations, a direct and simple computational algorithm for the gradient approach is utilized. This algorithm can be implemented in the existing finite element code.

5.1. Numerical Integration

For numerical procedures employed in the analysis of elasto-viscoplastic problems, it is necessary to integrate the constitutive equations governing the material behavior. In this section, the numerical integration of the nonlocal geometrically nonlinear thermo-viscoinelatic model presented in section 2 is developed. Let t_o, t_1, ... , t_n, $t_{n+1} = t_n + \Delta t$, ... be convenient time instances along the time interval over which the dynamic response of the body is sought. Consider the time step $\Delta t = t_{n+1} - t_n$: at $t = t_n$ where all quantities are known, which are the converged values of the previous step, and the solution must be computed at t_{n+1} for a given body load increment, $\Delta \boldsymbol{b}$, and surface load increment, $\Delta \boldsymbol{t}$.

The dynamic evolution of a hypoelastic-thermo-viscoplastic and thermo-viscodamaged body of volume V and surface S is governed at time step $n+1$, by the constitutive relations presented in the previous section and by the following momentum, initial, and compatibility relations:

$$\boldsymbol{L}^T \boldsymbol{\tau}_{n+1} + \rho_o \boldsymbol{b}_{n+1} = \rho_o \dot{\upsilon}_{n+1} \text{ in } V; \; \boldsymbol{t}_{n+1} = \boldsymbol{\tau}_{n+1} \boldsymbol{n} \text{ on } S_t \tag{214}$$

$$\boldsymbol{u} = \boldsymbol{u}_o, \quad \boldsymbol{\upsilon} = \boldsymbol{\upsilon}_o \quad \text{at } t = t_o$$

$$\boldsymbol{l}_{n+1} = \nabla \boldsymbol{\upsilon}_{n+1} = \boldsymbol{C} \boldsymbol{\upsilon}_{n+1} \text{ in } V$$

$$\boldsymbol{u}_{n+1} = \tilde{\boldsymbol{u}} \text{ on } S_u; \; \boldsymbol{\upsilon}_{n+1} = \tilde{\boldsymbol{\upsilon}} \text{ on } S_\upsilon; \; T_{n+1} = \tilde{T} \text{ on } S_T$$

where $(\bullet)_{n+1} = (\bullet)_n + \Delta(\bullet)$ is the additive decomposition of each of the internal variables. For algorithmic convenience, the authors have shifted to matrix vector notation in this section. $\boldsymbol{L}$ is the differential operator, $\boldsymbol{b}$ and $\boldsymbol{t}$ are the body force and the surface traction vectors, respectively, $\boldsymbol{u}$ is the three-component displacement vector, and $\boldsymbol{n}$ denotes the outward normal to the surface S. The boundaries S_u, S_υ, and S_T are parts of the boundary where the displacement $\tilde{\boldsymbol{u}}$, the velocity $\tilde{\boldsymbol{\upsilon}}$, and the temperature $\tilde{T}$ is prescribed, respectively. It is clear that $S_t \cup S_u \cup S_\upsilon \cup S_T = S$ and $S_t \cap S_u = 0$.

In the context of the finite element method, the discrete problem can be obtained via a spatial displacement–based projection of the semidiscrete (i.e. discrete in space and continuous in time) problem into a finite dimensional subspace of admissible continuous shape functions. Consequently, in the following sections the procedure for solving the derived set of governing equations using the finite element method is described thoroughly. In order to integrate the set of constitutive equations, a return mapping algorithm is developed in the subsequent sections.

5.2. Return Mapping Algorithm

Considering a given configuration of known set of positions $\boldsymbol{X}$ at time t_n, the problem is now to update all state variables to a new configuration defined by its respective set of positions $\boldsymbol{x}$ (which are supposed to be known) at time t_{n+1}. This situation typically arises in a nonlinear finite element problem where the new positions $\boldsymbol{x}$ are determined from the discretized version of the momentum equation, $(214)_1$

In this section a semi-implicit stress integration algorithm for rate-dependent problems (Voyiadjis, et al., 2006) is recalled. This stress update algorithm treats the rate-independent and rate-dependent problems in a unified way. Moreover, fully nonlocal coupled viscoplastic-viscodamage constitutive equations with a two-step predictor-corrector structure is presented: hypoelastic predictor and coupled viscoplastic-viscodamage corrector (Zhu and Cescetto, 1995). In the model presented in this paper, there exists two coupled surfaces, f and g, and for each iteration f and g should be corrected simultaneously (Figure 12). The different steps of the integration algorithm are detailed below.

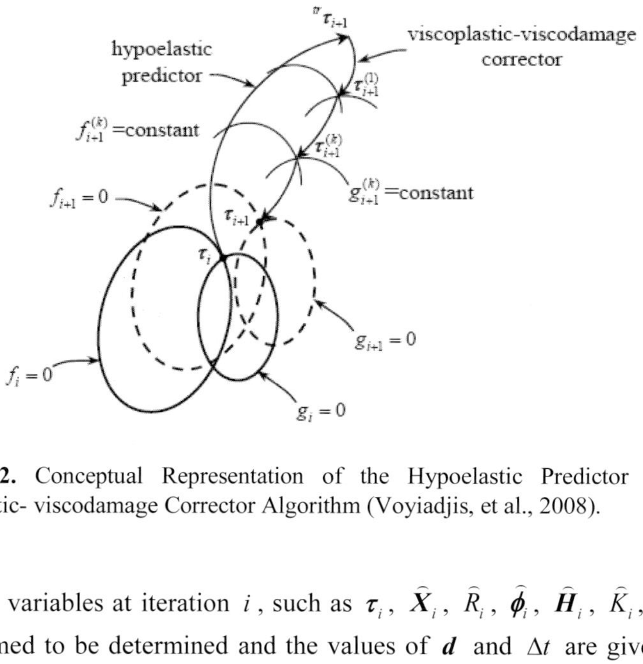

Figure 12. Conceptual Representation of the Hypoelastic Predictor and Coupled Viscoplastic- viscodamage Corrector Algorithm (Voyiadjis, et al., 2008).

If the variables at iteration i, such as $\boldsymbol{\tau}_i$, $\widehat{\boldsymbol{X}}_i$, $\widehat{R}_i$, $\widehat{\boldsymbol{\phi}}_i$, $\widehat{\boldsymbol{H}}_i$, $\widehat{K}_i$, $\widehat{\boldsymbol{Y}}_i$, and T_i, are assumed to be determined and the values of $\boldsymbol{d}$ and Δt are given, then $\boldsymbol{\tau}_{i+1}$ that satisfies the discretized constitutive equations can be obtained. In the following, a hypoelastic predictor and coupled viscoplastic-viscodamage corrector is proposed. In the first step, the hypoelastic predictor problem is solved with initial conditions that are the converged values of the previous iteration i while keeping the irreversible variables frozen. This produces a trial stress state, $\boldsymbol{\tau}^{tr}$, which, if outside the viscoplastic surface f and the viscodamage surface g is taken as the initial conditions for the solution of the viscoplastic-viscodamage corrector problem. The scope of this second step is to restore the generalized consistency condition by returning back the trial stress to the viscoplastic surface f and the viscodamage surface g simultaneously as conceptually represented in Figure 12.

However, one of the major challenges while integrating the constitutive equations in finite deformation context is to achieve the incremental objectivity, i.e. to maintain correct rotational transformation properties all along a finite time step. A procedure that has now become very popular is first to rewrite the constitutive equations in a corotational moving frame. Therefore, Voyiadjis et al. (2006) showed that by assuming that the variables of the model at iteration i and the displacement field $\boldsymbol{u} = \boldsymbol{x}_{i+1} - \boldsymbol{x}_i$ at iteration $i+1$ are known, the trial elastic stress can then be given in the corotational frame by

$$^{tr}\boldsymbol{\tau}_{i+1} = \boldsymbol{R}\left(\boldsymbol{\tau}_i + \square_i : \boldsymbol{E}\right)\boldsymbol{R}^T \tag{215}$$

where $\boldsymbol{R}$ is the polar decomposition ($\boldsymbol{F} = \boldsymbol{R}.\boldsymbol{U}$) rotation matrix and $\boldsymbol{E}$ is the (incremental) logarithmic strain tensor between the reference configuration and the current one, which is given by

$$\boldsymbol{E} = \ln \boldsymbol{U} = \frac{1}{2}\ln \boldsymbol{U}^2 = \frac{1}{2}\ln\left(\boldsymbol{F}^T\boldsymbol{F}\right) \tag{216}$$

In the above procedure, it is essential to realize that $\boldsymbol{F} = \boldsymbol{R}.\boldsymbol{U}$ are incremental tensors; in the case of rigid body motion, $\ln \boldsymbol{U} = \boldsymbol{0}$, thus the stress tensor will be updated exactly by the relation $\boldsymbol{\tau}_{n+1} = \boldsymbol{R}\boldsymbol{\tau}_n\boldsymbol{R}^T$ whatever the amplitude of the rotation. The rotation tensor $\boldsymbol{R}$ is directly and exactly computed from the polar decomposition and $\boldsymbol{R}$ only needs to be evaluated once per time step. All kinematic quantities are based on the deformation gradient $\boldsymbol{F}$ over the considered time step, a quantity that is readily available in nonlinear finite element codes.

5.3. Coupled Viscoplastic-Viscodamage Corrector algorithms

If $f\left({}^{tr}\tau_{i+1}, \widehat{\boldsymbol{X}}_i, \widehat{R}_i, \dot{\hat{p}}, T_i, \widehat{\boldsymbol{\phi}}_i\right) \le 0$ and $g\left({}^{tr}\widehat{Y}_{i+1}, \widehat{\boldsymbol{H}}_i, \widehat{K}_i, \dot{\hat{p}}_i, T_i, \widehat{\boldsymbol{\phi}}_i\right) \le 0$, the process is clearly undamaged elastic and the trial stress is in fact the final state. On the other hand, if $f > 0$ and $g > 0$, the Kuhn-Tucker loading/unloading conditions, Eqs. (11), are violated by the trial stress which now lies outside f and g (see Figure 12). Consistency, is restored by a generalization of the classical return mapping algorithm to rate-dependent problems (Voyiadjis et al., 2006). Since the objective rates reduce to a simple time derivative due to the fact that the global configuration is held fixed, the coupled viscoplastic-viscodamage corrector problem may then be rephrased as:

$$\dot{\boldsymbol{\tau}} = -\square : \left(\boldsymbol{d}^{vp} + \boldsymbol{d}^{vd}\right) - \boldsymbol{A} : \dot{\boldsymbol{\phi}} - \square : \alpha_t \dot{T}\mathbf{1} \tag{217}$$

The hypoelastic predictor and coupled viscoplastic-viscodamage corrector step yields the final stress as:

$$\boldsymbol{\tau}_{i+1} = {}^{tr}\boldsymbol{\tau}_{i+1} - \square_i : \left(\boldsymbol{d}^{vp} + \boldsymbol{d}^{vd}\right) - \boldsymbol{A}_i : \dot{\boldsymbol{\phi}} - \square_i : \alpha_t \dot{T}\mathbf{1} \tag{218}$$

where $\dot{T}$ is calculated from the heat balance equation, In Equations. (11-12), $\dot{\lambda}^{vp}$ and $\dot{\lambda}^{vd}$ are calculated using the nonlocal computational algorithm presented in (Abu Al-Rub and Voyiadjis, 2005, Voyiadjis and Abu Al-Rub, 2006). The temperature evolution can be expressed as follow

$$\dot{T} = \frac{1}{\rho_o c_p} J^e P \mathbf{1} : \boldsymbol{d} + Q_1^{tp}\, \dot{\lambda}^{vp} + Q_1^{td}\, \dot{\lambda}^{vd} \tag{219}$$

where Q_1^{tp} and Q_1^{td} are obtained from the previous iteration i and are given by

$$Q_1^{tp} = \frac{1}{\rho_o c_p} \Upsilon \boldsymbol{\tau}' : \frac{\partial f}{\partial \boldsymbol{\tau}}, \quad Q_1^{td} = \frac{1}{\rho_o c_p} \Upsilon \boldsymbol{\tau}' : \frac{\partial f}{\partial \boldsymbol{Y}} \tag{220}$$

Moreover, one requires the satisfaction of the generalized viscoplasticity consistency condition $\dot{f}$ at the end of iteration $i+1$, such that

$$\frac{\partial f}{\partial \bar{\boldsymbol{\tau}}} : \overset{\nabla}{\bar{\boldsymbol{\tau}}} + \frac{\partial f}{\partial \bar{\bar{\boldsymbol{X}}}} : \overset{\nabla}{\bar{\bar{\boldsymbol{X}}}} + \frac{\partial f}{\partial \bar{\bar{R}}} \dot{\bar{\bar{R}}} + \frac{1}{\Delta t} \frac{\partial f}{\partial \dot{\bar{\bar{p}}}} \dot{\bar{\bar{p}}} + \frac{\partial f}{\partial T} \dot{T} = 0 \tag{221}$$

Since the local iteration process is applied within the time step $t + \Delta t$ (i.e. at step $n+1$) and the updated Lagrangian formulation is used then $\dot{\bar{\bar{p}}} = \ddot{\bar{\bar{p}}}\Delta t$ is used in obtaining Eq.(221). Substitution of the rate equations from (15), (16), (217), and (219) into Eq. (221) yields the following expression:

$$\left(\frac{\partial f}{\partial \boldsymbol{\tau}} : \boldsymbol{C} + \mathbf{Z}^p\right) : \boldsymbol{d} + Q_1^p \dot{\lambda}^{vp} + Q_2^p \nabla^2 \dot{\lambda}^{vp} + Q_3^p \dot{\lambda}^{vd} + Q_4^p \nabla^2 \dot{\lambda}^{vd} = 0 \tag{222}$$

where $\mathbf{Z}^p$ and Q_k^p ($k = 1, \ldots, 4$) are obtained from the previous i iteration and are given by:

$$\boldsymbol{Z}^{p} = -\frac{\partial f}{\partial T}.\left(J^{e}P + T\alpha_{t}\right)\mathbf{1} \tag{223}$$

$$\begin{aligned} Q_1^p &= Q_1^{tp} - \frac{\partial f}{\partial \boldsymbol{\tau}} : \left(\boldsymbol{C} : \frac{\partial f}{\partial \boldsymbol{\tau}} + \boldsymbol{A} : \frac{\partial f}{\partial \boldsymbol{Y}} \right) + \left[\frac{\partial f}{\partial p} + \frac{1}{\Delta t}\frac{\partial f}{\partial \dot{p}} \right] \frac{\left(1 - k_1 \bar{\bar{R}}\right)}{1 - \hat{r}} + \\ & \left[\frac{\partial f}{\partial \nabla^2 p} + \frac{1}{\Delta t}\frac{\partial f}{\partial \nabla^2 \dot{p}} \right] \times \left[\frac{\nabla^2 r}{\left(1-\hat{r}\right)^2}\left(1 - k_1 \bar{\bar{R}}\right) - \frac{k_1}{1-\hat{r}} \nabla^2 \bar{R} \right] - \\ & \frac{1}{a_2}\frac{\partial f}{\partial \bar{\boldsymbol{X}}} : \hat{\boldsymbol{M}} : \left(\frac{\partial f}{\partial \bar{\boldsymbol{X}}} + k_2 \bar{\bar{\boldsymbol{X}}} \right) - \frac{2}{a_2 \ell^2}\frac{\partial f}{\partial \nabla^2 \bar{\boldsymbol{X}}} : \\ & \left[\nabla^2 \hat{\boldsymbol{M}} : \left(\frac{\partial f}{\partial \bar{\boldsymbol{X}}} + k_2 \bar{\bar{\boldsymbol{X}}} \right) + k_2 \hat{\boldsymbol{M}} : \nabla^2 \bar{\boldsymbol{X}} \right] + \frac{\partial f}{\partial \boldsymbol{\phi}} : \frac{\partial f}{\partial \boldsymbol{Y}} \end{aligned} \tag{224}$$

$$\begin{aligned} Q_2^p &= \left[\frac{\partial f}{\partial \nabla^2 p} + \frac{1}{\Delta t}\frac{\partial f}{\partial \nabla^2 \dot{p}} \right] \frac{\left(1 - k_1 \bar{\bar{R}}\right)}{1-\hat{r}} - \frac{2}{a_2 \ell^2}\frac{\partial f}{\partial \nabla^2 \bar{\boldsymbol{X}}} : \hat{\boldsymbol{M}} : \left(\frac{\partial f}{\partial \bar{\boldsymbol{X}}} + k_2 \bar{\bar{\boldsymbol{X}}} \right) + \\ & \frac{\partial f}{\partial \nabla^2 \boldsymbol{\phi}} : \frac{\partial f}{\partial \boldsymbol{Y}} \end{aligned} \tag{225}$$

$$Q_3^p = Q_1^{td} - \frac{\partial f}{\partial \boldsymbol{\tau}} : \left(\boldsymbol{C} : \frac{\partial g}{\partial \boldsymbol{\tau}} + \boldsymbol{A} : \frac{\partial g}{\partial \boldsymbol{Y}} \right) + \frac{\partial f}{\partial \boldsymbol{\phi}} : \frac{\partial g}{\partial \boldsymbol{Y}} \tag{226}$$

$$Q_4^p = \frac{\partial f}{\partial \nabla^2 \boldsymbol{\phi}} : \frac{\partial g}{\partial \boldsymbol{Y}} \tag{227}$$

Similarly, the generalized viscodamage consistency condition $\dot{g}$, equatios (11-12), needs to be satisfied. Since the viscodamage driving forces $\hat{\boldsymbol{Y}}$ is a function of $\boldsymbol{\tau}$ and $\hat{\boldsymbol{\phi}}$ one can then express $\dot{g}$ as follows:

$$\dot{g} \equiv \frac{\partial g}{\partial \boldsymbol{\tau}} : \overset{\nabla}{\boldsymbol{\tau}} + \frac{\partial g}{\partial \hat{\boldsymbol{\phi}}} : \overset{\nabla}{\hat{\boldsymbol{\phi}}} + \frac{\partial g}{\partial \hat{\boldsymbol{H}}} : \overset{\nabla}{\hat{\boldsymbol{H}}} + \frac{\partial g}{\partial \hat{K}}\dot{\hat{K}} + \frac{1}{\Delta t}\frac{\partial f}{\partial \dot{\hat{p}}}\dot{\hat{p}} + \frac{\partial g}{\partial T}\dot{T} = 0 \tag{228}$$

Similary substitution rate equations into equation (228) it yields the following expression:

$$\left(\frac{\partial g}{\partial \boldsymbol{\tau}}:\boldsymbol{C}+\mathbf{Z}^{d}\right):\boldsymbol{d}+Q_{1}^{d}\dot{\lambda}^{vp}+Q_{2}^{d}\nabla^{2}\dot{\lambda}^{vp}+Q_{3}^{d}\dot{\lambda}^{vd}+Q_{4}^{d}\nabla^{2}\dot{\lambda}^{vd}=0 \tag{229}$$

where $\mathbf{Z}^{d}$ and Q_{k}^{d} ($k=1,\ldots,4$) are obtained from the previous i iteration and are given by:

$$\boldsymbol{Z}^{d}=-\left(\frac{\partial g}{\partial T}-\frac{\partial g}{\partial \boldsymbol{\tau}}:\alpha_{t}\mathbf{1}\right).\left(J^{e}P+T\alpha_{t}\right)\mathbf{1} \tag{230}$$

$$Q_{1}^{d}=Q_{1}^{tp}-\frac{\partial g}{\partial \boldsymbol{\tau}}:\left(\boldsymbol{C}:\frac{\partial f}{\partial \boldsymbol{\tau}}+\boldsymbol{A}:\frac{\partial f}{\partial \boldsymbol{Y}}\right)+\frac{\partial g}{\partial \boldsymbol{\phi}}:\frac{\partial f}{\partial \boldsymbol{Y}} \tag{231}$$

$$Q_{2}^{d}=\frac{\partial g}{\partial \nabla^{2}\boldsymbol{\phi}}:\frac{\partial f}{\partial \boldsymbol{Y}} \tag{232}$$

$$Q_{3}^{d}=Q_{1}^{td}-\frac{\partial g}{\partial \boldsymbol{\tau}}:\left(\boldsymbol{C}:\frac{\partial g}{\partial \boldsymbol{\tau}}+\boldsymbol{A}:\frac{\partial g}{\partial \boldsymbol{Y}}\right)+\frac{\partial g}{\partial r}\left(1-h_{1}\widehat{K}\right)-\frac{2h_{1}}{a_{3}\ell^{2}}\frac{\partial g}{\partial \nabla^{2}r}\nabla^{2}K+\frac{1}{\Delta t}\frac{\partial g}{\partial \dot{p}}$$
$$-\frac{1}{a_{4}}\frac{\partial g}{\partial \boldsymbol{H}}:\left(\frac{\partial g}{\partial \boldsymbol{H}}+h_{2}\widehat{\boldsymbol{H}}\right)-\frac{2h_{2}}{a_{4}\ell^{2}}\frac{\partial g}{\partial \nabla^{2}\boldsymbol{H}}:\nabla^{2}\boldsymbol{H}+\frac{\partial g}{\partial \boldsymbol{\phi}}:\frac{\partial g}{\partial \boldsymbol{Y}} \tag{233}$$

$$Q_{4}^{d}=\frac{\partial g}{\partial \nabla^{2}r}\left(1-h_{1}\widehat{K}\right)+\frac{1}{\Delta t}\frac{\partial g}{\partial \nabla^{2}\dot{p}}-\frac{2h_{2}}{a_{4}\ell^{2}}\frac{\partial g}{\partial \nabla^{2}\boldsymbol{H}}\left(\frac{\partial g}{\partial \boldsymbol{H}}+h_{2}\widehat{\boldsymbol{H}}\right)+\left.\frac{\partial g}{\partial \nabla^{2}\boldsymbol{\phi}}\right|^{T}\frac{\partial g}{\partial \boldsymbol{Y}} \tag{234}$$

Due to the higher order of the governing equations, equations. (223) and (229), in the inelastic region, considerable difficulties are experienced with their numerical implementation. The consistency condition of inelasticity is no longer an algebraic equation but is a differential one because of $\nabla^{2}\dot{\lambda}^{vp}$ and $\nabla^{2}\dot{\lambda}^{vd}$, which needs to be calculated along with $\dot{\lambda}^{vp}$ and $\dot{\lambda}^{vd}$. Another complication is the higher-order boundary conditions that are necessary from the mathematical point of view and have to be prescribed on the moving elasto-inelastic boundary (Figure 12). These internal boundaries are not always easy to interpret physically. In the following, the robust numerical technique that have been developed in (Abu Al-Rub and Voyiadjis, 2005, Voyiadjis and Abu Al-Rub, 2006) is used to calculate the Laplacian terms $\nabla^{2}\dot{\lambda}^{vp}$ and $\nabla^{2}\dot{\lambda}^{vd}$.

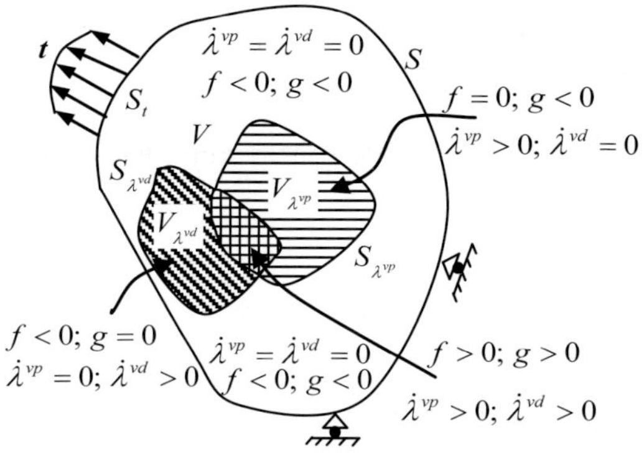

Figure 13. Schematic Representation of Elastic, Viscoplastic, and Viscodamage Boundaries (Voyiadjis et al., 2008).

6. Verification and calibration of the proposed theory

Verification of the proposed constitutive model requires that the modeling and numerical algorithms implemented are correct. In this section one point material model is used to calibrate the developed constitutive model by specifying various material constants, or constitutive parameters, for each material. Values for these parameters are obtained by fitting the predicted material behavior to the laboratory data for various load paths presented in the previous section. Some of the already established material constants are obtained from the literature by Yun and Palazotto (2007). The values of the material constants for Steel 1080 and Steel VascoMax are presented in Table 3. Finite element simulations are performed by implementing the proposed viscoplastic constitutive model in the commercial finite element code ABAQUS [6.8.3]. The material model is implemented computationally as a user defined subroutine (VUMAT) for the dynamic analysis in the FEA code (ABAQUS). User subroutines provide an extremely powerful and flexible tool for the analysis. The equations of motion are solved with the help of a finite-element package (ABAQUS), and the constitutive law by a solver for ordinary differential equations. The relevant constitutive information is passed to ABAQUS by a subroutine VUMAT which has to be supplied by the user. Starting from an equilibrium state at time, t_n ABAQUS performs an (incremental) loading as well as with the time increment Δt, and an initial guess $\Delta\varepsilon^n$, for the strain increment. The user subroutine VUMAT has to supply ABAQUS with a new Cauchy stress. Tensor $\sigma(t_n + \Delta t)$ is updated according to the constitutive law as well as with the derivative of stress with respect to the strain increment.

In the simulations, a uniaxial tension specimen is loaded with prescribed velocities and with different meshes in order to verify the appropriate material parameters for this numerical simulation. The results are compared to the experimentally obtained curves. In this regard a simple rectangular geometry is used for the uniaxial tension problem subjected to low and high velocities (Figure 14).

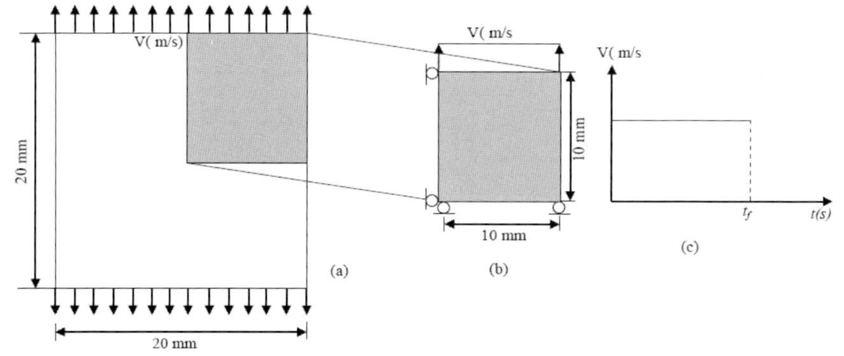

Figure 14. Problem Description for a Simple Uniaxial Tension Specimen

Due to symmetry, only a quarter of the problem is considered and solved with the mesh of one axisymetric element and boundary conditions (see Figure 14). The material parameters used in the simulation are listed in Table 3. In this simulation the effect of strain rate on the stress-strain curve is also investigated.

6.1. Calibration of the metalic material constants

The main goal of this approach is to determine the material constants and simulate the strain rate and all conditions exactly as they occur in the experiment. In the following figures the obtained results of the strain - stress relationship according to the constitutive model presented in the previous sections are presented.

Table 3: Material Constants for Steel 1080 and VascoMax for Presented Constitutive Model (Abu Al-Rub and Voyiadjis, 2005).

Material constant	**Steel 1080**	**Steel VascoMax**
Yong's Modulus	$E = 203GPa$	$E = 180GPa$
Poisson Ratio	$\upsilon = 0.3$	$\upsilon = 0.3$

Density	$\rho = 7800 kg/m^3$	$\rho = 8000 kg/m^3$
The Initial Yield Strength	$\sigma_y = 440 MPa$	$\sigma_y = 1.25 GPa$
Isotropic Hardening Modulus	$r = 12$	$r = 1$
Associated with the Kinematic Hardening Rule	$C = 12 \cdot 10^9$	$C = 2 \cdot 10^9$
	$\gamma = 40$	$\gamma = 6$
	$\beta = 0.1$	$\beta = 0.316$
Associated with the Isotropic Hardening Rule	$q_m = 300 \cdot 10^6$	$q_m = 170 \cdot 10^6$
	$q_0 = 10 \cdot 10^6$	$q_0 = 7 \cdot 10^6$
	$\mu = 10$	$\mu = 10$
Melting Temperature	$T_{melt} = 1670 \text{K}^0$	$T_{melt} = 1685 \text{K}^0$
Reference Temperature	$T_{ref} = 295 \text{K}^0$	$T_{ref} = 295 \text{K}^0$
Thermal Softening Exponent	$T_{soft} = 0.5$	$T_{soft} = 0.5$
Rate Sensitivity Exponent	$m = 1$	$m = 1$
Relaxation time/Viscosity Parameter	$T_{relax} = 0.005$	$T_{relax} = 0.005$
Specific Heat	$C_p = 460 J/kg \cdot K$	$C_p = 452 J/kg \cdot K$

The implemented theory is extremely adequate in characterizing the behavior of the material used in this research. Figure 15 presents the comparison of the uniaxial loading of a steel 1080 specimen as well as the computational simulation using the model by Abu Al-Rub and Voyiadjis (2005) and Voyiadjis and Abu Al-Rub (2003) (solid line). Significant improvement and precision in simulating the experimental approach is due to the gradient approach (Voyiadjis and Deliktas, 2009) and elaborated conditions for the yield surface. Figure 16 represents the same type of experiment; however, in this case the tested material is Steel VascoMax. Again the comparison of experiment with the computational model is very successful. Figures 17 and 18 compare the model and experimental results for the cyclic loading for both Steel 1080 and Steel VascoMax, respectively. Similar to the uniaxial case, the theory with incorporated damage and the gradient theory simulate well the experimental results for the cyclic loading. The comparison of both computational stress-strain relationships, presented in Figures 17-18, show the very good accuracy of the calibrated constitutive model and experiments conducted. There is relatively very small difference between the experimental and computational results. This way, the calibrated constitutive model is validated by the comparison to the experimental approach and therefore can be used for more sophisticated applications concerning friction and wear.

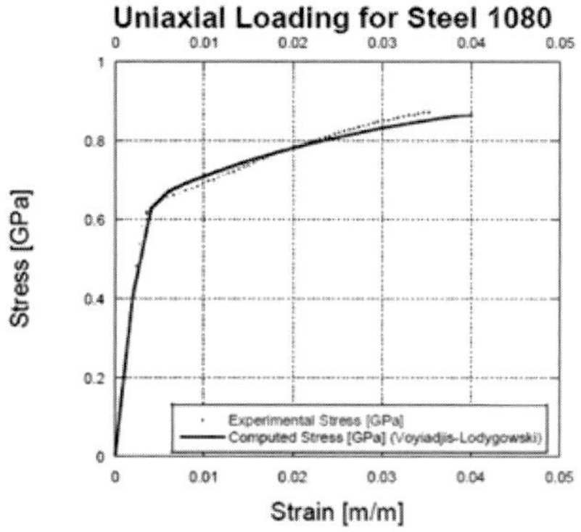

Figure 15. Comparison of Experimental and Computational Results for the Uniaxial Loading of Steel 1080 using the Proposed Constitutive Model (Voyiadjis, et al., 2010)

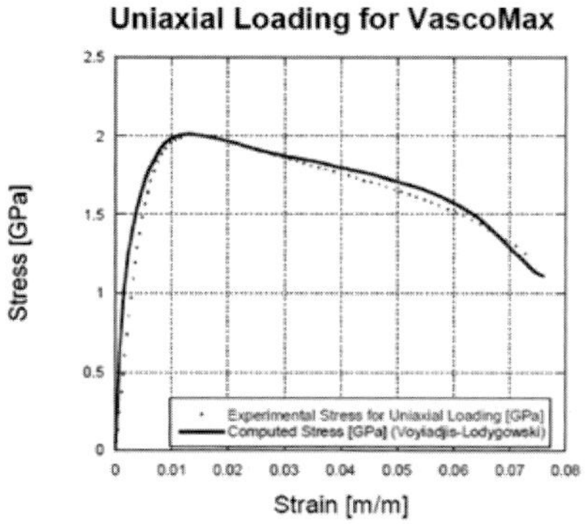

Figure 16. Comparison of Experimental and Computational Results for the Uniaxial Loading of VascoMax using the Proposed Constitutive Model (Voyiadjis, et al., 2010)

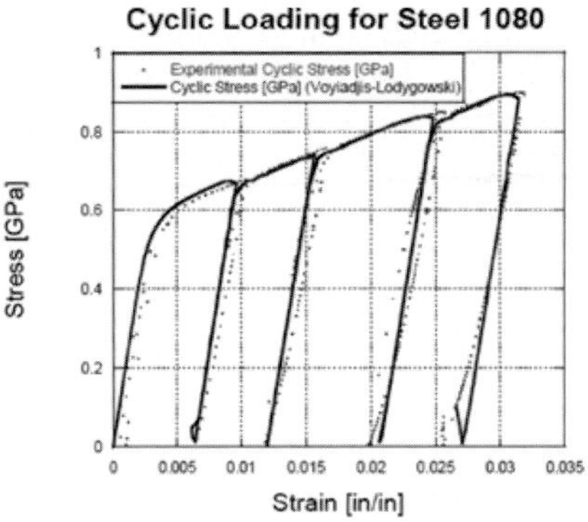

Figure 17. Comparison of Experimental and Computational Results for the Cyclic Loading of Steel 1080 using the Proposed Model (Voyiadjis, et al., 2010)

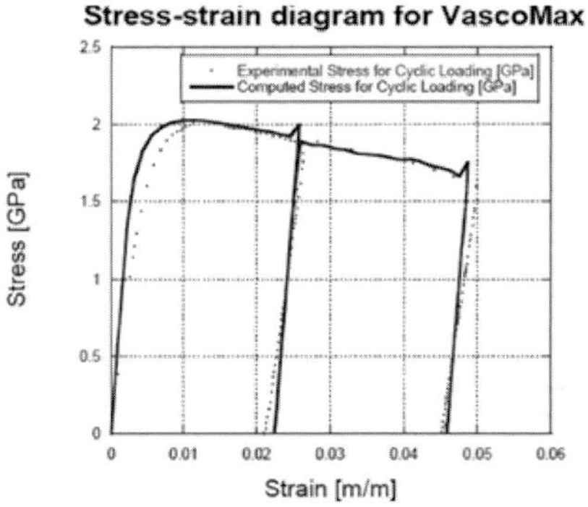

Figure 18. Comparison of Experimental and Computational Results for the Cyclic Loading of Steel 1080 using the Proposed Model (Voyiadjis, et al., 2010)

6.2. Calibration of the composite material constants

The flow chart of teh algorithm that is used for the calculations of stress and user dependent solution variables within the user material subroutine VUMAT is show in Figure 19.

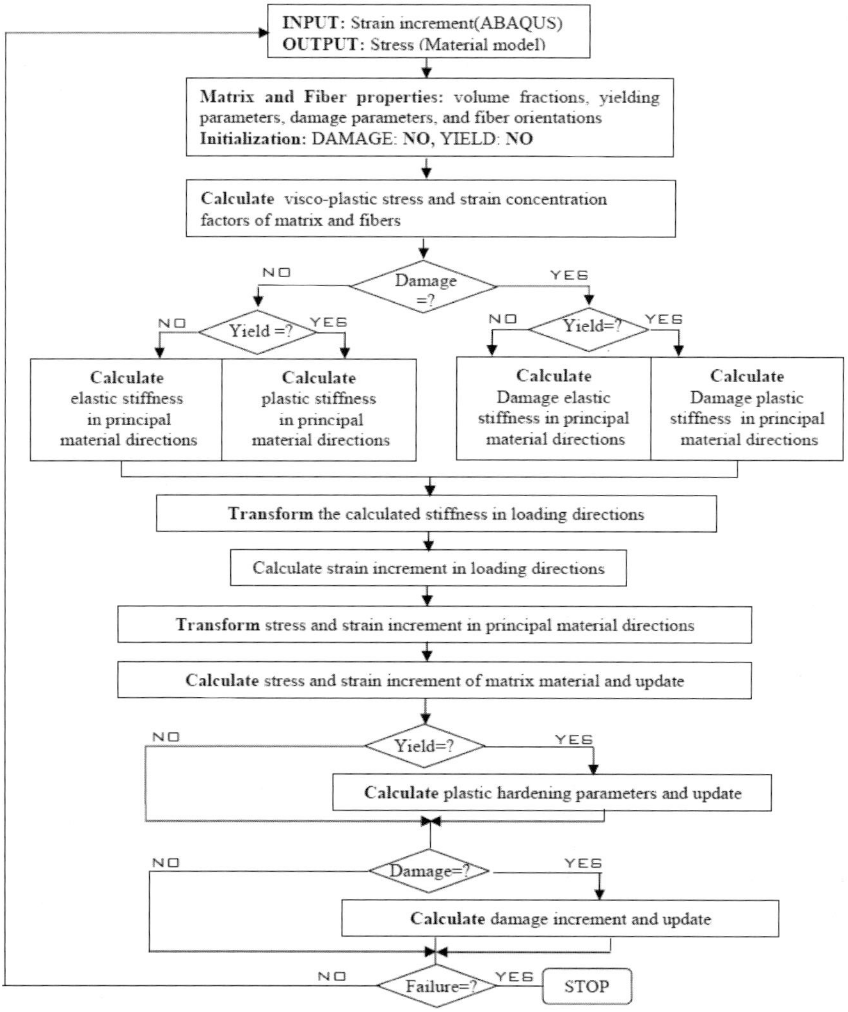

Figure 19. Flow Chart Representation of the Numerical Integration Algorithm for the Constitutive Model of the Single Lamina (Deliktas, et al., 2009).

The material parameters used in the simulation are listed in Table 4. The material constants used for constitutive model of the titanium based matrix material (TIMEAL 21S) are obtained from previous of works of Voyiadjis and coworkers (Thiagarajan and Voyiadjis, 2000, Voyiadjis, et al., 2008, Voyiadjis and Deliktas, 2000, Voyiadjis and Park, 1995, Voyiadjis and Thiagarajan, 1996)

Table 4. Material Constants of the Proposed Constitutive Model (Deliktas, et al., 2009)

Materials	Matrix	Fiber
Elastic constants	$E = 70.0$ GPa $\upsilon = .2$	$170.0GPa$ 0.29
Viscoplastic constants	$\sigma_y = 450MPa$ $b = 16.9$ $H^{vp} = 193.8MPa$ $L^{vp} = 5.85MPa$ $Q^{vp} = 300MPa$ $Q^d = 30MPa$ $n_1 = 1.0$ $n_2 = 2.7$ $k_1 = k_2 = k_3 = k_4 = 0.424$ $t^{vp} = 0.005s$	
Damage Constants	$\xi_1 = 0.92$ $\xi_2 = 0.9$ $\xi_3 = 0.9$ $\zeta_1 = 0.62$ $\zeta_2 = 0.62$ $\zeta_3 = 0.6$ $\upsilon_1 = 0.00$ $\upsilon_2 = 0.001$ $\upsilon_3 = 0.001$	

The first computer simulation is performed to show the micromechanical capability of the proposed theory wherein the matrix and the fiber are treated as two constituents with their respective constitutive models and the material behavior of the composite is obtained by using the Moi-Tanaka averaging (Mori and Tanaka, 1973) based on the Eshelby homogenization scheme . This is shown in Figure 20.

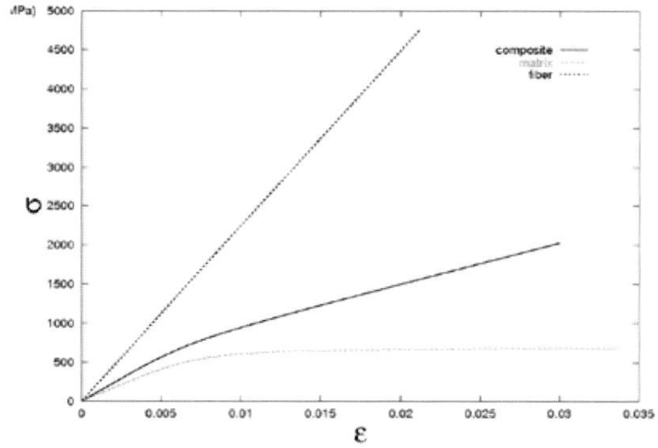

Figure 20. Uniaxial Tension Behavior of each Phases of the Composite Material (Deliktas, et al., 2009)

In these simulations the effects of the different model parameters are investigated. It is clear that the coupled viscoplastic damage model introduces the notion of the rate dependent and delayed material response. These responses are controlled by the relaxation time t^{vp} and reference temperature. Figure 21 shows the response to simple tension for different values of the relaxation time and the reference temperature.

In Figure 21 it can be seen that as the relaxation time t^{vp} increases the material response hardens its behavior. On the other hand, the uniaxial tension behavior of the proposed theory is investigated under different strain rate loadings. Experimental data for titanium based matrix material (TIMETAL 21S) have been reported by Arnold et al. (Arnold, et al., 1994) and Arnold (Arnold, 1990) and is used here for comparison purposes. Figure 22 represents the model simulation along with the experimental curves for the three strain rate loadings.

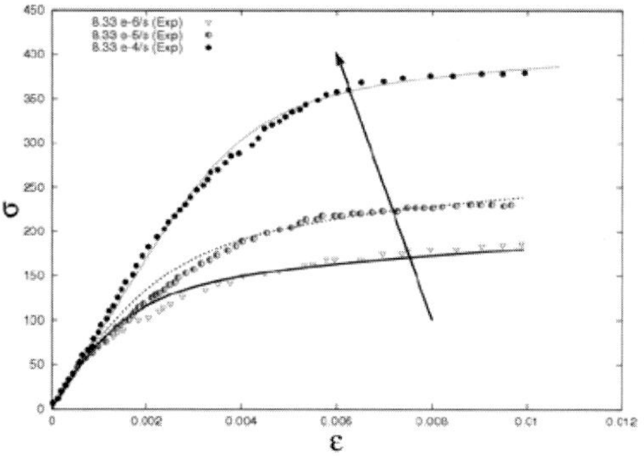

Figure 21. Uniaxial Tension Behavior under the Control of Relaxation Time Parameter $t^{vp} = 0, 0.001, 0.005$ (Deliktas, et al., 2009).

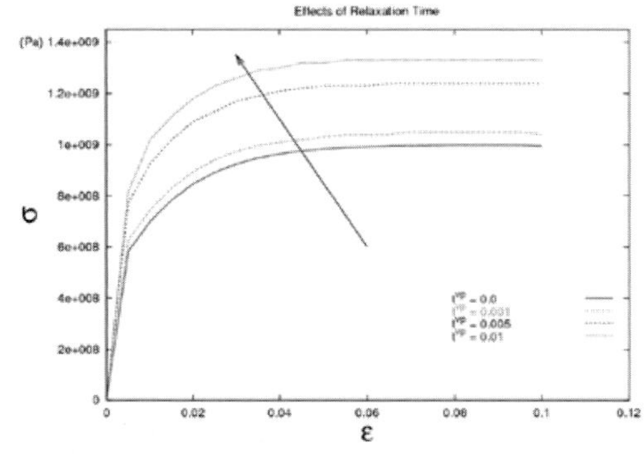

Figure 22. Model Predictions of the Proposed Model and Experimental data by Arnold et al. (1994) for the Uniaxial Tension Behavior under the Control of Different Strain Rates $\dot{p} = 0.00000833,\ 0.0000833,\ 0.000833$ (Deliktas, et al., 2009).

It is observed that at lower strain rates the material shows softer response as compared to the higher strain rates. Although the higher strain rate is well simulated, it is observed that the model predictions at the lower strain rates are sliglty different than those of experimental results. The initial yielding is observed to decrease with decreasing strain rates in the experiments as well as in the model predictions.

The cyclic response of the material is also predicted by the proposed theory in order to show the capability of the proposed theory in characterizing the kinematic and isotropic response of the material. In Figure 23 the cyclic behavior of the material is shown. The material response in a stress-controlled environment is obtained using the developed material model rouitine. The uniaxial stress is made to oscillate between $\pm$ 650 MPa and the corresponding response is plotted in Figure 23. The response of the model is stabilized after a single cycle which conforms well with the experimenta observations in the case of symmetrical cyclic loading (zero-mean stress or no ratcheting).

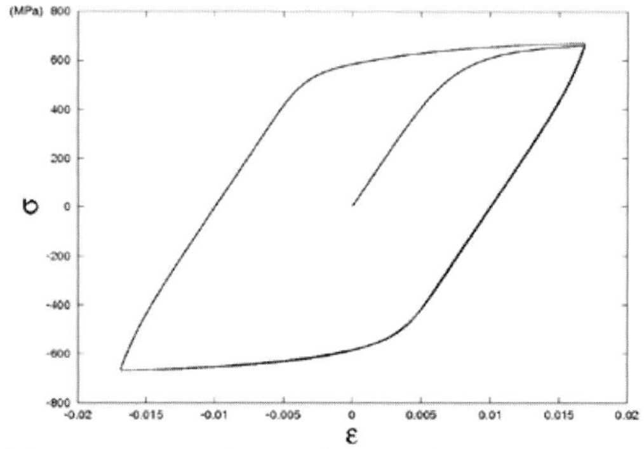

Figure 23. Material Response under Cyclic Loading (Deliktas, et al., 2009).

These simulations indicate that the proposed theory will able to predict the material characteristic behavior in the expected trend.

7. Application of the theory in studying high velocity contact problems

7.1 Simulation of experimental work on the study of friction between metallic surfaces

The experimental work on the study of friction between metallic surfaces was conducted by the authors in collaboration with the National Engineering School of Metz (ENIM), France. A modified Hopkinson's bar experiment shown schematically in Figure. 24(a), developed by Philippon at al. (2004), simulates high velocity contact between two surfaces. The experimental device, presented in Figure 25(b), has been designed to investigate dry friction between materials A (Steel 1080) and B (Steel VascoMax) which are mainly made up of two distinct parts. A dynamometer ring allows applying a known apparent normal force F_N on the contact surfaces between specimens A and B, and a thin tube, on which a set of two strain gauges is glued, that measures the tangential force due to friction. The ring is mounted on the thin tube (load sensor) and rigidly fixed by two screws. Specimen B is placed between two specimens A fixed symmetrically in the inner surfaces of the dynamometer ring. The value of the apparent normal force F_N exerted by this ring is calibrated by the thickness of specimen A and the width of specimen B.

In this experiment, an ultra high speed projectile is fired inside the tube of a gas gun towards the specimen B mounted between specimens A. During the impact it is assumed that projectile passes the entire kinetic energy on specimen B and hence the impact velocity becomes the initial velocity of the specimen B sliding between specimens A (Philippon, et al., 2003). During this sliding process the variation of contact and normal forces are recorded with the use of the set of oscilloscopes. The contact force exerted on the dynamometer (in Figure 24(b)), can be predicted by measuring the change in length of the friction device using strain gauges. Moreover, the latter (along with the sliding surface) are processed using SEM to characterize the topography of the before/after surfaces and consequently the size of contacting asperities. In this experiment, roughness measured through the SEM is characterized by the software package (Vision32) provided by Veeco Meteorology Group. The characterization of the sled shoe and the track surface indicates the roughness of 4-8 μm and it is postulated that damage phenomena are generated from dislocation generation at between crystallographically misoriented single crystals (at the microscale). However, the roughness of the prepared and machined specimens is significantly smaller (0.4-0.8 μm). In order to further characterize the sliding interfaces, the experimental study as previously described is conducted (see Figs. 24(a) and 24(b)) and again the measurement procedure in the same manner is performed.

The set of simulations is performed for different values of normal force and sliding velocity capturing the development of surface temperature, plastic deformation as well as the frictional force. The different mesh sizes are also considered to show the lack of mesh dependency due to the non-local gradient formulation adopted in this work according to the proposed theory. The simulations are carried out for the set for sliding velocities between 10m/s and 200m/s. The registration of the frictional force is done by defining the group of nodes on each specimen A and summing the occurring reaction forces afterwards to obtain the final tangential force. This force is further used to calculate the friction coefficient which will be presented in a forthcoming paper. Hence, in this work one concentrates on comparing the recorded computational temperature and its evolution leaving the friction coefficient determination for later development.

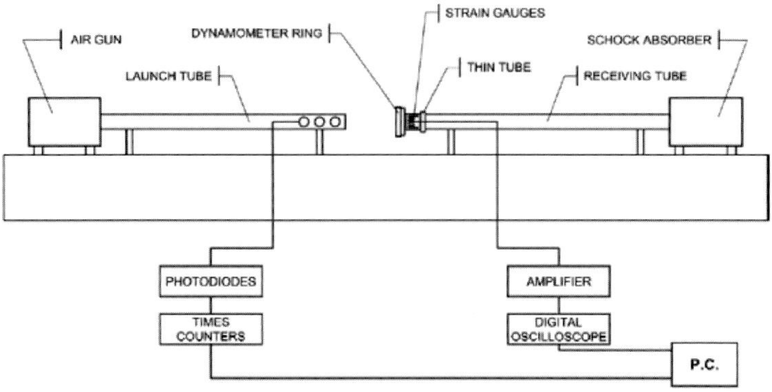

Figure 24(a). Schematic Description of the Gas Gun Experiment, Developed by Phillipon et.al. (2004) in ENIM, which is Used to Measure Contact Forces during the High Velocity Contact between Polycrystallines.

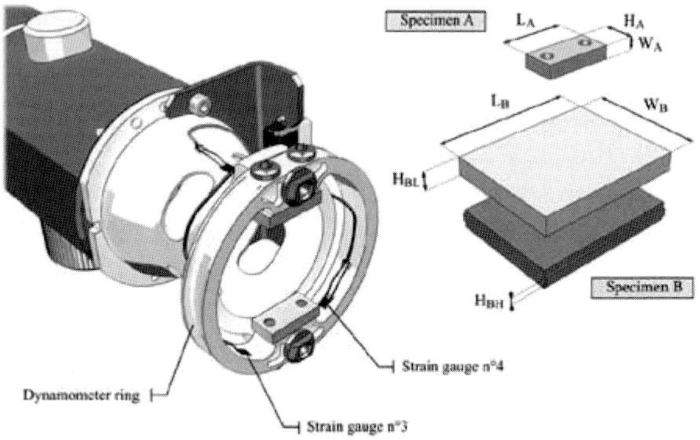

Figure 24(b). Friction Device Used for Experimental Measurement of Contact Forces with Specimen Description (Phillipon et.al., (2004).

In this work the authors concentrate on temperature as the major variable in the analysis of the high velocity contact problem. It is analyzed in the following manner. The set of surface nodes on the inner part of specimen A is created. It is important to note that not all inner nodes are in contact with specimen B and not all of them experience the temperature changes. However, it is the most precise way to determine the global temperature change of the inner part of specimens A. The temperature is recorded for each node on the previously mentioned surface and finally the average temperature is calculated. These results are presented in Figures 25 and 26. For the sake of completeness the same analysis is conducted for specimen B (Figure 26).

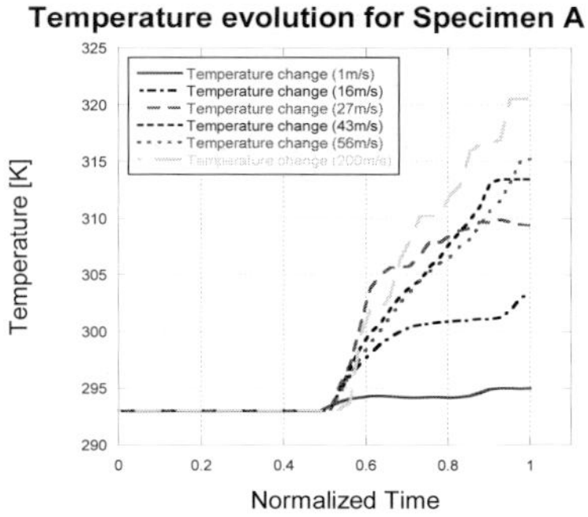

Figure 25. Temperature Evolution as an Average from the entire Inner Surface of Specimen A (Voyiadjis, et al., 2010).

As is clearly visible in Figure 25 the increase of sliding velocity causes the increase of the average temperature. This increase is in good agreement with the theoretical approach presented by Bowden and Hughes (1939), and Bowden and Persson (1960). Despite the fact that the overall increase is substantial, it is not linear with respect to the velocity unlike the increase of maximum temperature occurring on specimen A (Figure 25). It can be clearly seen that the particular increase of velocity for higher range of velocities does not give the same increase of temperature (i.e. almost quadrupling the velocity does not result in quadrupling the temperature).

Despite the fact that the overall increase is substantial, it is not linear with respect to velocity. It must be noted that unlike the increase of maximum temperature occurring on specimen A (Figure 26) occurring to be almost linear, the increase of the average temperature with respect to velocity is logarithmic and converging to some critical temperature. It can be clearly seen that the particular increase of velocity for higher range of velocities does not give the same increase of temperature (i.e. almost quadrupling the velocity does not result in quadrupling the average temperature). This phenomenon is captured in Figure 27 and stays in a good correlation with the theories and experimental results presented in vast variety of sources such as Oden and Martins (1984), Lim and Ashby (1989), Philippon, et al. (2003) and others. It is theoretically clear and also confirmed by formulations presented in this paper that due to dry friction the generated heat will logarithmically increase reaching some particular level. The further part of this research concentrates on experimental valuation of these theoretical approaches as well as temperature variation model verifying the experimental results and will be published in a close future.

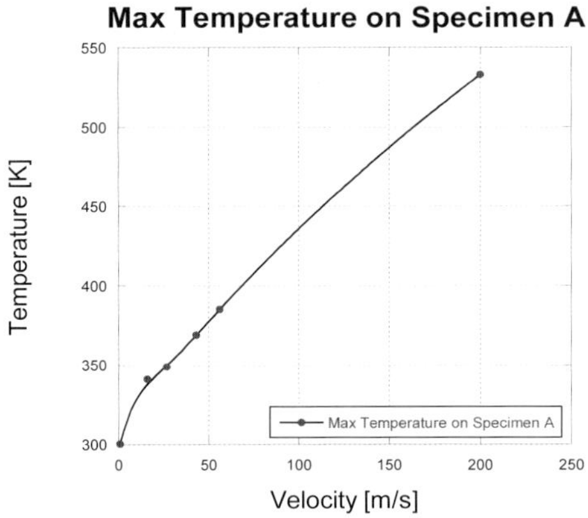

Figure 26. Maximum Temperatures Occurring During the Sliding Process on the inside Surface of Specimen A for Particular Velocity (Voyiadjis, et al., 2010).

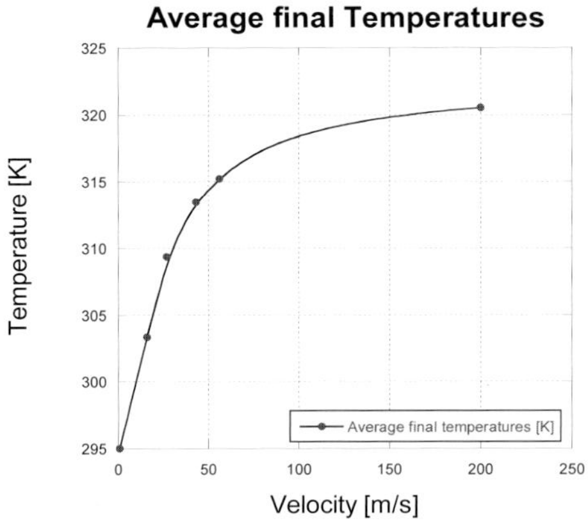

Figure 27. Average Final Temperatures of the inside Part of Specimen A (Voyiadjis, et al., 2010).

For the sake of completeness Figure 27 presents the evolution of the temperature on the contact surface (L_B x H_{BL} on Figure 24(b)) of specimen B.

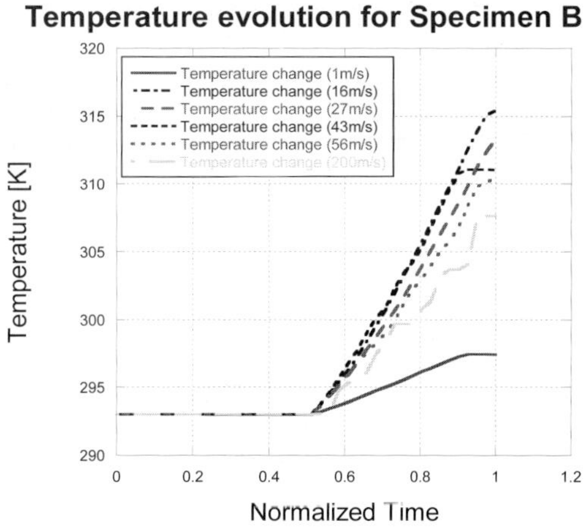

Figure 28. Temperature Evolution as an Average from the Entire Sliding Surface of Specimen B (Voyiadjis, et al., 2010).

In this case however, due to large increase of sliding velocity and the fact that the entire sliding surface of specimen B is not the entire time in contact during the sliding process (see Figures 29 and 30) the temperature does not increase with the increase of the sliding velocity. It is assumed in the theoretical formulations as well as in computational approach that 90% of the entire plastic energy is converted into the heat. Due to increase of the sliding velocity the shear bands do not manage to open. Hence, the generated plastic energy decreases in comparison with the plastic energy generated for the lower sliding velocities resulting in smaller temperature increase. It clearly implies that this phenomenon can be interpreted as adiabatic process and will be considered by the authors, along with already mentioned temperature variation model, in the following publications. Moreover this fact invokes a very good heat resistance of high strength steel VascoMax and provides valuable information for the usage of steel VascoMax, in particular, the application to the shoe construction in the Holloman High Speed Test Track and Air Force base in Ohio, partially supporting this research.

Figures 29 and 30 present three stages of the friction simulation with the different sliding velocities. This approach utilizes the use of coarse and fine mesh approach showing the mesh independent solution.

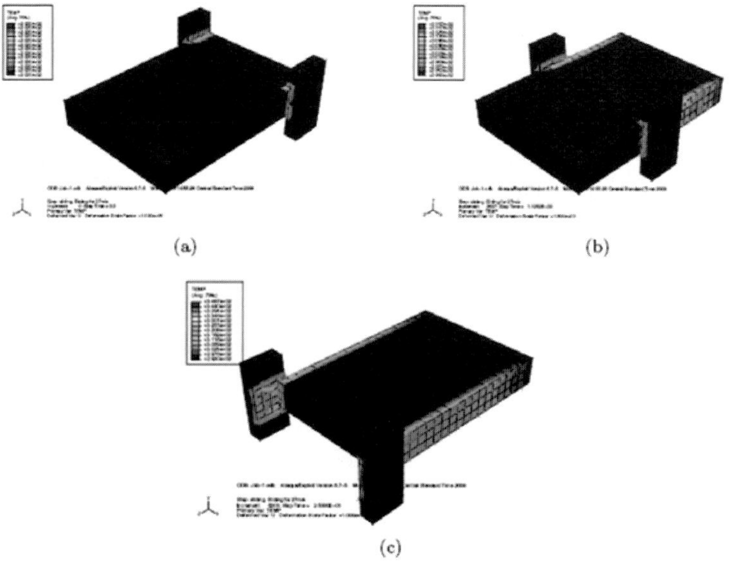

Figure 29. Temperature Evolution on both Sliding and Fixed Surfaces of Specimens A and B for Coarse Mesh and Velocity 27m/s (Voyiadjis, et al., 010).

Initial stage (Figures 29(a) and 30(a)) present the application of all boundary conditions just before applying the sliding velocity to the specimen B. Figures 29(b) and 30(b) present the development of the temperature as one of the mainly considered variables. Last Figures 29(c) and 30(c) show the moment when the contact between the specimens is lost. Figure 30 with the coarse mesh presents the results for the sliding velocity of 27m/s whereas Figure 30 presents the fine mesh for the sliding velocity 56m/s. These results clearly indicate that the solution is not mesh dependant and friction force analysis clearly indicates convergence with the experimental results conducted by Philippon, et al. (2004). The two velocities were simulated for both the coarse and fine meshes originally without any substantial difference in the obtained results.

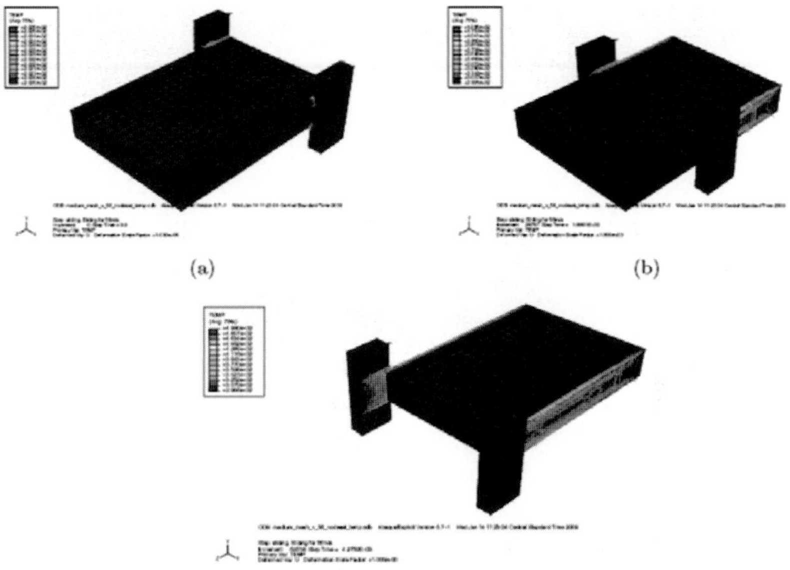

(a) (b) (c)

Figure 30. Temperature Evolution on both Sliding and Fixed Surfaces of Specimens A and B for Fine Mesh and Velocity 56m/s (Voyiadjis, et al., 2010).

7.2 Shear localizations in strip tension

In the first example of this section, local viscoplastic fromulation is considered only for a shear localization problem in order to study the adiabatic shear band localizations in tensile deformation for several bcc metals (Abed and Voyiadjis, 2007, Voyiadjis and Abed, 2006).

A simple uniaxial plane strain tension problem, with 10mm x 20mm dimensions as shown in Figure 31, is considered in this case. The theoretical formulations and computational algorithms of the proposed elasto-viscoplastic constitutive models are tested by studying the mesh sensitivity problem using different mesh configurations. In this regard, the considered problem is modeled and meshed with five different mesh discretizations (Mesh1: 10×20, Mesh2: 15×30, Mesh3: 20×40, Mesh4: 25×30, Mesh5: 30×60 elements) using the four-node plane strain element with reduced integration.

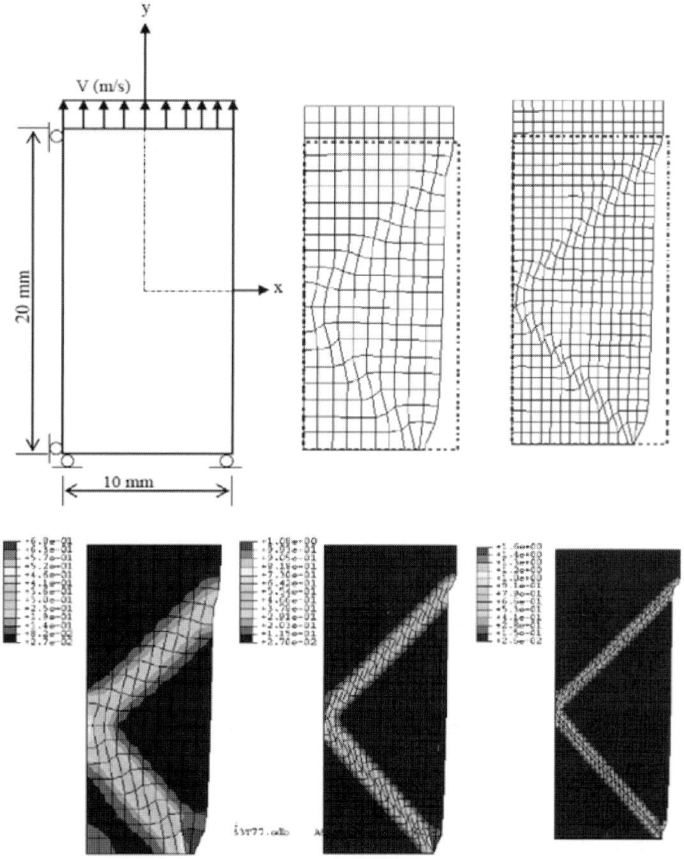

Figure 31. Illustration for a Subjected to Certain Velocity with a Specified Boundary Condition and Deformed (solid lines) and Undeformed (dashed lines) Patterns at 70μs Time Step (2.1mm elongation) (Voyiadjis and Abed, 2006)

In this example, the differences between the rate-independent and rate-dependent plasticity constitutive equations was investigated by applying constant velocities at certain initial temperatures using the five considered meshes. For this type of problem, shear bands may be developed and formed in different directions as there is no specific imperfection or defects induced in the sample. Deformations of a homogenous body with uniform initial conditions become inhomogeneous because of the interaction with the incident waves reflected from the boundaries and corner. This may happen at different instants in the considered metals due to the differences in their material properties. Moreover, it may also encounter at different instants within the same metal if different velocities (different waves) are considered. However, this problem should not be affected by the mesh refinement. (i.e., by using different mesh configurations).

It is obvious from these patterns that the observed width of shear bands is determined by the element size such that the deformation is localized along a line of integration points. The von Mises stresses at the localized regions over the total elongation of the specimen (3.3mm) clearly show the mesh dependency of the considered rate-independent analysis as shown in Figure 32.

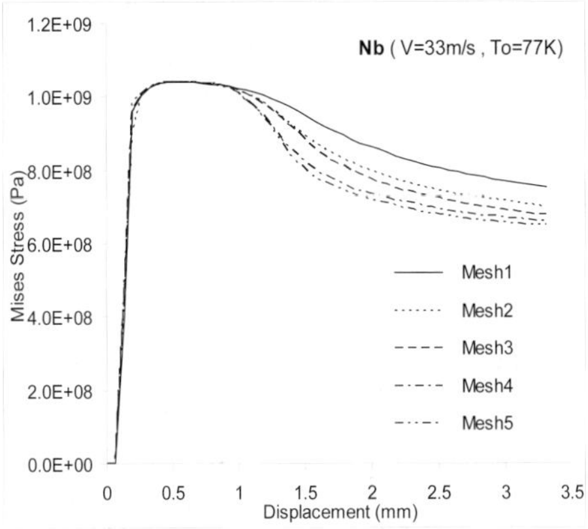

Figure 32. Mesh-dependent Results for Niobium using the Mises-displacement Curves. (Voyiadjis and Abed, 2006)

In order to show that the rate dependent model regularize nonlocal problem and provide mesh independent results the proposed viscoplastic model was used by by applying the velocity (V=33m/s) and initial temperature (77K) over a total time step of 100μs (3.3mm total displacement) using three mesh discretizations (Mesh1, Mesh2, and Mesh3). The corresponding deformed displacement patterns are illustrated in Figure 33(a) at a time step of 83μs (2.7mm displacement). The stress (Mises)-displacement curves over the total time step is also plotted in Figure 33(b). By comparisons with the results obtained from the rate-independent analysis, one observes a finite width of shear bands that is independent of the finite element size. Moreover, the Mises-displacement curves (Figure 33(b)) show almost identical results for all of the considered meshes as compared to those obtained for the rate-independent analysis.

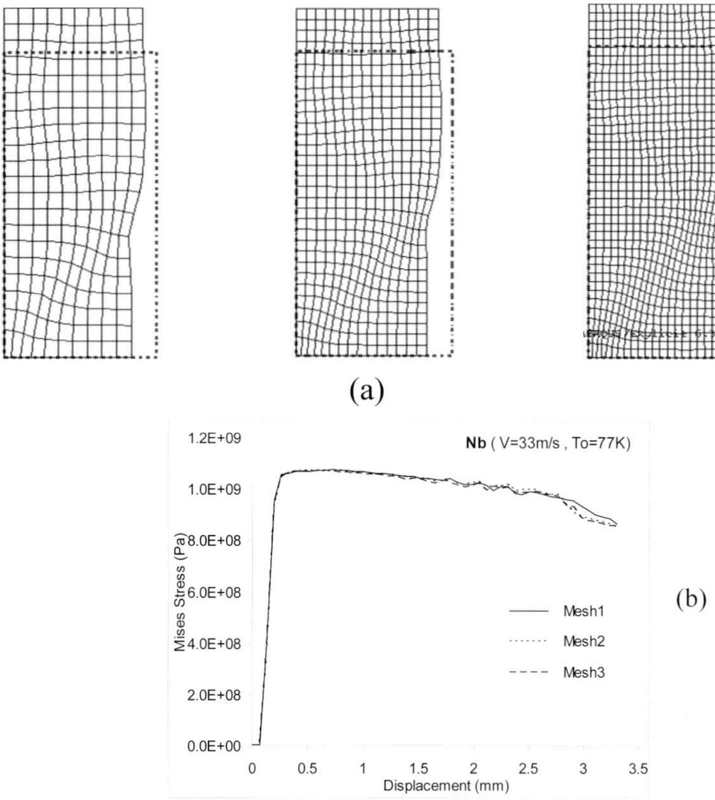

Figure 33. Mesh-independent Results for Niobium at V=33m/s (1650s-1 strain rate) and To=77K. (a) Deformed (solid lines) and Undeformed (dashed lines) Patterns at 83μs Time Step (2.7mm elongation). (b) Mises-displacement Curves (Voyiadjis and Abed, 2006) .

7.3 Shear localizations in cylindrical hat-shaped samples and comparisons with experimental results

Numerical simulations of the dynamic deformation response of tantalum hat-shaped specimens, shown schematically in Figure 34, are performed in this study under prescribed conditions. Comparisons of these results to experiments conducted by three different authors (Bronkhorst, et al., 2006, Nemat-Nasser, et al., 1998, Perez-Prado, et al., 2001) are also made. The hat-shaped sample geometry is commonly used to study shear localization growth. The shape indicates an axi-symmetric condition with a shear zone limited between the upper (hat) and lower (brim) portions. The proposed finite element simulations of the proposed samples assume axi-symmetry condition in which a four-node axi-symmetric type of elements is used for the problem mesh geometry.

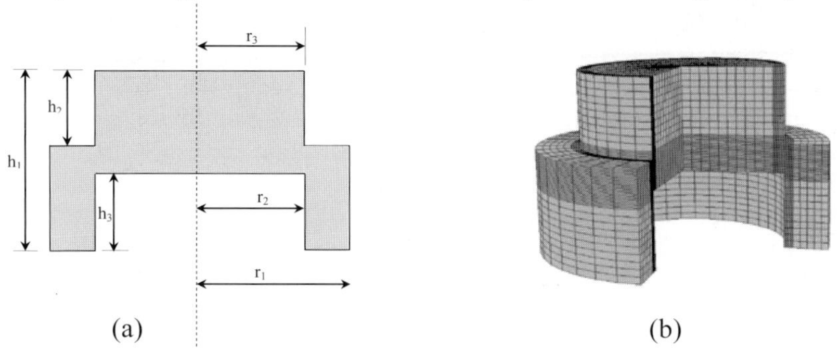

Figure 34. Cylindrical Hat-shaped Specimen. (a) A Schematic Diagram (dimensions given in Table 1). (b) A Three-dimensional Shape Corresponding to 3/4 of the Sample (Voyiadjis and Abed, 2006).

The finite element simulations of the above-mentioned hat-shaped sample, subjected to compressive loading at the ambient initial temperatures and three different strain rates (for the three different authors), are performed. Results for the adiabatic shear stress versus displacement of the three considered mesh density simulations compared with corresponding experimental data are given in Figure 35.

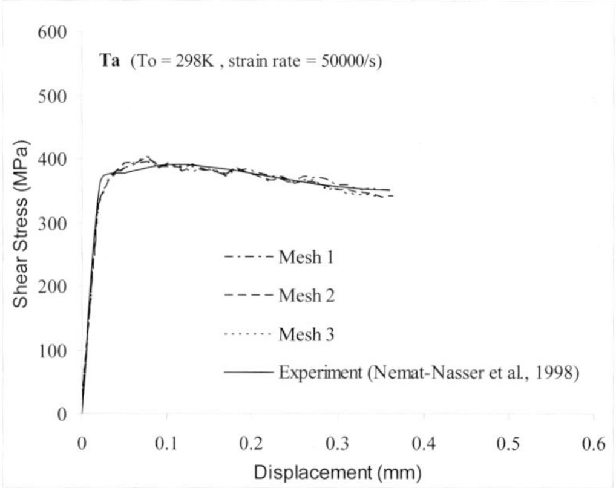

Figure 35. Shear Stress vs Displacement Curves for Tantalum under Adiabatic Condition. Comparison between Experiment (Nemat-Nasser, et al., 1998) and Simulation Results using Three Different Mesh Configurations (Voyiadjis and Abed, 2006).

In addition to the good comparisons between the results from simulations and experiments, it is interesting to also note that the stress-displacement curves of the adiabatically tested specimen and model simulations share a common feature, that is, the shear stress first reaches a peak value, and then decreases with increasing displacement. Simulation results of the equivalent plastic strain contours for the tantalum adiabatic simulation at the end of the considered displacement is shown in Figure 36(a). Shear localization corresponding to pure tantalum deformed at the above mentioned loading condition compares well with the optical micrographs taken at the forced localized shear region as illustrated in Figure 36(b). The width of the shear region increases with increasing shear strain within a given specimen.

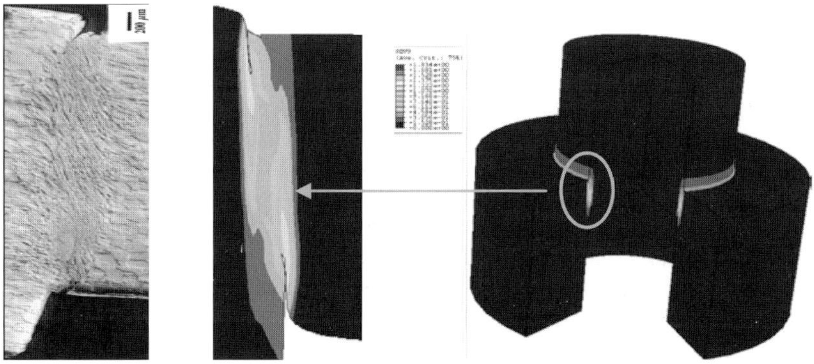

(a) (b)

Figure 36. (a) Shear Localization Results from Experimental Observations ((Perez-Prado, et al., 2001)) and Model Simulation. (b) Contour Plots (Voyiadjis and Abed, 2006).

7.4 A blunt projectile impacting a target

The objective of this numerical example is to investigate if the proposed constitutive equations are able to describe the structural response to projectile impact damage when different failure modes are expected to occur. This is done by conducting numerical simulations of the experimental tests presented by Borvik et al. (2005) for a blunt projectile made of hardened Arne tool steel impacting a circular plate made of Weldox 460 E steel. The model material constants of the target material of Weldox 460 E steel and for the projectile material of hardened Arne tool steel are listed in Table 5.

Table 5. Material Constants of the Target Material of Weldox 460 E Steel and for the Projectile Material of Hardened Arne Tool Steel.

Target Material of Weldox 460 E steel			
$E = 200\,GPa$	$a_1 = 400\,MPa$	$h_1 = 0$	$r_0 = 0$
$\nu = 0.33$	$a_2 = 20\,GPa$	$h_2 = 0$	$\gamma = 1.7$
$\overline{Y}_o = 490\,MPa$	$a_3 = 0$	$\Upsilon = 0.9$	$m = 0.94$
$T_r = 295\,K$	$a_4 = 0$	$\ell = 5\,\mu m$	$n = 1$
$T_m = 1800\,K$	$k_1 = 0.1\,MPa^{-1}$	$\alpha_t = 1.1\times10^{-5}\,/\,K$	$\phi^c_{eq} = 0.30$
$\rho_o = 7850\,kg\,/\,m^3$	$k_2 = 15\,GPa^{-1}$	$c_p = 452\,J\,/\,kg.K$	$P_{cutoff} = 160\,GPa$
$\eta^v = 0.01\,s$	$\eta_r = 0$	$c_v = 266\,J\,/\,kg.K$	
Projectile Material of Hardened Arne Tool Steel			
$E = 204\,GPa$	$\nu = 0.33$	$\rho_o = 7850\,kg\,/\,m^3$	$Y_o = 1900\,MPa$
$E_t = 15\,GPa$	$\varepsilon_f = 2.15\%$		

In these simulations a 4-node 2D axisymmetric element with one integration point and a stiffness based on hourglass control is used. Plot of the initial configuration, showing a part of the target plate and the blunt projectile just prior to impact, is shown in Figure 37.

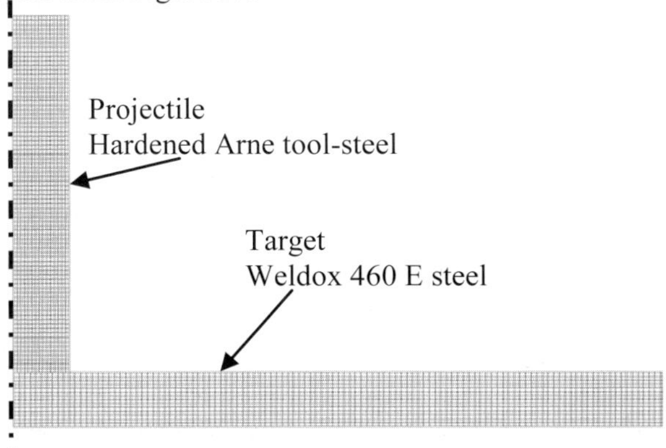

Figure 37. Finite Element Mesh Plot of the Axisymmetic Initial Configuration just before Impact.

The target plate has a nominal thickness of 12mm and a diameter of 500mm, while the nominal length and diameter of the hardened projectile are 80mm and 20mm, respectively. In each run, the target plate is fully clamped at the support, while the projectile is given an initial velocity similar to the one used in the corresponding experiment conducted by Borvik et al. (2002). In order to reduce the computational time, which is affected both by the element size and number, the mesh was somewhat coarsened towards the boundary. Owing to this coarsening the total number of elements in the target plate is less than about 10,000 in the simulations. Contact was modeled using an automatic 2D single surface penalty formulation available in ABAQUS/Explicit. A friction coefficient of 0.05 is assumed between the projectile and the target. Time increments of order $10^{-8}\,s$ are used to satisfy the stability criteria.

Borvik et al. (2002) simulations indicate the problem involving shear localization and plugging for blunt projectiles to be mesh size sensitive. However, the numerical solution using the present model is mesh size independent and converges monotonically towards a limit solution when the number of elements over the target thickness becomes sufficiently large. Therefore, the numerical results improve as the element size is reduced until it stabilizes at some value; i.e. the mesh size dependency is not pathological. This is expected since the width of the shear band is known to be in the order of 10-100 μm. A direct comparison between the numerical and experimental residual velocity curves for blunt projectile is shown in Figure 38.

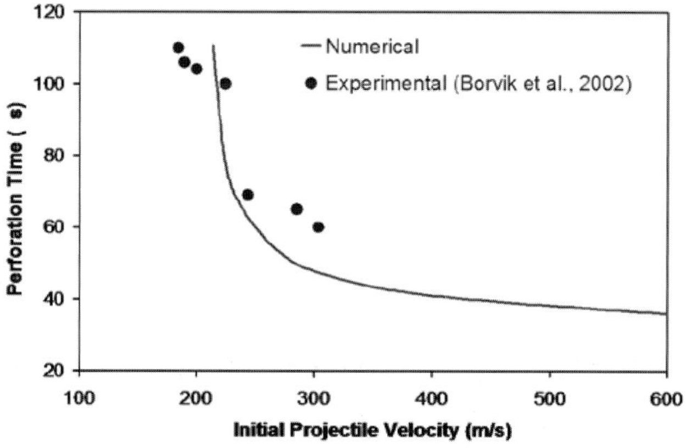

Figure 38. A Comparison between Numerical and Experimental Results by a Blunt Projectile for the Initial Impact Velocity versus Residual Projectile Velocity (Voyiadjis et al., 2008).

It is noteworthy that in Figure 37 one cannot simply obtain the perforation time by knowing the initial and residual velocities of the projectile and the target thickness. For example, for 303.5m/s impact velocity the residual velocity from

Fig. 10 is 199.7m/s. By linear kinematics and knowing that the target thickness is 12mm, one can simply calculate the perforation time to be 116 μs , whereas experimentally it is reported as 60 μs which deviates from the calculated one. High-speed camera images showing perforation of the target plate at impact velocities close to the ballistic limit are shown in Figure 39.

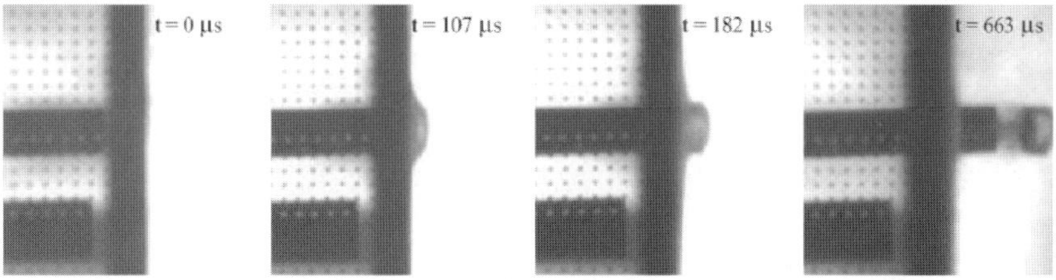

Figure. 39. High-speed Camera Images Showing Perforation of the Target Plate at Impact Velocities Close to the Ballistic Limits with a Blunt Projectile; after Borvik et al. (Borvik, et al., 2005).

Numerical plots showing perforation of the target plate by a blunt projectile at impact velocity close to the ballistic limit of 210m/s are shown in Figure 40. The contours of accumulated viscoinelastic strain are plotted on the deformed mesh. It can be seen that limited inelastic deformation occurs outside the localized shear zone. These plots clearly demonstrate that the numerical model qualitatively captures the overall physical behavior of the target during penetration and perforation. Notice also that in these plots, only a part of the complete target plate is shown.

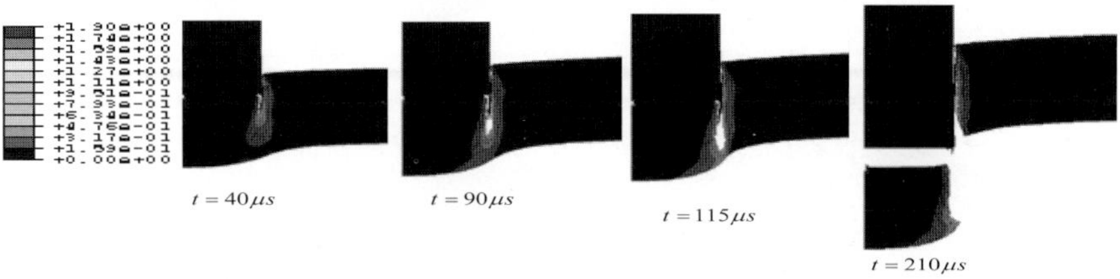

Figure 40. Perforation of the Target Plate by a Blunt Projectile of Initial Impact Velocity of 210 m/s Plotted as Contours of Accumulated Inelastic Strain and at Different Times. The Green Region Indicates an Accumulated Damage between 0.25 and 0.30 (Voyiadjis et al., 2008).

7.5. A blunt projectile impacting a composite plate

The proposed model is used to simulate a hemispherically tipped cylindrical steel projectile, 5.5mm long and 7.5mm diameter, impacting and penetrating into a 50x50x5mm laminated metal matrix composite plate subjected to an initial velocity of 1000m/s. Table 6 shows the geometrical dimensions and material parameters that are used for the projectile.

Table 6. Material and Dimensional Properties of the Projectile.

Material Properties	Youngs's modulus, E : 210 E 9Pa Poisson ratio: 0 .28 Density : 7850kg/m^3
Dimensions of the projectile	Radius : 7.5mm Length: 5.5mm

Numerical analysis of the impact stage provides considerable kinetics and kinematic information of the penetration and perforation processes. Generally, strong shock wave-material interaction is generated and propagated along both the projectile and the targets that can lead to fracture global inelastic strains. Phenomenologically, the penetration can be viewed as a process to generate a cone shaped macrocrack in the material, in which the kinetic energy of the penetrator is dissipated. In Figure 41 the sequence of stages of impact can be seen.

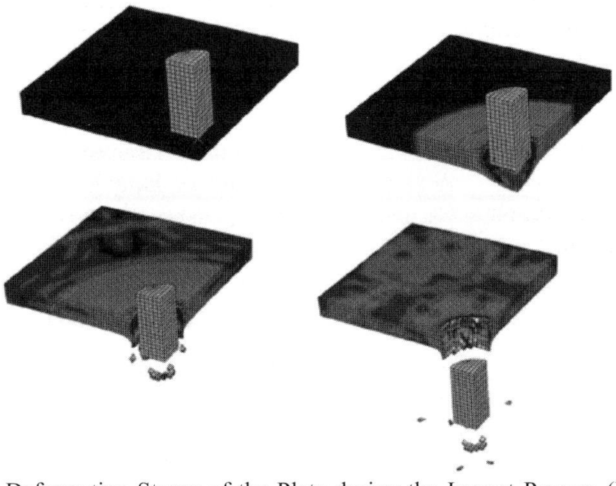

Figure 41. Deformation Stages of the Plate during the Impact Process (Deliktas, et al., 2009).

Global bending and dishing is observed in the third stage. Numerical plots of the perforation of the target plate by the projectile shows the limited inelastic deformation that occurs outside of the localized shear zone. These plots clearly indicate that the numerical model qualitatively captures the overall physical behavior of the target during penetration and perforation. In the third and fourth stages of deformation of the plate (a and b) it can be seen that the elements in the impacted area are significantly distorted. However, stable results are nevertheless obtained. This distortion caused delay in the damage evolution process and, consequently the erosion of the failed elements. This simulation also indicates that the initiation and propagation of the microdefects depend on both the amplitude and distribution of the inelastic strain/stress in the vicinity around the microdefects. This implies that the evolution of the material inelasticity is practically a nonlocal process. Hence, such nonlocality in the material response requires such a micromechanical approach that incorporates material length scale parameter into the classical constitutive relations.

8. Conclusions

The authors presented the micro-mechanical damage model that accounts for the nonlocal microscopic interactions between material points (i.e. to take into account the influence of an internal state variable at a point on its neighborhood) in the simulation of metal/composite impact and severe contact stress problems. This is achieved by introducing the contributions of damage and its corresponding gradients in the virtual power relations as measures of micro motion of damage within the bulk. By using these internal state variables together with displacement and temperature, the constitutive model is formulated with state laws based on the free energies and the complimentary laws based on the dissipation potentials.

The presented nonlocal microdamage model is based on the enhanced gradient-dependent theory which is successful in explaining the size effects encountered at the micron scale and in preserving the well-posed nature of the initial boundary value problem that governs the solution of material instability triggering strain localization. This is modeled through coupling continuum damage with strain gradient plasticity.

In this work the mechanics of small damage are also presented using a consistent mathematical and mechanical framework based on the equations of damage mechanics. In this regard, the new scalar damage variable is investigated in detail. Two different approaches to damage characterization in composite materials are illustrated. It is shown that both the overall and local approaches to damage yield exactly the same results for elastic composite materials. The investigation in this work has been carried out for seeking a physical basis is sought for the damage tensor $[M]$ that is used to link the damage state of the material with effective undamaged configuration. The approach presented here provides for a strong physical basis for this missing link. In particular, the authors have made an important link between the damage tensor and fabric tensors

Computaional aspects of the presened theory were discussed. Numerical integration algoriths, verification and validation process of the theory have been discussed. In this sense one point material model is first used to calibrate the developed constitutive model by specifying various material constants, or constitutive parameters, for each material. Values for these parameters are obtained by fitting the predicted material behavior to the laboratory data for various load paths presented in the previous section. The finite element simulations are also performed by implementing the presented model in the commercial finite element code ABAQUS [6.8.3] as a user defined subroutine (VUMAT) in order to perform following simulations i) Simulation of experimental work on the study of friction between metallic surfaces ii) the experimental work on the study of friction between metallic surfaces was conducted by the authors in collaboration with the National Engineering School of Metz (ENIM), France iii) a projectile impact damage for different failure modes iv) shear localizations in cylindrical hat-shaped samples and v) a projectilel impacting a composite plate

Results of these analyses lead one to remark the following points:

- The proposed theory is verified successfully in the simulation of the material response with respect to hardening and softening behaviors under uniaxial and cyclic loading cases. The reliable and robust simulation capability of the model shows that the theory is extremely adequate in characterizing the behavior of steel 1080 and Steel VascoMax, respectively. Accurate material model parameters that are needed for the simulation of the experimental work in the study of friction between metallic surfaces is conducted by the authors in collaboration with the National Engineering School of Metz (ENIM), France.
- The simulations of the above experimental set up are performed with the finite element commercial software ABAQUS according to the proposed theory and coded as a VUMAT subroutine. The set of simulations is performed for different values of normal force and sliding velocity capturing the development of surface temperatures, plastic deformations as well as the frictional forces. Different mesh sizes are also considered to show the lack of mesh dependency due to the non-local gradient formulation adopted in this work according to the proposed theory. The

simulations are carried out for a set of sliding velocities between 10m/s and 200m/s.

- The increase of sliding velocity causes the increase of the average temperature. This increase is in good agreement with the theoretical approach presented by Bowden and Hughes (1939), and Bowden and Persson (1960). Despite the fact that the overall increase is substantial, it is not linear with respect to the velocity unlike the increase of maximum temperature occurring on the specimen. It can be concluded that the particular increase of velocity for higher range of velocities does not give the same increase of temperature (i.e. almost quadrupling the velocity does not result in quadrupling the temperature).
- Numerical analysis of the impact stage provides considerable kinetics and kinematic information of the penetration and perforation processes. Generally, strong shock wave-material interaction is generated and propagated along both the projectile and the targets that can lead to fracture global inelastic strains.

This model provides a potential feature for enabling one to relate the non-local continuum plasticity and damage of the bulk material to friction and wear at the contact interfaces. The findings of this research effort is invaluable in providing a material model and numerical procedure that will be used within a hydrocode to better facilitate the design components of severe contact stress applications.

References

Abed, F. H., and Voyiadjis, G. Z., 2007. Adiabatic shear band Localizations in BCC metals at high strain rates and various initial temperatures. *International Journal for Multiscale Computational Engineering* 5, 325-349.

Abu Al-Rub, R. K., and Voyiadjis, G. Z., 2003. On the coupling of anisotropic damage and plasticity models for ductile materials. *International Journal of Solids and Structures* 40, 2611-2643.

Abu Al-Rub, R. K., and Voyiadjis, G. Z., 2004. Analytical and experimental determination of the material intrinsic length scale of strain gradient plasticity theory from micro- and nano-indentation experiments. *International Journal of Plasticity* 20, 1139-1182.

Abu Al-Rub, R. K., and Voyiadjis, G. Z., 2005. A direct finite element implementation of the gradient-dependent theory. *International Journal for Numerical Methods in Engineering* 63, 603-629.

Abu Al-Rub, R. K., and Voyiadjis, G. Z., 2006. A finite strain plastic-damage model for high velocity impact using combined viscosity and gradient localization limiters: Part I - Theoretical formulation. *International Journal of Damage Mechanics* 15, 293-334.

Abu Al-Rub, R. K., Voyiadjis, G. Z., and Bammann, D. J., 2007. A thermodynamic based higher-order gradient theory for size dependent plasticity. *International Journal of Solids and Structures* 44, 2888-2923.

Aifantis, K. E., and Willis, J. R., 2005. The role of interface in enhancing the yield strength of composite and polycrystal. *Journal of the Mechanics and Physics of Solids* 53, 1047-1070.

Aifantis, K. E., and Willis, J. R., 2006. Scale effects induced by strain gradient plasticity and interfacial resistance in periodic and randomly heterogeneous media. *Mechanics of Materials* 38, 702-716.

Allix, O. P., Ladeveze, P., Gilleta, D., and Ohayon, R., 1989. A Damage Prediction Method for Composite Structures. *International Journal of Numerical Methods in Engineering* 27, 271-283.

Anand, L., Gurtin, M. E., Lele, S. P., and Gething, C., 2005. A one-dimensional theory of strain-gradient plasticity: Formulation, analysis, numerical results. *Journal of the Mechanics and Physics of Solids* 53, 1789-1826.

Armstrong, R. W., and Zerilli, F. J., 1994. Dislocation Mechanics Aspects of Plastic Instability and Shear Banding. *Mechanics of Materials* 17, 319-327.

Arnold, S. M., 1990. Quntification of numerical stifness for a unified viscoplastic constitutive model. *Journal of Engineering Materials and Technology-Transactions of the Asme* 112, 271-276.

Arnold, S. M., Saleeb, A. F., and Castelli, M. G., 1994. A fully associative ' non-linear kinematic' unified viscoplastic model for tinanium based matrices. In: *NASA*, (Ed., Washington D.C, pp. 203=209.

Bammann, D. J., Chiesa, M. L., McDonald, A., Kawahara, W. A., Dike, J. J., and Revelli, V. D., 1990. Prediction of Ductile Failure in Metal Structures,. *AMD* 107, 7–12.

Batra, R. C., and Wei, Z. G., , 2006. Shear Bands Due to Heat Flux Prescribed at Boundaries. *International Journal of Plasticity* 22, 1-15.

Batra, R. C., and Chen, L., 1999. Shear Band Spacing in Gradient-Dependent Thermoviscoplastic Materials. *Computational Mechanics* 23, 8-19.

Batra, R. C., and Kim, C. H., 1988. Effect of material characterisctic length on the intiation, growth and band widh of adiabatic shear bands in dipolar materials. *Journal De Physique* 41-46.

Batra, R. C., and Kim, C. H., 1990. The interaction among adiabatic shear bands in simple and dipolar materials *International Journal of Engineering Science* 28, 927-942.

Bittencourt, E., Needleman, A., Gurtin, M. E., and Van der Giessen, E., 2003. A comparison of nonlocal continuum and discrete dislocation plasticity predictions. *Journal of the Mechanics and Physics of Solids* 51, 281-310.

Borvik, T., Clausen, A. H., Eriksson, M., Berstad, T., Hopperstad, O. S., and Langseth, M., 2005. Experimental and numerical study on the perforation of AA6005-T6 panels. *International Journal of Impact Engineering* 32, 35-64.

Borvik, T., Clausen, A. H., Hopperstad, O. S., and Langseth, M., 2004. Perforation of AA5083-H116 aluminium plates with conical-nose steel projectiles-experimental study. *International Journal of Impact Engineering* 30, 367-384.

Borvik, T., Dey, S., and Clausen, A. H., 2006. A preliminary study on the perforation resistance of high-strength steel plates. *Journal De Physique Iv* 134, 1053-1059.

Borvik, T., Hopperstad, O. S., Berstad, T., and Langseth, M., 2002. Perforation of 12 mm thick steel plates by 20 mm diameter projectiles with flat, hemispherical and, conical noses Part II: numerical simulations. *International Journal of Impact Engineering* 27, 37-64.

Bronkhorst, C. A., Cerreta, E. K., Xue, Q., Maudlin, P. J., Mason, T. A., and Gray, G. T., 2006. An experimental and numerical study of the localization behavior of tanatlum and stainless steel,". *International Journal of Plasticity* 22, 1304-1335.

Budiansky, B., and O'Connell, R. J., 1976. Elastic Moduli of a Cracked Solid. *International Journal of Solids and Structures*, 12, 81-97.

Camacho, G. T., and Ortiz, M., 1997. Adaptive Lagrangian modelling of ballistic penetration of metallic targets. *Computer Methods in Applied Mechanics and Engineering* 142, 269-301.

Cauvin, A., and Testa, R., 1999. Damage Mechanics: Basic Variables in Continuum Theories. *International Journal of Solids and Structures* 36, 747-761.

Celentano, D. J., Tapia, P. E., and Chaboche, J.-L., 2004. Experimental and Numerical Characterization of Damage Evolution in Steels. In: G. Buscaglia, and E. Dari, O. Z., (Eds.), *Mecanica Computacional*, Vol. XXIII, Bariloche, Argentina.

Cermelli, P., Fried, E., and Gurtin, M. E., 2004. Sharp-interface nematic-isotropic phase transitions without flow. *Archive for Rational Mechanics and Analysis* 174, 151-178.

Chaboche, J. L., 1981. Continuous Damage Mechanics – A Tool to Describe Phenomena Before Crack Initiation. *Nuclear Engineering and Design* 64, 233-247.

Chelluru, S. K., 2007. Finite Element Simulation of Ballistic Impact of Metal and Composite Plates, *Whichita State University*.

Chow, C., and Wang, J., 1987. An Anisotropic Theory of Elasticity for Continuum Damage Mechanics. *International Journal of Fracture* 33, 3-16.

Christiansen, E. L., and Friesen, L., 1997. Penetration equations for thermal protection materials. *International Journal of Impact Engineering* 20, 153-164.

Curran, D. R., Seaman, L., and Shockey, D. A., 1987. *Dynamic Failure of Solids. Physics Reports* Vol. 147.

Deliktas, B., Voyiadjis, G. Z., and Palazotto, A. N., 2009. Simulation of perforation and penetration in metal matrix composite materials using coupled viscoplastic damage model. *Composites Part B: Engineering* 40, 434-442.

Doghri, I., 2000. Mechanics of Deformable Solids-Linear, Nonlinear, Analytical and Computational Aspects. Springer-Verlag, Berlin.

Dorgan, R. J., and Voyiadjis, G. Z., 2003. Nonlocal dislocation based plasticity incorporating gradients of hardening. *Mechanics of Materials* 35, 721-732.

Dorgan, R. J., and Voyiadjis, G. Z., 2006. A mixed finite element implementation of a gradient-enhanced coupled damage-plasticity model. *International Journal of Damage Mechanics* 15, 201-235.

Dorgan, R. J., and Voyiadjis, G. Z., 2007. Functional forms of hardening internal state variables in modeling elasto-plastic behavior. *Archives of Mechanics* 59, 35-58.

Dorgan, R. J., and Voyiadjis, G. Z., 2007. Nonlocal coupled damage-plasticity model incorporating functional forms of hardening state variables. *Aiaa Journal* 45, 337-346.

Eftis, J., Carrasco, C., and Osegueda, R. A., 2003. A constitutive-microdamage model to simulate hypervelocity projectile-target impact, material damage and fracture. *International Journal of Plasticity* 19, 1321-1354.

Fleck, N. A., and Willis, J. R., 2009. A mathematical basis for strain gradient plasticity theory- Part I: Scalar plastic multiplier. *Journal of the Mechanics and Physics of Solids* 57, 161-177.

Fleck, N. A., and Willis, J. R., 2009. A mathematical basis for strain gradient plasticity theory- Part II: Tensorial plastic multiplier. *Journal of the Mechanics and Physics of Solids* 57, 1045-1057.

Fredriksson, P., and Gudmundson, P., 2005. Size-dependent yield strength and surface energies of thin films. *Materials Science and Engineering a-Structural Materials Properties Microstructure and Processing* 400, 448-450.

Fredriksson, P., and Gudmundson, P., 2007. Competition between interface and bulk dominated plastic deformation in strain gradient plasticity. *Modelling and Simulation in Materials Science and Engineering* 15, S61-S69.

Fremond, M., and Nedjar, B., 1996. Damage, gradient of damage and principle of virtual power. *Int. J. Solids Structures* 33, 1103, 1996.

Gudmundson, P., 2004. A unified treatment of strain gradient plasticity. *Journal of the Mechanics and Physics of Solids* 52, 1379-1406.

Gurtin, M. E., 2000. On the plasticity of single crystals: free energy, microforces, plastic-strain gradients. *Journal of the Mechanics and Physics of Solids* 48, 989-1036.

Gurtin, M. E., 2004. A gradient theory of small-deformation isotropic plasticity that accounts for the Burgers vector and for dissipation due to plastic spin. *Journal of the Mechanics and Physics of Solids* 52, 2545-2568.

Gurtin, M. E., and Anand, L., 2009. Thermodynamics applied to gradient theories involving the accumulated plastic strain:The theories of Aifantis and Fleck and Hutchinson and their generalization. *Journal of the Mechanics and Physics of Solids* 57, 405-421.

Gurtin, M. E., and Needleman, A., 2005. Boundary conditions in small-deformation, single-crystal plasticity that account for the Burgers vector. *Journal of the Mechanics and Physics of Solids* 53, 1-31.

Hansen, N. R., and Schreyer, H. L., 1994. A Thermodynamically Consistent Framework for Theories of Elastoplasticity Coupled with Damage. *International Journal of Solids and Strucutres* 31, 359-389.

Ireman, P., Klarbring, A., and Stromberg, B., 2003. A model of damage coupled to wear. *International Journal of Solids and Structures* 40, 2957–2974.

Johnson, G. R., and Cook, W. H., 1985. Fracture Characteristics of Three Metals Subjected to Various Strains, Strain Rates, Temperatures and Pressures. *Engineering Fracture Mechanics* 21, 31-48.

Ju, J. W., 1990. Isotropic and Anisotropic Damage Variables in Continuum Damage Mechanics. *Journal of Engineering Mechanics*, ASCE 116, 2764-2770.

Kachanov, L., 1958. On the Creep Fracture Time. Izv Akad, *Nauk USSR Otd Tech* (in Russian) 8, 26-31.

Kattan, P. I., and Voyiadjis, G. Z., 1993. Micromechanical Modeling of Damage in Uniaxially Loaded Unidirectional Fiber-Reinforced Composite Laminae. *International Journal of Solids and Structures* 30, 19-36.

Kattan, P. I., and Voyiadjis, G. Z., 1993. Overall Damage and Elastoplastic Deformation in Fibrous Metal-Matrix Composites. *International Journal of Plasticity* 9, 931-949.

Kattan, P. I., and Voyiadjis, G. Z., 1996. Damage-plasticity in a uniaxially loaded composite lamina: Overall analysis. International *Journal of Solids and Structures* 33, 555-576.

Kattan, P. I., and Voyiadjis, G. Z., 2001. Decomposition of damage tensor in continuum damage mechanics. *Journal of Engineering Mechanics*-Asce 127, 940-944.

Krajcinovic, D., 1996. Damage Mechanics, Elsevier. The Netherlands.

Ladeveze, P., and Lemaitre, J., Damage Effective Stress in Quasi-Unilateral Conditions. *The 16th International Cogress of Theoretical and Applied Mechanics*, Lyngby, Denmark., 1984.

Ladeveze, P., Poss, M., and Proslier, L., 1982. Damage and Fracture of Tridirectional Composites, Progress in Science and Engineering of Composites. Proceedings of the Fourth International Conference on Composite Materials. *Japan Society for Composite Materials*, pp. 649-658.

Lee, H., Peng, K., and Wang, J., 1985. An Anisotropic Damage Criterion for Deformation Instability and its Application to Forming Limit Analysis of Metal Plates. *Engineering Fracture Mechanics* 21, 1031-1054.

Lemaitre, J., 1984. How to Use Damage Mechanics. *Nuclear Engineering and Design* 80, 233-245.

Lemaitre, J., 1984. How to Use Damage Mechanics. *Nuclear Engineering and Design* 80, 233-245.

Lemaitre, J., and Chaboche, J. L., 1990. Mechanics of solid materials. Cambridge University New York.

Lemaitre, J., Morchois.Y, Monthule.A, Noppe, J. M., and Riviere, C., 1970. Influence of Fatigue Damage on Strength Characteristics of Materials. Recherche Aerospatiale, 274-&.

Loret, B., and Prevost, H., 1990. Dynamics Strain Localization in Elasto-(Visco-)Plastic Solids, Part 1: General Formulation and One-Dimensional Examples. *Computer Methods in Applied Mechanics and Engineering* 83, 247–273.

Lubarda, V., and Krajcinovic, D., 1993. Damage Tensors and the Crack Density Distribution. *International Journal of Solids and Structures* 30, 2859-2877.

Luccioni, B., and Oller, S., 2003. A Directional Damage Model. Computer Methods in Applied Mechanics and Engineering 192, 1119-1145.

Molinari, A., 1997. Collective Behavior and Spacing of Adiabatic Shear Bands. *Journal of the Mechanics and Physics of Solids* 45, 1551–1575.

Mori, T., and Tanaka, K., 1973. Average Stress in Matrix and Average Elastic Energy of Materials with Misfitting Inclusions. *Acta Metallurgica* 21, 571-574.

Murakami, S., 1983. Notion of Continuum Damage Mechanics and its Application to Anisotropic Creep Damage Theory. *Journal of Engineering Materials and Technology* 105, 99-105.

Needleman, A., 1988. Material Rate Dependent and Mesh Sensitivity in Localization Problems. *Computer Methods in Applied Mechanics and Engineering* 67, 69–85.

Nemat-Nasser, S., Isaacs, J. B., and Liu, M., 1998. Microstructure of high-strain, high-strain-rate deformed tantalum. *Acta Metallurgica*, 46: 46, 1307-1325.

Nichols, J. M., and Abell, A. B., 2003. Implementing the Degrading Effective Stiffness of Masonry in a Finite Element Model", *North American Masonry Conference*, Clemson, South Carolina, USA.

Nygards, M., and Gudmundson, P., 2004. Numerical investigation of the effect of non-local plasticity on surface roughening in metals. European Journal of Mechanics a-Solids 23, 753-762.

Park, T., and Voyiadjis, G. Z., 1997. Damage analysis and elasto-plastic behavior of metal matrix composites using the finite element method. *Engineering Fracture Mechanics* 56, 623-646.

Park, T., and Voyiadjis, G. Z., 1998. Kinematic description of damage. *Journal of Applied Mechanics-Transactions of the Asme* 65, 93-98.

Perez-Prado, M. T., Hines, J. A., and Vecchio, K. S., 2001. [Microstructural evolution in adiabatic shear bands in Ta and Ta-W alloys. *Acta Materialia*, 49: 2905-2917. 49, 2905-2917.

Philippon, S., Sutter, G., Dedourge, A., and Molinari, A., 2003. Etude experimentale du frottement. *International Journal of Mechanical Production Systems Engineering, High Speed Machining*, 57–60.

Philippon, S., Sutter, G., and Molinari, A., 2004. An experimental study of friction at high sliding velocities Wear 257, 777–784.

Rabotnov, Y., Creep Rupture", in , edited by 1968, Berlin, pp. . *Twelfth International Congress of Applied Mechanics*, Berlin, 1969, pp. 342-349.

Sidoroff, F., Description of Anisotropic Damage Application to Elasticity. *IUTAM Colloquium on Physical Nonlinearities in Structural Analysis*, Berlin, 1981, pp. 237-244.

Sluys, L. J., 1992. Wave Propagation, Localization and Dispersion in Softening Solids (Disertation), The Delft Univ. of Technology, Delft, Netherlands.

Steinberg, D. J., and Lund, C. M., 1989. A Constitutive Model for Strain Rates from 10^{-4} to 10^{6}/s. „*Journal of Applied Physics* 65, 1528–1533.

Thiagarajan, G., and Voyiadjis, G. Z., 2000. Directionally constrained viscoplasticity for metal matrix composites. *Journal of Aerospace Engineering* 13, 92-99.

Tjernlund, J. A., Gamstedt, E. K., and Gudmundson, P., 2006. Length-scale effects on damage development in tensile loading of glass-sphere filled epoxy. *International Journal of Solids and Structures* 43, 7337-7357.

Voyiadjis, G. Z., 1988. Degradation of Elastic-Modulus in Elastoplastic Coupling with Finite Strains. *International Journal of Plasticity* 4, 335-353.

Voyiadjis, G. Z., and Abed, F. H., 2006. A coupled temperature and strain rate dependent yield function for dynamic deformations of bcc metals. *International Journal of Plasticity* 22, 1398-1431.

Voyiadjis, G. Z., and Abed, F. H., 2006. Implicit algorithm for finite deformation hypoelastic-viscoplastici in fcc metals. International *Journal for Numerical Methods in Engineering* 67, 933-959.

Voyiadjis, G. Z., and Abu Al-Rub, R. K., 2005. Gradient plasticity theory with a variable length scale parameter. *International Journal of Solids and Structures* 42, 3998-4029.

Voyiadjis, G. Z., and Abu Al-Rub, R. K., 2006. A Finite Strain Plastic-Damage Model for High Velocity Impacts Using Combined Viscosity and Gradient Localization Limiters, Part II: Numerical Aspects and Simulations *International Journal of Damage Mechanics* 15, 335–373.

Voyiadjis, G. Z., and Abu Al-Rub, R. K., 2007. Nonlocal gradient-dependent thermodynamics for Modeling scale-dependent plasticity. *International Journal for Multiscale Computational Engineering* 5, 295-323.

Voyiadjis, G. Z., Abu Al-Rub, R. K., and Palazotto, A. N., 2003. Non-local coupling of viscoplasticity and anisotropic viscodamage for impact problems using the gradient theory. Archives of Mechanics 55, 39-89.

Voyiadjis, G. Z., Abu Al-Rub, R. K., and Palazotto, A. N., 2004. Thermodynamic framework for coupling of non-local viscoplasticity and non-local anisotropic viscodamage for dynamic localization problems using gradient theory. *International Journal of Plasticity* 20, 981-1038.

Voyiadjis, G. Z., Abu Al-Rub, R. K., and Palazotto, A. N., 2006. "On the Small and Finite Deformation Thermo-elasto-viscoplasticity Theory: Algorithmic and Computational Aspects," European Journal of Computational Mechanics (in press). *European Journal of Computational Mechanics* (Revue Européenne de Mécanique Numérique), 15, 945-987.

Voyiadjis, G. Z., Al-Rub, R. K. A., and Palazotto, A. N., 2008. Constitutive modeling and simulation of perforation of targets by projectiles. *AIAA Journal* 46, 304-316.

Voyiadjis, G. Z., and Almasri, A. H., 2008. A physically based constitutive model for fee metals with applications to dynamic hardness. *Mechanics of Materials* 40, 549-563.

Voyiadjis, G. Z., and Deliktas, B., 1997. Damage in MMCs using the GMC: theoretical formulation. *Composites Part B-Engineering* 28, 597-611.

Voyiadjis, G. Z., and Deliktas, B., 2000. A coupled anisotropic damage model for the inelastic response of composite materials. *Computer Methods in Applied Mechanics and Engineering* 183, 159-199.

Voyiadjis, G. Z., and Deliktas, B., 2000. Multi-scale analysis of multiple damage mechanisms coupled with inelastic behavior of composite materials. *Mechanics Research Communications* 27, 295-300.

Voyiadjis, G. Z., and Deliktas, B., 2009. Modeling strengthening and softening in inelastic nanocrytalline materials with reference to the triple junction and grain boundaries using Strain Gradient Plasticity. *Journal of the Mechanics and Physics of Solids*(Submitted).

Voyiadjis, G. Z., Deliktas, B., and Lodygowski, A., 2010. Non-local modeling of heterogeneous media to assess high velocity contact using coupled viscoplasticity damage model. (submitted to a journal).

Voyiadjis, G. Z., and Kattan, P. I., 1992. A Plasticity-Damage Theory for Large Deformation of Solids .1. Theoretical Formulation. *International Journal of Engineering Science* 30, 1089-1108.

Voyiadjis, G. Z., and Kattan, P. I., 2006. Damage mechanics with fabric tensors. *Mechanics of Advanced Materials and Structures* 13, 285-301.

Voyiadjis, G. Z., and Kattan, P. I., 2007. Evolution of fabric tensors in damage mechanics of solids with micro-cracks: Part I – Theory and fundamental concepts. *Mechanics Research Communications* 34, 145-154.

Voyiadjis, G. Z., and Kattan, P. I., 2007. Evolution of fabric tensors in damage mechanics of solids with micro-cracks: Part II - Evolution of length and orientation of micro-cracks with an application to uniaxial tension. *Mechanics Research Communications* 34, 155-163.

Voyiadjis, G. Z., and Kattan, P. I., 2007. Evolution of fabric tensors in damage mechanics of solids with micro-cracks: Part II – Evolution of length and orientation of micro-cracks with an application to uniaxial tension. *Mechanics Research Communications* 34, 155-163.

Voyiadjis, G. Z., and Kattan, P. I., 2009. Mechanics of small damage in fiber-reinforced composite materials. *Composite Structures* In Press, Corrected Proof, Available online.

Voyiadjis, G. Z., Kattan, P. I., and Taqieddin, Z. N., 2007. Continuum approach to damage mechanics of composite materials with fabric tensors. *International Journal of Damage Mechanics* 16, 301-329.

Voyiadjis, G. Z., and Park, T., 1995. Anisotropic Damage of Fiber-Reinforced Mmc Using Overall Damage Analysis. *Journal of Engineering Mechanics-*Asce 121, 1209-1217.

Voyiadjis, G. Z., and Park, T., 1995. Local and Interfacial Damage Analysis of Metal-Matrix Composites. *International Journal of Engineering Science* 33, 1595-1621.

Voyiadjis, G. Z., and Park, T., 1997. Anisotropic damage effect tensors for the symmetrization of the effective stress tensor. *Journal of Applied Mechanics-*Transactions of the Asme 64, 106-110.

Voyiadjis, G. Z., and Park, T., 1997. Local and interfacial damage analysis of metal matrix composites using the finite element method. *Engineering Fracture Mechanics* 56, 483-511.

Voyiadjis, G. Z., Taqieddin, Z. N., and Kattan, P. I., 2007. Micromechanical approach to damage mechanics of composite materials with fabric tensors. *Composites Part B-Engineering* 38, 862-877.

Voyiadjis, G. Z., and Thiagarajan, G., 1996. A cyclic anisotropic-plasticity model for metal matrix composites. *International Journal of Plasticity* 12, 69-91.

Yun, S., and Palazotto, A., 2007. Damage mechanics incorporating two back stress kinematic hardening constitutive models. *Engineering Fracture Mechanics* 74 2844–2863.

Zhu, Y. Y., and Cescetto, S., 1995. Fully Coupled Elasto-Visco-Plastic Damage Theory for Anisotropic Materials. *International Journal of Solids and Structures*, 32, 1607-1641.

Zukas, J. A., 1990. High Velocity Impact Dynamics. Wiley, New York.

PART II

Advances in Statistical Damage Mechanics (SDM): New Modeling Strategies

by

Antonio Rinaldi

Advances in Statistical Damage Mechanics (SDM): New Modeling Strategies

Antonio Rinaldi
NAST Centre & Department of Chemical Science and Technology
University of Rome "Tor Vergata", Rome, Italy
(formerly at MAE Dept., Arizona State University, Tempe, AZ)
antonio.rinaldi@uniroma2.it

Abstract: A handful of methodologies can be pursued in alternative to (and in combination with) continuum mechanics to develop advanced damage models that are intrinsically suited to address complex issues, such as strain localization phenomena and sample-size dependence of structural failure, which are intimately related to microscale phenomena. This chapter features concepts and solutions from Statistical Damage Mechanics (SDM), which seeks to devise non-conventional approaches to fracture and damage by means of discrete damage models accounting for the essential microscale property of a material or structural system. The ultimate goal is the development of multiscale methodologies that can reliably predict the materials macroscale response in consideration of the microstructural evolution caused by the damage process. The constitutive damage models in this overview are established from physically-based definitions of the damage parameter(s) that stem from a pervasive (bottom-up) statistical rationale aided by a vast arsenal of modeling tools (*e.g.* descriptive and inductive statistics, fractal theory, computer graphics, numerical methods).

1. Introduction: from Continuum to Discrete Modeling

The properties of an engineering material at the macroscale are always the result of the properties of its microstructure, the details of which are normally observable only at a microscale of much finer resolution length[1]. When an irreversible transformation of the microstructure takes place due to nucleation, propagation,

[1] Microscale is not to be intended metrically, as having a length scale of a micrometer, which will be instead referred to as micrometer scale to avoid confusion. The prefix *micro* will be used consistently in the chapter to indicate microscale quantities.

and coalescence of defects, particularly microcracks[2] and voids, in a material subject to applied loads and diverse external stimuli (during both manufacturing and exercise), such a damage process is expressed at the continuum scale by the progressive deterioration in its mechanical and functional properties. Accordingly, the main objective of damage mechanics (DM) is the determination of such dependence between the microstructural evolution from microcracking and the macroscale effects to predict material and structural properties crucial to engineering design and maintenance (*e.g.* damage tolerance[3] and residual life of aircraft and aerospace structural elements).

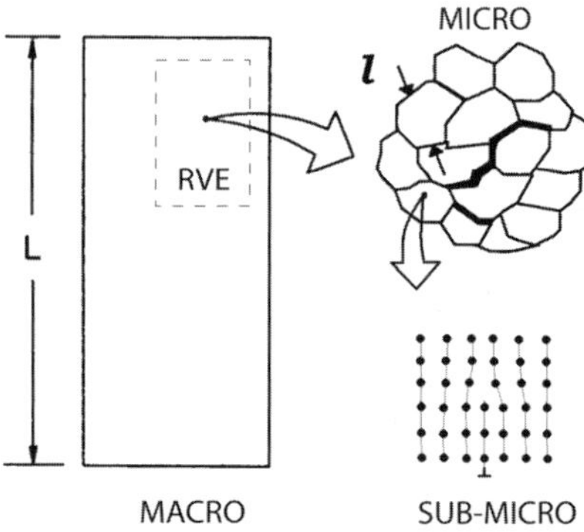

Figure 1. Hierarchy of length scales associated to a damage problem. The scope of the present multiscale study does not encompass for the sub-microscale.

[2] Microcracks "are an ubiquitous feature of engineering materials,[..] defined as flat defects with an atomically sharp tip. They will be modeled by a surface which is not penetrated by the interatomic bonds, *i.e.* as a pair of mating surface which do not adhere to each other. An internal surface will be, therefore, referred to as a microcrack if it: (a) can support a discontinuity in at least one component of the displacement vector, (b) has an atomically sharp tip and (c) is commensurable in size to the micro texture (size of the grain facet, distance between two adjacent fibers, etc.). A microcrack of the microstructural heterogeneity size "sees" the surrounding material as heterogeneous, anisotropic and often discontinuous. This consequence of the microcrack "smallness" is seldom acknowledged and almost never introduced into analyses" - Krajcinovic (1996).

[3] "Damage tolerance" – at least as defined in the aircraft industry - is the ability of a structure to sustain anticipated loads in the presence of fatigue, corrosion or accidental damage until such damage is detected through inspections or malfunctions and repaired, whereas "residual life" is the ability of the structure to sustain degradation from such sources as fatigue, accidental damage and environmental deterioration to the extent that they can be controlled by economically acceptable maintenance and inspection programs, Goranson (1993).

In the example depicted in Figure 1, the diffuse intergranular microcracking process ongoing in a polycrystalline material, although affecting the overall mechanical behavior and eventually leading to failure, would be typically hidden at the macroscale (the continuum level) if the resolution length L (defining the application) is much coarser than the average grain size l characteristic of the microstructure and of the microscale itself. As every theory or experiment has its own characteristic length that defines resolution and scope of the data, so in this case the microscale is the characteristic scale of the microcracking problem and there is generally no need to consider sub-micro scales (*e.g.* micrometer, nanometer, atomic, and sub-atomic scales, depending on the grain size), at least in first instance and provided that is not required by other inelastic processes like plasticity or creep[4]. In this view, damage is seen as a multiscale problem where a minimum of two length scales needs consideration for an accurate description. Unfortunately, if the gap between L and l is too large, the model resolution can hardly be set to l, which is still the typical situation encountered in the design of large structures (*e.g.* airplanes, cars, buildings, components, etc.) despites the enormous and continued increase in computational power. Then, the selection of the model resolution is driven by a trade-off between computational capabilities and desired accuracy, making it prohibitive (at present) to account directly for random texture and properties of extended material microstructures, and forcing instead the recourse to an effective continuum description where microscale defects are smeared out.

Continuum models, such as those discussed in previous chapters, represent in fact the mainstream in DM and often voice a top-down approach that seeks to "stay" on the macroscale rather than confront directly microstructural heterogeneity. Thus, a damage state is frequently described therein by a continuum damage parameter that is either chosen on a phenomenological basis or derived from micromechanics via one of the many *homogenization* or *coarse graining* techniques available. In particular, the "representative volume element" (RVE) is the basic tool of micromechanics to convert a disordered material into an equiv-

[4] According to Krajcinovic (1996), page 6, "A rational and rigorous description of the geometry of the microstructure (of a polycrystalline solid) on the atomic level must involve an appropriate specification of a particular random microstructure and a statistical characterization of the microstructural texture and a multitude of point, line and planar micro defects. The statistical characterization must emphasize the universal aspects and simultaneously blur away the inconsequential details. The subtleties, precision and rigor of the selected statistical characterization of the microstructural texture and the micro-defect population depends on the purpose of the analysis and the required resolution of the model".

alent continuum model and represents the smallest specimen volume of disordered matter (ref. Figure 1) that can be considered as statistically homogeneous (and, hence, in thermodynamic equilibrium – Krajcinovic and Rinaldi (2005)) under the action of nearly-uniform tractions at its boundaries. As such, the RVE size ξ sets a mesoscale (*i.e.* $l < \xi \leq L$) marking the lower threshold for the continuum description. In the RVE domain the local micro-properties can be simply averaged into effective properties, which yet depend on the specific averaging technique used to enforce the global redistribution of local effects of individual microcracks. The "dilute concentration" and "self-consistent" models in Krajcinovic (1996) are emblematic examples in this respect. Most notably by a damage localization viewpoint, the RVE may not even exist if the damage distribution ceases to be statistically homogeneous, as depicted in Figure 2. Then the relevant random micro-fields become dependent on position in space and are governed by the presence of cooperative phenomena[5] between existing defects and microcracks (Kreher and Pompe, (1989)). Early micromechanical models were usually adequate in the range of small damage density, but fell short indeed at the onset of damage localization, when the material heterogeneity increases to the point that one or few micro defects in a localized region exert a dominant influence on the macro-response of the solid. Henceforth, rate, mode, and pattern of material degradation depend on the random topology and morphology of the microstructure in an essential manner. The transition in structural behavior caused by damage localization is marked by the rise of several complex issues, such as the sample-size dependence of the material behavior, the increased data scatter in the mechanical response, and all sorts of numerical problems (*e.g.* mesh dependency). From that moment, direct consideration of the microscale cannot be denied any longer and the problem requires a true multiscale strategy. Of course, the presence of two length scales represents an intrinsic difficulty to continuum modeling and several sophisticated (analytical and numerical) stratagems have been developed to cope with the situation, as discussed at length previously in the book. Internal "length scale" parameters, for example, are used in non-local continuum to

[5] "A *cooperative phenomenon* is an effect which requires simultaneous interaction among many closely spaced defects. For example, the existing microcracks in a specimen subjected to compressive normal stresses in all three directions may not be able to grow. The mean-field stresses are sufficient to maintain order. As the system becomes more and more disordered by nucleation of additional microcracks the effect of the average ordering is diminished. The elastic energy release rate of a system of preferentially oriented microcracks may be increased by their direct interaction and eventually reach the critical magnitude. The range over which these defects grow as a result of their direct interaction may finally grow to the specimen size. The threshold of the short to long range correlation transition, or *localization* (emergence of the shear bands), is often reached in an "avalanche" mode" – Krajcinovic (1996), page 21.

transfer the microscopic evolution to the macroscale, although such phenomenological parameters may sometimes be difficult to justify or identify by a physical standpoint. "Integrated" approaches also exist, where coupled governing equations for microscale and macroscale are solved at once, such as the XFEM and the multifield theories (*e.g.* Moes et al (1999), Mariano and Stazi (2001)). Great effort, though, is being made to cast hierarchical and modular multiscale approaches that properly link together separate models for the two scales (*e.g.* Belytschko et al. (2008)), which favors the implementation of damage model at the microscale within commercial finite elements software.

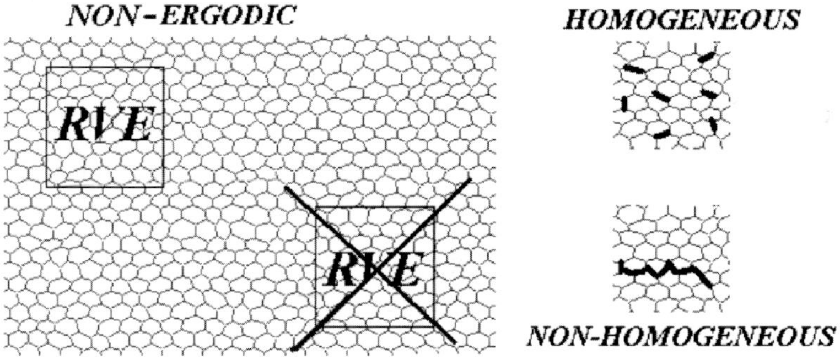

Figure 2. (Statistical) homogeneous *vs.* non-homogeneity of damage distribution and implication on the existence of an RVE of given size.

But continuum modeling is not the only option to address damage localization problems. Alternative, less conventional routes can be pursued. In particular, this last part of the book is dedicated to statistical damage mechanics (SDM) and microscale modeling. While an exhaustive treatment of SDM is out of scope here, the chapter will focus on recent advances and seminal ideas that can potentially drive or inspire the subject development in the short-term. By reversing the usual modeling perspective, the SDM approach stems from a bottom-up modeling rationale that sees the material as a system of interconnected microscale elements and seeks to obtain the macroscale properties straight from the direct consideration of the microstructural properties. SDM attempts to accurately reproduce the random microstructure of a material system by means of discrete models where inner disorder is accounted for through statistical distributions representative of the micro-properties of interest. Such discrete models rarely admit closed-form solutions due to the enormous number of degrees of freedom (*i.e.* number of equations), but need to be solved numerically by means of demanding computations. However, discrete models are applicable, regardless of whether the microstructure is statistically homogeneous or not. That appears unquestionably the most appealing characteristic of SDM modeling. The output data from this type of models is potentially very accurate and informative of the damage process in all regimes, which permits an unprecedented widespread application of statistical methods (see *e.g.* Hansen and Roux (2001)) and non-standard mathematical tools (*e.g.* fractal theory) to derive physically-based constitutive relations and damage theories that effectively reckon subtle aspects, such as sample-size effects, localization threshold, intrinsic variability of mechanical properties, and damage-induced anisotropy.

At the current stage it is hard to envision SDM as a full substitute for continuum models. SDM is rather a powerful synergic element of a hybrid strategy that includes continuum mechanics too. In fact, unless it became feasible to routinely solve systems of multibillions equations, SDM cannot realistically compete as a stand-alone, self-sufficient approach for engineering design of large structures. And even when that were possible, a "brute force" computational design approach would be impractical and erroneous. Instead, the constitutive models derived from SDM could fit well into the modular, multiscale logic mentioned above.

At large, the SDM strategy is really a compound endeavor requiring a series of diverse, multidisciplinary steps:

1- experimental characterization of the microstructure to obtain the necessary input data (in a statistical format);
2- computer representation of one (or of one class of) microstructure that captures both geometrical and physical properties;
3- simulations to produce accurate numerical predictions of the damage process;
4- analysis of the output data to characterize the model response via descriptive statistics (*e.g.* mean response, spread, distribution fitting);
5- identification of the damage parameter, derivation of approximate analytical solutions applicable in the continuum level models (*e.g.* scaling relations accounting for sample-size effects).

The SDM approach assumes that a virtually complete information of the microstructure be made available in terms of morphology, phases, and properties to build realistic microscale models. In reality, obtaining this type of data is by no means trivial and represents currently a formidable problem on its own, determining a gap between microscale modeling and characterization capabilities. Lately this issue has gained the renewed attention of the research community worldwide, both in the academia and in the industry, due to the significant advances in material science – especially in the areas of small scale microscopy and micro-mechanical probing (*e.g.* see MRS Bullettin (2008))[6]– that promise to bridge the current gap in the near future. In this overview, though, the microscale "characterization and representation" problem (*i.e.* the first two enlisted steps) is of secondary importance since the emphasis is on the damage simulation and modeling part (*i.e.* the last three enlisted steps) and can be overlooked. Instead, for argument sake, principles and key aspects of SDM modeling are presented on simple, ideal damage models consisting of parallel bundles (*i.e.* one dimensional models) and networks (*i.e.* higher dimensional models) made of spring-like elements, which none the less are representative of certain classes of real material and structural systems.

A pure damage process is revealed by the progressive loss of secant stiffness of the material and by a merely elastic unloading upon load removal (*e.g.* see damage in ceramic matrix composite monitored via ultrasonic evaluation by Audoin and Baste (1994)). In mathematical terms the corresponding constitutive equations of damage for the full 3D problem take up the following structure originally proposed by Kachanov (1958)

$$\bar{\sigma}_{ij} = \bar{C}_{ijkl}\left(\bar{D}_{ijkl}\right)\bar{\varepsilon}_{kl} = \bar{C}^{0}_{ijkl}\left(1-\bar{D}_{ijkl}\right)\bar{\varepsilon}_{kl} \quad \left(i,j,k,l=1..3\right), \tag{1}$$

where $\{\bar{\sigma}_{ij}, \bar{\varepsilon}_{kl}\}$ are the stress and strain second order tensors respectively, $\bar{C}^{0}_{ijkl}$ and $\bar{C}_{ijkl}$ are the macro-scale elastic components of the generalized Hooke's law in pristine and damaged conditions, and $\bar{D}_{ijkl}$ is an appropriate macroscopic damage (tensor-valued) parameter[7]. For one-dimensional systems or materials subject to a uniaxial loading, either tensile or compressive, a reduced scalar formulation is commonly adopted

$$\bar{\sigma} = \bar{E}(\bar{D})\ \bar{\varepsilon} = \bar{E}^{0}(1-\bar{D})\bar{\varepsilon}\ , \tag{2}$$

where $\{\bar{\sigma}, \bar{\varepsilon}\}$ are stress and strain along the loading axis, $\bar{E}^0$ and $\bar{E}$ are the Young's moduli in pristine and damaged conditions, and $\bar{D}$ is the scalar damage parameter. In either case, the modeling problem basically consists of determining

[6] In particular, the advent of nanotechnology has fostered improvements of electron (*e.g.* SEM, TEM) and probing (*e.g.* AFM, STM, SPM, etc.) microscopy, together with novel technologies, such as microtesting machines, nanoindenters, microtomography, and focused ion beam (FIB), for imaging, machining, and testing of materials down to the submicrometer scale. Software packages (*e.g.* MIMICS, SURFDRIVER , AMIRA, etc.) have been created in conjunction to facilitate the handling of the enormous amount of data provided by such innovative tools. Today, for instance, the virtual reconstruction of a microstructure from serial sectioning is still an involved task but not prohibitive any more, and a full automation is already at sight.

[7] Throughout the chapter, the bar sign over symbols, e.g. Eq. (1), shall indicate an "effective" macroscopic quantity, to distinguish it from its counterpart at the microscale.

the proper damage variable, which fully encapsulates the complexity of the damage process. SDM is particularly well-suited for this task and can lead to identify the correct parameter, *i.e.* $\bar{D}_{ijkl}$ or $\bar{D}$, for a given situation based on its physical bearing. In this regard, rational derivations of $\bar{D}$ are illustrated extensively in the discussion at several points.

The chapter is conceived as a small collection of case studies, arranged for increasing complexity and logically linked to show primarily how the intricacy of the damage process rises with the dimensionality of the system. Of course, to set the stage, one dimensional models (parallel bundles) are analyzed first, highlighting classical closed-form results and the rational definition of the damage parameter. Higher dimensional models of damage are examined next, featuring localization processes and sample size effects like in most real systems. In the latter case, emphasis is placed on the derivation of what appears to be the first rational model of damage developed to date for disordered networks. At last, sample size effects are analyzed, presenting one possible approach to derive scale-invariant relations based on fractal theory.

For the most part, the discourse will refer to quasi-brittle materials affected by sole damage (such as certain ceramics and concrete) and treatable with Eqs.(1-2). Yet, brief consideration is given to the coupling between damage and plasticity in ductile systems, for in that case cracking formation is rarely an isolated inelastic process and is almost invariably accompanied with other deformation mechanisms, such as plasticity, creep, corrosion. More specifically a novel one dimensional SDM model is discussed here for the first time, with the twofold intent to show the implications of the presence of more inelastic processes active simultaneously and the solutions offered by SDM.

2. One Dimensional Models: Parallel Bundles Systems (PBS)

A parallel bundle model consists of N links (springs or bars) of equal length connected in parallel, and is henceforth also called "parallel bar systems" (PBS). Each link is a two-force structural element characterized by at least an axial stiffness k_i that fully determines the mechanical response of the i-th bar in the elastic domain. If the bars are perfectly brittle it is necessary to prescribe the rupture threshold, whereas, if the bars are ductile, it is also necessary to prescribe the onset of yielding and the hardening law. The introduction of disorder is facilitated by the simplicity of the model. Typically, the disorder may be quenched so that probability distributions must be assigned to links stiffnesses, strengths, and yield thresholds. As a consequence of disorder, the nonlinear behavior during brittle or ductile deformation arises from the sequential random rupturing or yielding of bars, respectively[8].

A PBS can be either *loose* or *tight* depending on the rule by which the externally applied force is shared by extant links. In the former case the externally applied force is equally distributed through all extant links according to a *Global Sharing Rule* (GSR). Such a "democratic" sharing of loads in the system is enforced by admitting no load-transfer between neighbor links and providing at each end of the bundle a transversal rigid bus-bar that can move only in the direction of the applied force (longitudinal axis of the system, as in Figure 3). On the contrary, in the other case, a *Local Sharing Rule* (LSR) is assumed by letting adjacent bars to interact via some form of a shear-lag or frictional operative mechanism that causes extant links near a defect to carry a larger quota of the load. Analytical solutions are available in these two extreme cases for average load curve or other relevant statistics (*e.g.* see Sornette (1989), Kloster et al (1997)). For more general load sharing rules only numerical solutions are generally available, like for compliant bus-bars redistributing load non-homogenously near broken links (*e.g.* Touissant (2006)). By restraining the scope to loose PBS, two models are considered next for the cases of "pure damage" and "damage-plasticity".

2.1. Pure Damage: Quasi-Brittle PBS

Perhaps the simplest discrete model of damage evolution in a disordered material is the PBS analyzed by Krajcinovic and Silva (1982), made of an ensemble of N elastic bars of same size and stiffness connected by two rigid end-bars as in Figure 3 and subject to quasi-static monotonic axial elongation $\overline{u}$, *i.e.* the controlled variable. A statistical distribution $p_f(\overline{u})$ of critical force strengths[9] is used

---

[8] Historically, the PBS was first used in applied mechanics to estimate the rupture strength of threads, low twist yarns and cables. This rather simple and attractive artifice was later extensively applied to studies of ductile and brittle phenomena (Krajcinovic (1996), page 56, and references therein). As of today, the underlying rationale is still applied in ISO and UNI technical standards related to structural systems such as steel ropes and brakes.

[9] In the text the function $p_f(\overline{u})$ is referred to as strength distribution, while it is actually the distribution of the associated critical displacements $\hat{u}_i$ *(i*=1..*N)* of the bars. Due to the one-to-one correspondence between $\hat{u}_i$ and the force strength $\hat{f}_i$ (representing the actual physical property), it is indeed practically equivalent to consider either one of them. However, it is more convenient to think in terms of $\hat{u}_i = f(\hat{f}_i)$ because the displacement $\overline{u}$ is consistently taken as the controlled macro-variable in this discourse, like in displacement controlled tests. For elastic bars, $\hat{u}_i$ and $\hat{f}_i$ are linearly related through k_i as $u_i = f_i / k_i$ and express substantially the same variability. The reader be aware that later in this book chapter springs can be assigned a distribution either of critical displacements

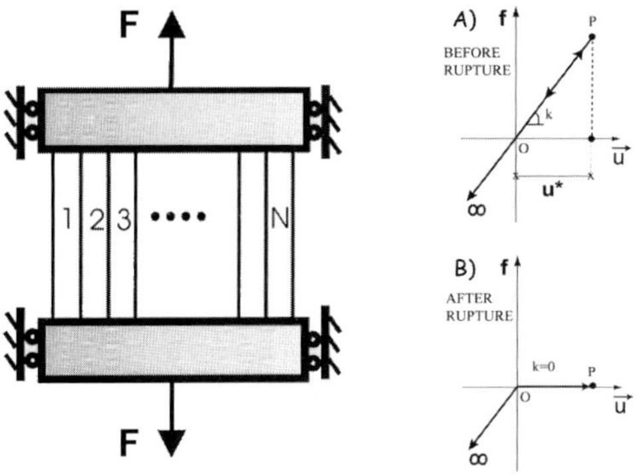

Figure 3. PBS model (left) and micro constitutive law of one bar (right), showing the tensile response prior and after rupture.

to introduce disorder such that bars fail randomly at different time during pulling. By definition, in this PBS there is no privileged interaction between near-neighbors links and load redistribution, caused by the rupture of one bar, affects equally the whole ensemble of survivor bars by means of the connective rigid bars. This model is appealing for its simplicity because constitutive relations in Eq.(2) are derivable analytically. More specifically, by assuming that prior to random failure the i-th link bears a load-share $f_i = k \cdot \overline{u}$, then the undamaged secant stiffness of the pristine system is $\overline{K}_o$ and the axial equilibrium renders the $\overline{F} - \overline{u}$ response

$$\overline{F} = \overline{K} \cdot \overline{u} = \sum_{i=1}^{N-n} f_i = \overline{K}_o \left(1 - \frac{n}{N}\right) \cdot \overline{u} = \overline{K}_o \left(1 - \overline{D}\right) \cdot \overline{u} \quad , \tag{3}$$

where $\overline{K} = \overline{K}_o[1-(n/N)]$ is the effective secant stiffness at any given $\overline{u}$. Hence, the damage parameter $\overline{D}$ emerges simply as the ratio of the number of broken bars n to N. Since damage is a stochastic process and the N strength values randomly sampled from $p_f(\overline{u})$ are unknown, the exact determination of $\overline{D}$ is generally impossible and statistical estimates must be pursued instead. If each rupture can be correlated to the displacement $\overline{u}_i^*$ that caused it during the process, then analytical estimates can be found for $\overline{D}$. In this specific PBS, this linkage is provided by the strength distribution. Since the displacement field is homogeneous at the microscale ($u_i = \overline{u}$)

$p_f(\overline{u})$ (*e.g.* the PBS in Section 2) or of critical strains $p_f(\overline{\varepsilon})$ (*e.g.* for the networks in Section 3), depending on whether the strength represents a force or a stress respectively.

and each bar has the same cumulative probability of failure $P(\overline{u})$, the mean estimate of $\overline{D}$ is simply given by the expected value (*i.e.* the linear operator $\langle \circ \rangle$) of the quenched distribution at $\overline{u}$

$$\langle \overline{D} \rangle = \frac{\langle n \rangle}{N} = \int_0^n \frac{dn}{N} = P(\overline{u}) = \int_0^{\overline{u}} p_f(u)du \,. \tag{4}$$

Interestingly enough, this definition does not depend on the specific distribution p_f and provides the rational connection between microscale disorder and macroscale structural degradation. Closed-form solutions are evidently possible depending on the integrability of p_f . For the uniform distribution $p_f = 1/\overline{u}_{\max}$ shown in Figure 4, the expected damage curve $< \overline{D}(\overline{u}) >$ is a straight line and the force response is a parabola according to

$$\langle \overline{F} \rangle = K_0 \left(1 - \frac{\overline{u}}{\overline{u}_{\max}} \right) \overline{u} \quad , \tag{5a}$$

$$\langle \overline{D} \rangle = \int_0^{\overline{u}} \frac{1}{\overline{u}_{\max}} du = \frac{\overline{u}}{\overline{u}_{\max}} \quad . \tag{5b}$$

The smoothness of this damage curve reflects that the damage process is based merely on random microcracks nucleation in this inherently 1D loose PBS model. In fact, unlike the higher dimensional models discussed later, no homogeneous-to-heterogeneous phase transition takes place in the PBS for the GLS rule prevents damage localization. As seen later, two-dimensional models are better suited to capture cooperative phenomena from microcracks interaction but the mathematical treatment is no longer as simple. Before moving on to higher dimensional damage models (Section 3), prior consideration will be given to the generalization of this PBS to ductile systems in Section 2.2.

The Eqs.(5) fulfill only partially the modeling goal. The statistical characterization of any stochastic process is generally incomplete without a measure of the variability accompanying the mean estimate. This fact should not be overlooked by scientists, engineers, and practitioners, who are sometimes unfamiliar with the broader statistical thinking and instinctively confine their attention just to the mean response. The consideration of the variability is indeed fundamental, especially when the coefficient of variation[10] gets larger, in order to appreciate the precision and appropriateness of the mean response. For example, the triangular grayed area OAB in the *F-u* space in Figure 4, which delimits the domain of all infinite PBS responses from any possible set of *N* strength values randomly sampled from p_f ,

[10] The coefficient of variation is the ratio CV% = Standard Deviation / Average*100.

offers visual evidence that individual responses can deviate greatly from the mean prediction (*i.e.* the parabola (5a) plotted therein) and more markedly when $\bar{u}$ increases. Thus, the sole mean response is misrepresentative of the actual damage process. To cope with this shortfall, a satisfactory statistical characterization of Eq.(3) could consist, for example, of the two statistics

$$\langle \bar{F} \rangle = \bar{K}_o \left(1 - \langle \bar{D} \rangle\right) \cdot \bar{u} \ , \tag{6a}$$

$$Var\left(\bar{F}\right) = \bar{K}_o^2 \bar{u}^2 \cdot Var\left(\bar{D}\right) = \bar{K}_o^2 \bar{u}^2 \cdot \sigma_{\bar{D}}^2 = \sigma_{\bar{F}}^2 \ , \tag{6b}$$

where the variance operator *Var(○)* is chosen arbitrarily for convenience sake but any other measure of spread (*e.g.* range, standard deviation, etc.) can be chosen instead in Eq.(6b)[11]. The latter equation depends on $Var\left(\bar{D}\right)$ that in this case can be computed explicitly.

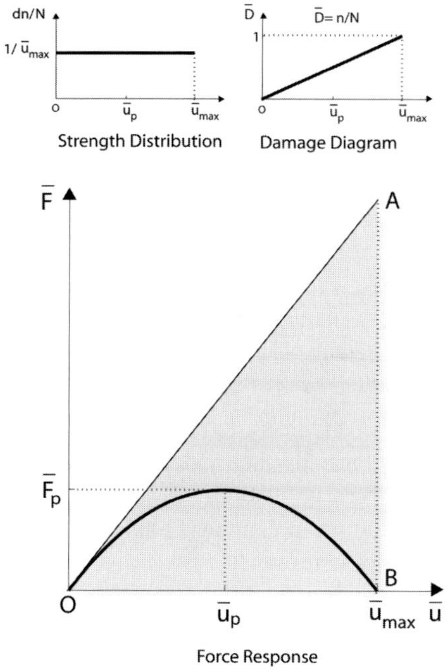

Figure 4. Average force response of the PBS model with uniform distribution of bar strengths under uniform tensile test. Strength distribution and damage curve shown on top. The grayed area AOB represents the domain of all possible PBS responses.

[11] Recall that if $y(x) = a + bx$ and x is a random variable, then $Var(y)=b^2 Var(x)$.

According to Eq.(4), $\bar{D}$ is indeed distributed as the number of broken bars n in a sample of N bars. Since each bar is independently and identically distributed (IID) according to p_f and has a failure probability $P(\bar{u}) = <\bar{D}>$, the N bars form what is known as a set of Bernoulli trials[12]. Under these circumstances n (and $\bar{D}$) is distributed as the binomial distribution[13]

$$p(n;\bar{u}) = \binom{N}{n} P(\bar{u})^n (1-P(\bar{u}))^{N-n} \quad , \quad n=0,1,..,N \tag{7}$$

and $Var\left(\bar{D}\right)$ can be readily expressed as an explicit function of $<\bar{D}>$

$$Var\left(\bar{D}\right) = \frac{Var(n)}{N^2} = \frac{P(\bar{u})[1-P(\bar{u})]}{N} = \frac{\langle\bar{D}\rangle\left(1-\langle\bar{D}\rangle\right)}{N} \quad . \tag{8}$$

Therefore, after the due substitutions in Eq.(6b), the mean constitutive model of this simple PBS and its variability assume the ultimate form

$$\langle\bar{F}\rangle = \bar{K}_o\left(1-\langle\bar{D}\rangle\right)\cdot\bar{u} \quad , \tag{9a}$$

$$Var\left(\bar{F}\right) = \bar{K}_o^2\bar{u}^2 \cdot \frac{\langle\bar{D}\rangle\left(1-\langle\bar{D}\rangle\right)}{N} \quad . \tag{9b}$$

which provide the final solution to this problem. Such notable equations are independent on the choice of p_f, are completely characterized by $<\bar{D}>$, and render closed-form solutions if $<\bar{D}>$ is an analytical form. In addition, they highlight that the mean response of the PBS is sample-size independent, but the variability (*i.e.* $Var(\bar{F}) = f(\bar{u};\bar{D},N)$) is not and diminishes with N. During the damage process, the variance is the result of three competing factors (*i.e.* $\bar{u}^2$, $\bar{D}$, $1-\bar{D}$) but tends to zero in the limit of zero damage ($\bar{D}\to 0$) and at failure ($\bar{D}\to 1$), and has its maximum at $<\bar{D}> = 0.5$ [14].

[12] In the statistical jargon, each bar in the sample is a Bernoulli trial that can be either a "success" with probability P, or a "failure" with probability $1-P$ (ref. Montgomery et al. (2001)). Actually, by a damage perspective the success is associated to the "rupture" event while the failure to the "no rupture" outcome. The failure probability is $P(\bar{u}) = <\bar{D}>$.

[13] The (discrete) binomial distribution $p(n)$ has 2 parameters $N \geq 0$ and $0 < P < 1$ such that

$$p(n) = \binom{N}{n} P^n (1-P)^{N-n} \quad , \quad n=0,1,..,N$$

and has mean $<n> = NP$ and variance $Var(n) = NP(1-P)$.

[14] Computed from $\partial Var\left(\bar{F}\right)/\partial\langle\bar{D}\rangle\Big|_{\bar{u}} = 0$ at fixed $\bar{u}$.

This example put clearly in evidence that the characterization of the variability of a process is generally a task far more complex than evaluating the mean value. Unlike the expected value operator $\langle \circ \rangle$, the $Var(\circ)$ operator is nonlinear and, hence, harder to handle. The functional determination of $Var(\bar{F})$ as stated by Eq.(6) is relatively_simple for this PBS just because Eq.(3) is linear in the only random variable $\bar{D}$. As ascertained later, that is rarely the case for more complex damage models, where modeling variability is a demanding task performed numerically most often than not. Such numerical evaluation represents a specialized topic on its own, falling outside the present scope and requiring substantial statistical expertise. Thus, while stressing the importance of quantifying variability case by case, this discourse about SDM is essentially restricted to the mean behavior.

2.2. Damage and Plasticity: Generalized PBS

The above loose PBS was built on the assumption that only damage was the active mechanism, which ultimately means that at the microscale the bars are allowed to behave elastically until a brittle failure (*i.e.* microcracking) occurs. But it is reasonable to question what would happen if their behavior were instead more complicated and other inelastic mechanism would pair with damage. In particular, it is interesting to gain a perspective of whether or not a closed-form solution would still be possible for such augmented problems. To date not many researchers in SDM have engaged with the detailed study of this sort of coupled problems and much work is necessary in that direction.

This section deals with a modified scenario where the previous PBS does not only undergo damage but also plasticity by allowing the bars to yield at the microscale. The modeling objective is to develop a generalized solution of the full coupled problem that yet would render the results from Section 2.1 as a special case. For tractability sake and with no loss of generality, the ductility of the bars is here accounted for by assuming the simple bilinear micro-constitutive relation shown in Figure 5, which is fully determined by four parameters: the elastic constant k, the hardening rate k^*, the strength, and the yield point. It is further assumed that k and k^* have pre-determined and identical values for the entire ensemble, while strength and yielding threshold are random variables sampled from the two quenched distributions $p_f(\bar{u})$ and $p_p(\bar{u})$ respectively[15].

[15] Analogously to the convention established in footnote # 9 about the strength distribution $p_f(\bar{u})$, the yielding threshold is conveniently expressed in terms of displacement and, hence, the yield distribution $p_p(\bar{u})$ does not refer to the yielding force f_i^* $(i=1..N)$ but rather to the associated yielding displacements u_i^* of the bars. Again, due to the one-to-one correspondence between u_i^* and the force strength f_i^* (representing the actual physical property), it is equivalent to consider either one of the two random variable. Therefore in this second PBS model, $p_f(\bar{u})$ and $p_p(\bar{u})$ are the distribution of the random

In this scheme, at the PBS level, damage and plasticity represent two distinct stochastic processes, the former being associated to the rupturing of the bars and ultimately controlled by the strength distribution $p_f(\overline{u})$, and the latter being linked to the yielding in the bars and to $p_p(\overline{u})$. Nonetheless, this conceptual partition of the coupled problem does not imply a true separation between the two fundamental inelastic processes, which can be readily shown to have different rank and to be linked hierarchically. In fact, by considering that all N bars have finite strength and are doomed to break eventually - regardless of plasticity - for the PBS to fail and that plasticity may only occur in survival bars, it is clear that damage and plasticity are related in a hierarchical order, with the latter being subdued to damage. In other words, damage can take place in the PBS without plasticity and can even hinder plasticity completely, while the converse is not true. This intuitive fact shall reflect in the structure of the constitutive model about to be developed.

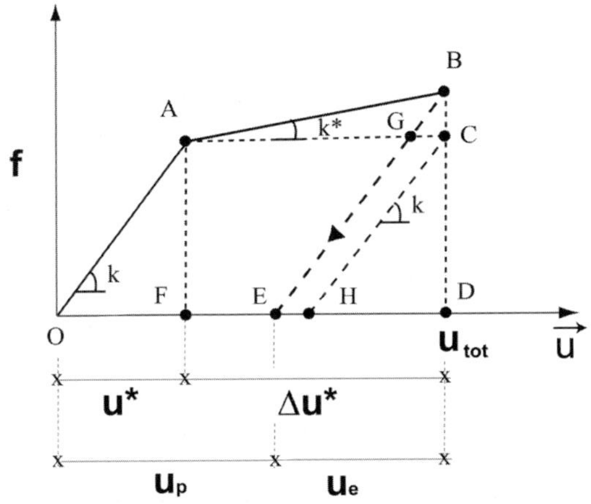

Figure 5. The bilinear micro-constitutive law of an elasto-plastic bar in the generalized PBS for ductile systems. Also a yield threshold and a constant hardening rate k^* are assigned to each bar, besides the usual k and strength. While k and k^* are equal for the entire ensemble, strength and yielding are random variables sampled from statistical distributions.

According to the previous bottom-up approach, also in this case the response of the PBS to monotonic tensile elongation $\overline{u}$ is the sum of the contributions from the

---

micro-variables $\widehat{u}_i^*$ and u_i^*. The expressions $\widehat{u}_i^* = f(\widehat{f}_i)$ and $u_i^* = g(f_i^*)$ are easy to determine but irrelevant to the discussion. Just note that it is no longer $\widehat{u}_i = \widehat{f}_i / k_i$ unless $\widehat{u}_i^* < u_i^*$ for the i-th bar.

unbroken bars, but each of them may have yielded or not. Then, if the scalar q represents the total number of broken bars at a given $\bar{u}$, the system response is

$$\bar{F} = \sum_{i=1}^{(N-q)} f_i = F_E + F_P \quad . \tag{10}$$

where the sum of all i-th contributions gets subdivided into two components F_E and F_P , the former cumulating the purely elastic responses from un-yielded bars and the latter accounting for the plastic response from "yielded" ones. If p indicates the number of yielded links, based on the assumed bilinear model in Figure 5, F_E and F_P can be calculated as

$$F_E(\bar{u}) = \sum_{i=1}^{(N-q)-p} k_i \bar{u} \quad , \tag{11a}$$

$$F_P\left(u_1^*,..,u_p^*;\Delta u_1^*,..,\Delta u_p^*\right) = \sum_{i=1}^{p}\left(k_i u_i^* + k_i^* \Delta u_i^*\right) \quad , \tag{11b}$$

where each random pair $\{u^*, \Delta u^*\}_i$ consists of the yielding displacement and the plastic displacement of the i-th bar and obeys the constraint

$$u_i^* + \Delta u_i^* = \bar{u} \, . \tag{12}$$

Then, by plugging Eqs.(11) back into (10) and recalling that k and k^* are assumed equal to all bars, the exact constitutive model in terms of microscale quantities is

$$\bar{F} = \sum_{i=1}^{(N-q)-p} k\,\bar{u} + \sum_{i=1}^{p} k\,u_i^* + \sum_{i=1}^{p} k^* \Delta u_i^* \quad . \tag{13}$$

This form of the PBS response serves the sole purpose of highlighting the structure of the problem but for practical purposes is intractable. Its exact computation would indeed require the knowledge of a large number of microscale degrees of freedom (*dof*) that are normally "hidden" to the engineer or scientist and - even if they were accessible - are too many and random in nature. More precisely, the total response $\bar{F}(\bar{u};u_i^*,..,u_p^*)$ depends on "p+1" *dof* since the knowledge of p *dof* (say, for instance, u_i^*) would suffice to determine F_P given the p constraint Eqs.(12) between u_i^* and Δu_i^* . Therefore, besides the sole controllable variable $\bar{u}$, also p random micro-parameters would be necessary, with p ranging up to N if all bars were to yield and potentially being an exceedingly large. Besides, q and p are themselves random variables that depend on the load history and vary from one PBS to another.

This is the typical scenario encountered in statistical mechanics, where ensembles in thermodynamic equilibrium characterized by an unmanageable number of *dof* are conveniently described in a statistical fashion in terms of average behavior and intrinsic variability. Likewise, here we pursue a statistical constitutive model representative on average of a PBS ensemble and dependent just on a handful of macroscale parameters (possibly just $\bar{u}$). To this purpose, strict knowledge of one PBS realization is irrelevant, while adequate statistical knowledge of the PBS microstructure valid any realization (*i.e.* p_f and p_p) is necessary to derive meaningful statistics. The expected response $< F >$ can be obtained from averaging all random variables in Eq.(13) (generally not correlated with each other) as

$$\langle \bar{F} \rangle = \left[\left(N - \langle q \rangle \right) - \langle p \rangle \right] k\, \bar{u} + \langle p \rangle k \left\langle u^* \right\rangle + \langle p \rangle k^* \left\langle \Delta u^* \right\rangle \quad . \tag{14}$$

The average quantities on the right hand side can be estimated from the quenched distributions $p_f(\bar{u})$ and $p_p(\bar{u})$. In the most general case, these two distributions need not to be related, and $p_f(\bar{u})$ and $p_p(\bar{u})$ can be in principle selected independently, provided that strength and yielding threshold are independent variables. Under this assumption and owing to the homogeneous displacement field at the microscale, the average numbers of broken and yielded bars can be estimated as

$$\langle q \rangle = N \int_0^{\bar{u}} p_f \; du \quad , \tag{15a}$$

$$\langle p \rangle = \left(N - q \right) \int_0^{\bar{u}} p_p \; du \quad . \tag{15b}$$

As anticipated in the initial comments, the latter parameter associated with plasticity depends on the former associated with damage. In analogy to the brittle PBS, Eqs.(15) can then be rewritten more meaningfully as

$$\langle q \rangle = \left\langle \bar{D}_f \right\rangle N \quad , \tag{16a}$$

$$\langle p \rangle = \left\langle \bar{D}_p \right\rangle \left(1 - \left\langle \bar{D}_f \right\rangle \right) N \quad , \tag{16b}$$

with the introduction of the two intensive thermodynamic quantities $\bar{D}_f$ and $\bar{D}_p$ that represent (macro) *generalized damage parameters* for each of the two ongoing inelastic processes and are defined as

$$\left\langle \bar{D}_f \right\rangle = \int_0^{\bar{u}} p_f \; du = \frac{\langle q \rangle}{N} \quad , \tag{17a}$$

$$\left\langle \bar{D}_p \right\rangle = \int_0^{\bar{u}} p_p \; du = \frac{\langle p \rangle}{N - \langle q \rangle} \quad . \tag{17b}$$

The substitution of Eqs.(16) into Eq.(14) renders

$$\bar{F} = N\left[1-\bar{D}_p+\bar{D}_f\bar{D}_p-\bar{D}_f\right]k\,\bar{u}+ \\ +N\,\bar{D}_p\left(1-\bar{D}_f\right)k^*\left\langle\Delta u^*\right\rangle+N\,D_p\left(1-D_f\right)k\left\langle u^*\right\rangle \tag{18}$$

where the average symbol $\langle\circ\rangle$ is implied for $\bar{F}$, $\bar{D}_f$, and $\bar{D}_p$ without endangering clarity, and so will be omitted hereafter for readability.

The other two averages $<u^*>$ and $<\Delta u^*>$ in Eq.(18) can also be estimated from the knowledge of $p_p(\bar{u})$. In fact the former is simply the first moment of the distribution while the latter is found by difference from Eq.(12), as

$$\left\langle u^*\right\rangle = \int_0^{\bar{u}} u\cdot p_p du \quad , \tag{19a}$$

$$\left\langle \Delta u^*\right\rangle = \bar{u} - \left\langle u^*\right\rangle \ . \tag{19b}$$

The definitions (17) and (19) fully specialize Eq.(18) that provides the actual average constitutive equation of the generalized PBS. In comparison to model (9a) for the brittle PBS, the structure of the equation is significantly more complex, but closed-form solutions are still possible depending on the integrability of both p_f and p_p. This statistical rational model can be further rearranged to reflect the partition in Eq.(10) according to

$$\bar{F} = \overbrace{N\left[1-\bar{D}_p+\bar{D}_f\bar{D}_p-\bar{D}_f\right]k\,\bar{u}}^{F_E} + \overbrace{N\,\bar{D}_p\left(1-\bar{D}_f\right)\left[k^*\left\langle\Delta u^*\right\rangle+k\left\langle u^*\right\rangle\right]}^{F_P} \tag{20}$$

and finally

$$\bar{F} = \overbrace{N\left(1-\bar{D}_p\right)\left(1-\bar{D}_f\right)k\,\bar{u}}^{F_E} + \overbrace{N\,\bar{D}_p\left(1-\bar{D}_f\right)\left[k^*\left\langle\Delta u^*\right\rangle+k\left\langle u^*\right\rangle\right]}^{F_P} \tag{21}$$

that has a clear-cut statistical meaning. In fact, the factor $\left(1-\bar{D}_p\right)\left(1-\bar{D}_f\right)$ in F_E corresponds to the fraction of bars that are both unbroken and not yielded, and exhibit a mere elastic behavior $f_i = k\,\bar{u}$. On the other hand, the factor $\bar{D}_p\left(1-\bar{D}_f\right)$ in F_P accounts for the fraction of bars that are both unbroken and yielded. Statistically speaking, $\left(1-\bar{D}_p\right)\left(1-\bar{D}_f\right)$ and $\bar{D}_p\left(1-\bar{D}_f\right)$ are conditional probabilities associated to two types of alternative events (*i.e.* elastic response *vs.* plastic response) that the $N\left(1-\bar{D}_f\right)$ survivor bars can take up. Also, Eq.(21) translates in mathematical terms the hierarchal dependence pointed out earlier, with damage being dominant and plasticity subordinate to it. In fact, plasticity can occur only before complete failure (*i.e.* prior than $\bar{D}_f = 1$), where the damage process is entirely ultimated and plasticity gets "switched off " along.

Similarly to the former case, the result for the generalized PBS has been achieved under rather general requisites, independently of a specific choice of distributions p_f and p_p. Only the absence of correlation between them was assumed so far and, consequently, the model can account for an enormous variety of system responses. As depicted in Figure 6, three types of scenario can occur, depending on the mutual "positioning" of p_f and p_p, shown along with the macro response. Case A depicts a fully ductile response of the PBS before damage, resulting from disjoint distributions such that all bars have a yielding threshold much smaller than the strength (*i.e.* p_p precedes p_f on the $\bar{u}$ axis) and damage takes place only after bars have yielded. By a macroscopic standpoint, the plastic process has the effect of lowering the forward tangent stiffness $\bar{K}_T = \partial\bar{F} / \partial\bar{u}$ while not altering the backward secant stiffness $\bar{K}_S = \bar{F} / \bar{u}_e$ (with $\bar{u}_e$ being the elastic component of the displacement $\bar{u}$). In the limit of bars having infinite strengths (*i.e.* $p_f(\bar{u})$ sitting at the infinity of the $\bar{u}$ axis or sufficiently away from $p_p(\bar{u})$), damage would be suppressed, leaving an indefinitely ductile PBS affected just by plasticity. Correspondingly, by setting $\bar{D}_f = 0$ in Eq.(21), the general model would specialize as

$$\bar{F} = \overbrace{N\left(1-\bar{D}_p\right)k\,\bar{u}}^{F_E} + \overbrace{N\,\bar{D}_p\left[k^*\left\langle\Delta u^*\right\rangle + k\left\langle u^*\right\rangle\right]}^{F_P} \quad . \tag{22}$$

On the other extreme, when the sampled yielding thresholds are much larger than the strengths (*i.e.* p_f precedes p_p) , only damage would happen as all bars break before yielding can occur, as portrayed in Case C. Evidently, unlike plasticity, the damage process alters both $\bar{K}_T$ and $\bar{K}_S$. A reduced model for pure damage is obtained by setting $\bar{D}_p = 0$ in Eq.(21) as

$$\bar{F} = \overbrace{N\left(1-\bar{D}_f\right)k\,\bar{u}}^{F_E} \quad , \tag{23}$$

which unsurprisingly retrieves Eq.(9a) for the brittle PBS. This confirms that the generalized model indeed comprehends the previous PBS as a special case, as desired. In between these two situations, when the spans of p_p and p_f overlap, the PBS exhibits a mixed response shown in Case B, which is a compound of damage and plasticity and forces to use the full model equation (21).

Of course, as stated in Section 2.1, the model (21) for the mean response should be accompanied by another equation describing the variability of the generalized PBS, such as the variance

$$Var\left(\bar{F}\right) = f(\bar{u}, \bar{D}_p, \bar{D}_f, N) \quad . \tag{24}$$

This equation is given in functional form because it cannot be expressed explicitly as easily as Eq.(6b) or Eq.(9b) for Eq.(21) can be easily proved to be non linear in the two random damage variables $\overline{D}_p$ and $\overline{D}_f$. An analytical solution of this variability problem is not yet available and, as anticipated, is not attempted here.

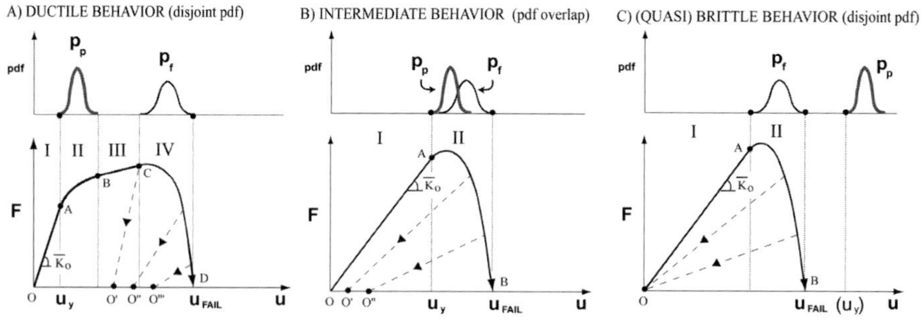

Figure 6. Three possible cases covered by the generalized PBS.

2.2.1. Correlation Between p_p and p_f

The model derivation must be properly modified if the two distributions p_f and p_p are correlated. Similar situations are easily envisioned in nanotechnology applications dealing with CNTs, metal wires, rods, and so on. The adjustments need to be evaluated case by case. For illustrative purposes, consider the hypothetical scenario depicted in Figure 7 of a PBS made of perfect single crystal metal bars. It is further assumed that the material has cubic symmetry (*e.g.* Ni, Cu, Au, etc.) and that all bars have the *c*- material axis in common and normal to the bundle plane. Then, the other two ones, *i.e.* the *a-b*, can freely rotate in plane randomly in the range θ = 0°-90°. Figure 7(A) shows the material axes for θ = 45° and θ = 0°. Then, strength and yield threshold of a bar do not only depend on θ but are also mutually related since the <010> and <100> planes represent both the slip planes and the cleavage planes in this type of materials. Assuming that the strength $\sigma_f^{<100>}$ and τ_{CRSS} are the strength and critical resolved shear stress of these planes, the axial load necessary to cause failure or yielding in bar are reported in Figure 7(B) as a function of θ and normalized to $\sigma_f^{<100>}$ and τ_{CRSS} . It is intuitive that a bar has maximum failure probability when the yielding probability is minimum and vice versa, regardless of the specific choice of p_f and p_p . Interestingly, no plasticity would occur if θ=0° for the assumed single slip mechanism is expected to be hindered by a zero shear stress on the slip plane. By a modeling standpoint, the rupturing of the weaker bars during the damage process would imply the depletion of the lower tail of p_p , potentially slowing down plasticity. Then a conditional distribution $p_p(\overline{u}; p_f, q)$ should be adopted to account for such evolution.

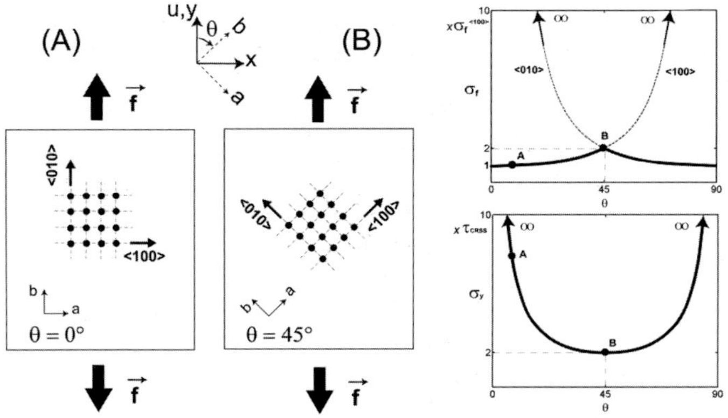

Figure 7. An ideal PBS made of single crystal metal bars with cubic symmetry and random orientation in the *a-b* plane. Schemes A) and B) (left) indicate the slip plane orientation for θ=45° and θ=0°; the load necessary to break or yield a bar is reported in the graphs (right) as a function of θ as normalized to $\sigma_f^{<100>}$ and τ_{CRSS} .

2.3. Final Thoughts About PBS Applications

The generalized model can have several applications. It can serve to solve a direct design problem to achieve a PBS-like system with controlled statistical properties, by optimizing the microstructural properties. Conversely, it can be used in the inverse problem (in conjunction with statistical methods such as ordinary least squares, maximum likelihood estimates, or alike) to deduce the microstructural properties, otherwise hard to access directly, starting from the observed macro response. Furthermore, nanotechnology nowadays provides novel major opportunities for discrete models application and re-launches the importance of models such as this one. For instance, as of today, it is very hard to access the mechanical properties of individual 1-D structural elements of micro- and nano-scale size, such as CNT, wires, tubes, rods, which very often come in bundles. A bundle of such hard-to-handle elements, *e.g.* the system of parallel polymer fibers on display in Figure 8(A), could instead be more comfortably tested in certain cases to infer the properties of a single constituent via inverse analysis aided by an *ad hoc* SDM model. Testing a single sub-micron fiber would require a rather involved and costly micrometric set-up such as the one in Figure 8(B) by Jaeger et al. (2009).

Other foreseeable areas of nanotechnology and nanoelectronics that would benefit for PBS models are for example the design and manufacturing of self-assembled nanosystems or Micro-Electro-Mechanical-Systems (*i.e.* MEMS) and Nano-Electro-Mechanical-Systems (*i.e.* NEMS). To this end, the generalized

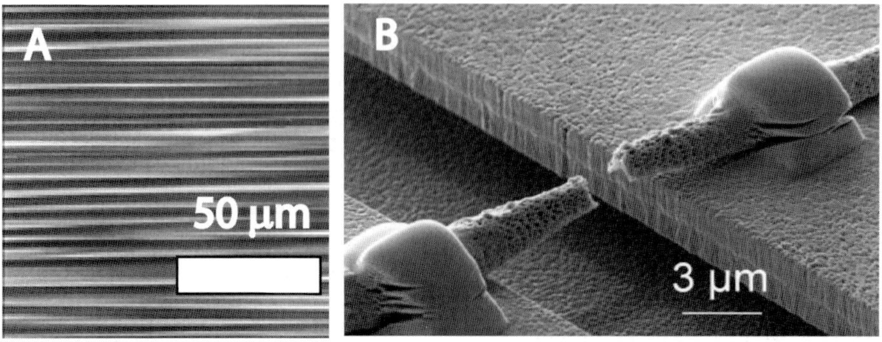

Figure 8. (A) An example of PBS made of a real bundle of aligned electrospun microfibers used for biomedical and tissue engineering applications. (B) A torn single fiber after a tensile test in the micro/testing machine (credit: © Fraunhofer IWM - Jaeger et al. (2009)).

model presented here provides a flexible and tailorable template to devise custom models accounting for two or more inelastic deformation processes simultaneously.

3. Higher Dimensional Models: Spring Networks

Lattice networks of springs, trusses, beams, or fuses are the logical multidimensional extension of the PBS model in the 2D and 3D cases. These higher dimensional models are intrinsically subject to local load redistribution associated to damage formation and, hence, are suitable for the study of related cooperative phenomena. Lattices are an extremely flexible tool to represent finely interconnected structural and material systems. They can reproduce actual reticular structures (*e.g.* truss structures) but also provide a simple representation of the complexity and disorder of the microscopic texture of a polycrystalline material[16]. As such, these simple models have a rather general scope.

Historically, percolation lattices were amongst the first ones used to study damage in the attempt to model the damage evolution via percolation theory (*e.g.* Born and Huang (1954), Scher and Zallen (1970), Jagota and Scherer (1993)). In there, defects are randomly nucleated until the percolation threshold is reached and a

[16] Lattice models are also reminiscent of a larger class of materials with microstructure, such as cellular solids, or multiphase superalloys with cellular precipitates (*e.g.* some duplex steels rich in Cr and Nb). The damage process in those materials often depends on the damage of a characteristic microstructural network that represents the weaker structural feature and may consists of grain boundaries, thin membranes, of brittle second phases depending on the specific case. Then, modeling such network rather the entire solid may suffice. Other innovative nanostructured materials, such as graphene sheets, can be modeled by 2D or 3D lattice models as well.

cluster of coalesced defects spans the entire lattice length. This procedure fails to consider the interaction among microcracks. Hansen et al. (1990) considered fuse lattices consisting of a resistor network to understand the damage process using an electrical analogy. A voltage is assigned to the network and a resistor burns out as soon as the current value exceeds an assigned maximum threshold. The voltage is increased until the lattice becomes a dielectric signifying material breakdown. The major advantage of the electric lattice over the mechanical one is that the former deals with scalar fields (voltage and current) rather than vector ones (displacement and force), which is relevant for computability. Although the model succeeds in accounting for the local interaction between defects, the electrical analogy provides just qualitative insight and a mechanical network is required when either accurate models of real microstructures, or multiaxial loading schemes, or frictional effects, or plasticity, or creep are sought. The mechanical network of springs investigated here has been extensively employed in the study of damage by many authors (*e.g.* Hansen et.al. (1989), Krajcinovic and Basista (1991), Delaplace, et al. (1996), Mastilovic and Krajcinovic (1999), Van Mier (2000)). An example of random lattice mimicking a polycrystalline microstructure is depicted in Figure 9. The sites of the lattice are associated to the grains and are connected to each other according to a desired topological scheme by links transferring grain-boundary forces.

The lattice can inherit either a brittle or a ductile response depending on the micro-constitutive relation characterizing the behavior of its link, just like the PBS.

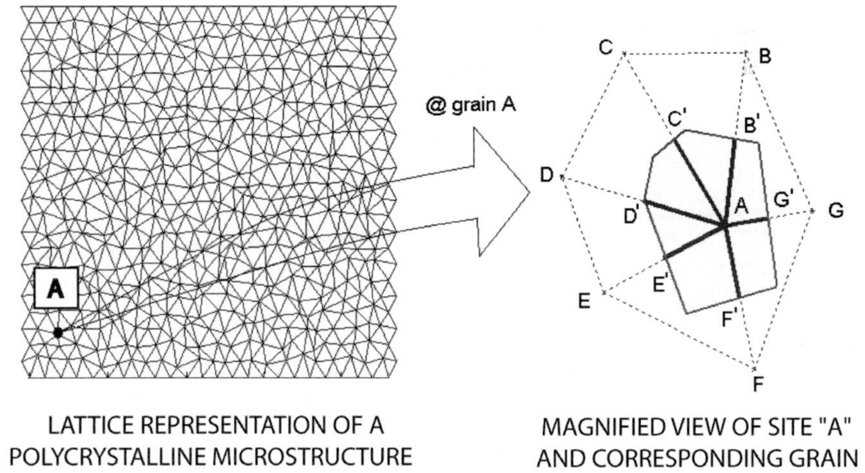

Figure 9. Lattice representing a disordered polycrystalline microstructure. The grain at lattice site A is shown on the right; other adjacent grains are at sites B to G (not plotted).

However, the scope of the discussion about the lattice is limited to brittle systems, subject solely to the damage process. More specifically, an extensive and detailed examination of a selected spring network is presented with the twofold goal, on the one hand, to highlight key general aspects of its damage process common to many real quasi-brittle materials, and, on the other one, to elucidate some novel SDM modeling routes. After defining the lattice model, numerical results from simulations of both tensile and multiaxial tests of many random samples are presented, mostly from a phenomenological angle. Such numerical data are, then, used as a basis to develop analytical and statistical damage models. The discussion devoted to modeling aspects consists of two main blocks. The first, main one is dedicated to establishing the rational foundation of the damage parameter for the springs network system. Departing from the analysis of the "energetics" of the lattice, an innovative methodology delivers a rational damage model for the uniaxial case, along with some analytical approximation of it. Instead, the second and last block of the discussion addresses the problem of sample-size effects on damage response. One modeling approach is covered stemming from fractal theory. Then, the sequence of topics featured in the present network analysis is orderly

1. lattice representation and model definition;
2. simulations results from tensile and multiaxial tests;
3. physical foundations of $\bar{D}$;
4. sample-size effects.

3.1. Lattice Representation and Model Definition

One practical way to build lattice models is to generate them as a Delaunay triangular network from a Voronoi tessellation approximating a polycrystalline microstructure (*e.g.* ref. Espinosa and Zavatteri (2003)). Nowadays an increasing number of commercial software is available to generate optimal 2D and 3D Voronoi representations of real microstructures from their micrographs and other experimental data (ref. footnote #6). Alternatively, for mathematical and conceptual studies like this one, an artificial Voronoi tessellation generated with a computer[17],

[17] The computer generation is based on a random process to distribute a cloud of points, *i.e.* grains, in the lattice domain. The rationale of such a distribution can be controlled, not only to decide the number of grains in the microstructure, but also their relative size and density. For example, seeding points according to a Poisson process yields a uniform density of grains over the lattice domain. Several pieces of statistical-graphical software (also freeware) are available to perform this task in a fast and automated fashion. It should be noticed that not all random clouds admit a Delaunay/Voronoi construction, which also implies that not all microstructures are easily and rigorously described by a Voronoi graph. For example, real grains are not regular polyhedrons (polygons in 2D) and do not always have convex shapes (concavities cannot be directly reproduced with the basic

as in Figure 10, is enough. The corresponding Delaunay lattice, that is the topological dual representation of the Voronoi froth, can be obtained by a Legendre transformation according to a triangulation procedure explained at length in literature (*e.g.* Zallen (1983), Okabe et al. (1999)).

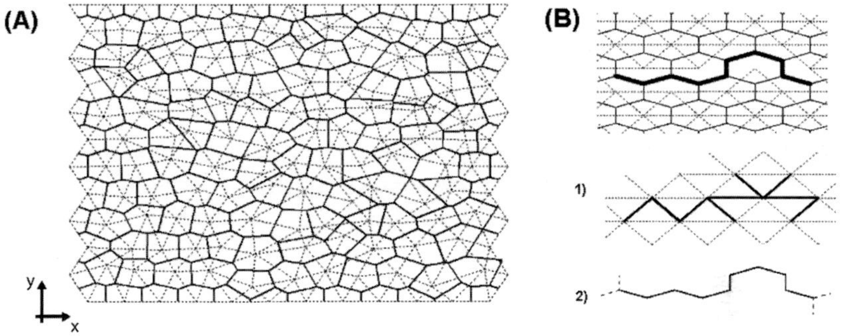

Figure 10. A) Sample Voronoi froth and Delaunay network of an imaginary microstructure. B) Damage (microcracks) representation in the two representations.

Notably, each Voronoi grain is a polygon (polyhedron) with measurable area (volume), geometry, and number of sides (facets), but it reduces to a simple point-particle connected with nearest neighbors in the Delaunay lattice[18]. The average grain size defines the resolution length l of the Voronoi graph, whereas the average link length defines the resolution length ℓ of the dual Delaunay model[19].

Voronoi construction made of convex tiles only). More advanced and complex tessellation techniques are available to overcome some of these issues. However, when pursuing an exact representation of a real microstructure is not the main intention – like in this essay – the focus can more fruitfully shift onto accounting for the major statistics of a given microstructure (*e.g.* distributions of grain size and grain orientation, two-point or higher order correlation functions, etc.) to create Voronoi tessellations that reflects such statistics. Different types of tessellations and their construction to achieve the "best" approximation represent an active research topic in the computer graphics research community at present.

[18] Therefore the Delaunay representation does not convey information about grain geometry but provides a lumped description of mechanical inter-granular interactions and topology of the material on the microscale. The advantage is a sensible reduction of *dof* with respect to, say, a FEM model based on the Voronoi representation (Rinaldi et al. (2008)).

[19] It is better to define distinct resolution lengths for both advocated representations, since Delaunay and Voronoi models can be used separately and for different purposes. For instance, the following analysis of the network model requires the sole consideration of the Delaunay lattice. Nonetheless, for all that said, especially about the connection to expe-

The Delaunay points are defined such that a link between two of them in the lattice is orthogonal to the dual grain-boundary in the Voronoi tessellation. The coordination number z_i of the generic i-th lattice site is the number of near-neighbors connected by a link to that site and is equal to the number of sides of the corresponding Voronoi grain.

The relevant mechanical properties of the microstructure are lumped in the links connecting pairs of lattice sites[20]. In this study, the links are linear springs of equal stiffness k (k = 100 is assumed) that break permanently once the tensile force reaches the assigned random strength, as in the PBS from Section 2.1 (ref. Figure 3)[21]. A broken spring, though, is still active (and hyper-elastic) in compression to account for crack closure[22]. Each of these failures corresponds to the formation of one new microcrack in the system that increases of damage. This specific model can evidently reproduce only a damage process that evolves by inter-granular micro-cracking. Figure 10(B) illustrates the graphical representation of such microscale damage in both the Voronoi and Delaunay schemes. Interestingly, the Voronoi version is self-evident because it allows readily assessing the adjacency between contiguous microcracks clustering to form larger inter-granular cracks unlike in the

rimental data, understanding the relation between these dual constructions is of self-explanatory importance.

20 The transposition of a given tessellation of solid grains into a network of one dimensional springs is a non-trivial matter that imposes a trade-off between loss of microstructural information and reduction of *dof* (*e.g.* Monette and Anderson (1994)). Rinaldi et al. (2008) explored this issue and formulated a discretization procedure to lump random continuum grain properties, such as geometry, number of boundaries, and elasticity of each crystal domain, into the lattice. In particular it was found that the mechanical properties of the lattice links can selected to account for the crystal orientation and actual geometry of the associated grains. Hence, by means of their algorithm or by a similar one it is possible to automatically identify a disordered Delaunay lattice – with naturally arising distributions of stiffness and length – starting from a Voronoi tessellation that is fully characterized (experimentally or virtually) both by a mechanical and geometrical standpoints.

21 The use of linear springs appears justified by the primacy of the short-range interactions in the solid phase (in analogy to the principle of local action underlying the Hooke's law in continuum mechanics - Billington and Tate (1981)) and is appropriate in the context of quasi-brittle materials withstanding small deformations and limited plasticity (*e.g.* ceramics with plastic strains $\varepsilon_{pl} < 0.1\%$ - ref. Davidge (1979)).

22 According to Krajcinovic (1996) "A microcrack subjected to a substantial compressive normal stress in the direction of its normal (combined with a nearly vanishing shear stress in its own plane) is a paradigm of a passive microcrack. The status of a microcrack may change if the stresses normal to its bedding plane change their sign during a deformation process".

Delaunay lattice[23]. Grain boundary cracks and other imperfections smaller than the resolution length (*e.g.* crack dislocations, glassy pockets, etc.) are "invisible" to the model and can be accounted only indirectly through the randomness of the mechanical parameters of the lattice. Consequently, since the formation of a single microcrack cannot be monitored, it is considered to propagate instantaneously from zero to its final size (*i.e.* the grain facet/boundary) when the corresponding spring breaks.

In this lattice, geometrical disorder adds up to the mechanical disorder already present in the PBS. The evolution and eventual localization of damage are stochastic processes that depend crucially on the resultant of both types of microscale disorders. The length and the tensile strength of the springs are the random microscale variables associated to the geometrical and mechanical disorder respectively. For convenience sake, they are regarded here as independent random variables, such that their quenched distributions can be chosen independently from one another. This way, it is possible to consider each type of disorder separately from the other, in particular to examine ordered lattices with perfect geometry but yet endowed with mechanical disorder[24] (*e.g.* in Section 3.3). Of course, each disorder type is a source of variability sufficient alone to give rise to damage localization. For these reasons, it is now necessary to specify the statistical information of the lattice model examined later. The statistical distributions of lengths and strengths below must be regarded as common to any lattice regardless of its size, unlike otherwise specified.

1) Geometrical Disorder. The network model consists of two-dimensional Delaunay lattices that are first generated from a honeycomb Voronoi tessellation reproducing a microstructure with perfect hexagonal grains (stage 1) and, then, distorted by displacing the nodes - everywhere in the lattice but on the side - from

[23] Many models in the literature do not contemplate the Voronoi tessellation but just the Delaunay lattice. This is well the case, for example, when the lattice represents an actual reticular structure rather than a microstructure. However, the consideration of the Voronoi froth may be necessary in the latter case to assess microcracks adjacency and coalescence, as remarked in the text. In fact, two adjacent broken springs are not necessarily two adjacent grain boundaries (microcracks), as shown in Figure 10(B) for the 2D case. Two contiguous springs correspond to two adjacent grain boundaries *only and only if* the dual Voronoi elements have a common node.

[24] This mechanical network is a highly idealized model but reproduces the damage process of real quasi-brittle materials to a sufficient degree to be useful as a benchmark system for the development of SDM tools and novel modeling ideas at this stage. Presently it is secondary whether the random variables are sampled from independent (and somewhat arbitrary) distributions, with no reference to any specific material or real experimental data. More detailed information about the microstructure is needed to correlate mechanical and morpho-geometrical disorder, which yet remains a fundamental research task to be pursued in the future to calibrate SDM models for engineering (ref. footnote # 20).

the initial position through a controlled random perturbation (stage 2) to impart geometrical disorder and simulate the random texture of a real microstructure. If N is now the number of nodes (*i.e.* grains) sitting on the lattice side, a $N\times N$ lattice indicates a network square in N (*i.e.* having same number of nodes on all four sides) that in the "stage 1" unperturbed configuration has N^2 nodes and a number of springs N_0

$$N_0 \cong 3N\cdot(N-1)-N/2 \quad . \tag{25}$$

(Be aware of the different notation from the PBS where N was defined differently). After the geometrical perturbation (*i.e.* in stage 2), the numbers of nodes and springs vary randomly but on average are well approximated by the above estimates. Figure 10(A) shows an instance of a 12×12 disordered lattice. This generation procedure is applied to obtained spring lengths ℓ_i (i=1..N_0) with average size $\langle\ell\rangle=1$ and normally distributed in the range $\alpha\langle\ell\rangle\leq\langle\ell\rangle\leq(2-\alpha)\langle\ell\rangle$, where $\alpha(0<\alpha\leq 1)$ is the characteristic parameter of the geometrical disorder related to the distribution of grain sizes. The values $\alpha=0.1$ is used to enforce geometrical disorder, while $\alpha=1$ corresponds to lack of it and renders a perfectly triangular lattice with hexagonal grains (z=6). Irrespectively of the disorder, a $N\times N$ lattice is a $L_X\times L_Y$ rectangle with $L_X\cong N-1$ and $L_Y\cong\sqrt{3}/2\cdot L_X$ (representing the macroscopic length scales along each coordinate axes). The scope is limited to 2D models but results and concepts hold seamlessly for the 3D case, which mostly entails heavier computations.

2) Mechanical Disorder. According to the same rationale adopted for the PBS (ref. footnote #9), rather than actual random strengths, the springs are assigned the corresponding critical strain $\varepsilon_i*=u_i*/\ell_i$ (i=1..N_0). (note the change in notation with the PBS where the critical displacement $\hat{u}_i$ is now replaced with u_i*). The latter ones are sampled from a uniform distribution $p_f(\bar{\varepsilon})=1/(\varepsilon_{MAX}-\varepsilon_{MIN})$ with $\varepsilon_{MIN}=0$ and $\varepsilon_{MAX}=0.01$. The defects smaller than the grain boundary are accounted indirectly through such distribution. This choice is arbitrary but convenient for a number of reasons. For instance, it simplifies the analytical calculations later on and, also, allows comparing the lattice with the brittle PBS in Figure 4 having equivalent mechanical disorder.

3.2. Simulations Results from Tensile and Multiaxial Tests

The lattice model defined above is suitable to study the damage process under a variety of simulated loading conditions. The discussion focus is about simulations of both uniaxial and multiaxial mechanical tests conducted quasi-statically in displacement-controlled conditions. By a structural viewpoint, the spring network is a statically indeterminate and globally stable structure, the response of which to an applied load is obtained from the resolution of a system of $2N\times 2N$ equations, corresponding to the number of *dof* in the lattice (each node has 2 *dof* in 2D) and is expressed in terms of nodal forces and displacements (the latter being the primary variable). The solution requires the application of proper boundary conditions (BCs), which both constrain the system against rigid body motion and enforce the kinematic requirements of the given test configuration. If a lattice is constrained along a side to be either fixed or subject to a certain displacement, its nodes are connected to a rigid bar free to translate of any controlled amount (including $\bar{u} = 0$) in any direction[25], similarly to the PBS. Once the proper BCs are identified, a mathematical solver proceeds to impose them in the system of equations to calculate the deformed configuration of the lattice and the damage. Both direct solvers (*e.g.* direct stiffness method, finite elements method) and iterative solvers (*e.g.* molecular dynamics) exist, each method offering advantages and disadvantages and being best-suited for certain applications. For the quasi-static simulations inhere both methods can be applied but direct solvers are certainly preferable until memory storage requirements allow it. Not to digress, the interested reader can find a description of the solution methods in Appendix 1. Instead, this section offer a critical dissertation about the results of those simulations, at first for the tensile test and then for more general multiaxial cases.

Tensile tests are simulated by fixing one lattice side and displacing the opposite one, as shown in Figure 11(A) for the x direction. For a quasi-static loading in displacement-controlled mode the lattice response, when sufficiently magnified, appears like a saw-toothed profile consisting of a series of equilibrated states, as qualitatively depicted in Figure 11(B).

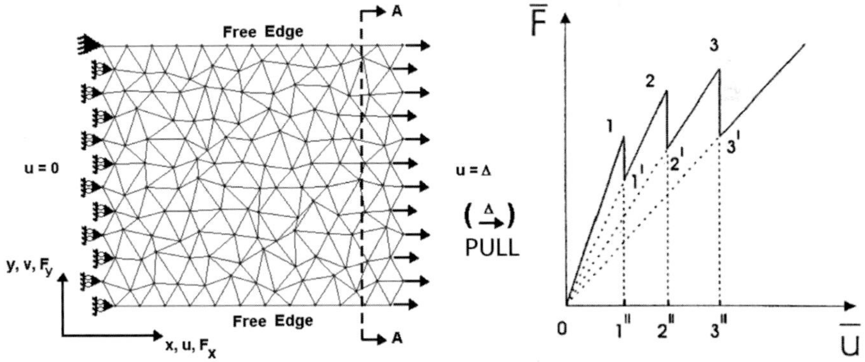

Figure 11. Simulations of uniform tensile tests and saw-toothed response $\bar{F}$ *vs.* $\bar{u}$.

25 In the simulations of displacement-controlled tests only geometrical or essential BCs (*i.e.* nodal displacements) are assigned, as opposed to natural BCs or mixed BCs entailing assignment of nodal forces too.

Any point on the elastic loading segments $\{\overline{01}, \overline{1'2}, \overline{2'3}\}$ represents an equilibrated state. The segments $\{\overline{11'}, \overline{22'}, \overline{33'}\}$ of the loading pattern are, instead, fictitious irreversible transformations between the equilibrated states corresponding to springs breaking. Points on the latter segments have no physical meaning because such transformations are virtually instantaneous, which holds true for the PBS too. Analogously to the PBS, the primary data collected during the damage process are the force response $\overline{F} - \overline{u}$ and the cumulative curve of broken springs $n - \overline{u}$. However, glaring differences emerge when comparing the tensile responses from the two models.

The sample results in Figure 12 correspond to a 192×192 lattice and highlight a non-symmetric force response that after the peak exhibits a softening phase[26] characterized by a sudden drop in the load bearing capability and by an increased variability in the system response. Most significantly, the curve of cumulative ruptures $n - \overline{u}$ ceases to be a straight line, like in Figure 4, and takes on a characteristic knee-shaped profile hiding a profound physical meaning that needs to be well understood before defining the damage parameter. In fact it is evident that the damage parameter cannot longer be obtained by a simple normalization of n over the number of springs like in Eq.(3), because at failure $n/N_0 << 1$ and only a small fraction of elements is involved in the entire damage process. The "*saturation knee*", already investigated by Hansen et al. (1989), is a notable feature of the $n - \overline{u}$ data that preludes to the softening phase in the force response and will be used later in modeling it.

In order to characterize the mean and spread of these output data scatter, as many random lattices as possible should be considered per each size N. Yet, the runtime of a simulation increases non-linearly with N and fewer replicates can progressively be collected. Table 1 summarizes the tensile test simulations investigated in this work, indicating the replicates per N, along with the numbers of nodes and springs[27].

---

[26] Convention: "hardening phase (or regime or response)" is the part of the force response preceding the peak and "softening phase (or regime or response)" the remaining post-peak data. The former has $d\overline{F}/d\overline{u} > 0$ (or $d\overline{\sigma}/d\overline{\varepsilon} > 0$) and the latter has $d\overline{F}/d\overline{u} < 0$ (or $d\overline{\sigma}/d\overline{\varepsilon} < 0$). Hence, the term "hardening" here has no bearing on the classical connotation from plasticity, *i.e.* the post-yield plastic hardening in ductile materials.

[27] The N range is chosen to cover at least an order of magnitude difference between minimum and maximum lattice size, which is important for the accuracy, precision, and robustness of the regression analysis. A range as extended as possible, compatibly with the computer power, should be always pursued. The size values were selected here as multiple of 12 and 24 arbitrarily to initially compare results against Hansen et al (1989), whose maximum lattice size was N=*24* back then. Lattices with N=*12* were discarded as proved too small to insure statistical homogeneity and robustness versus BCs.

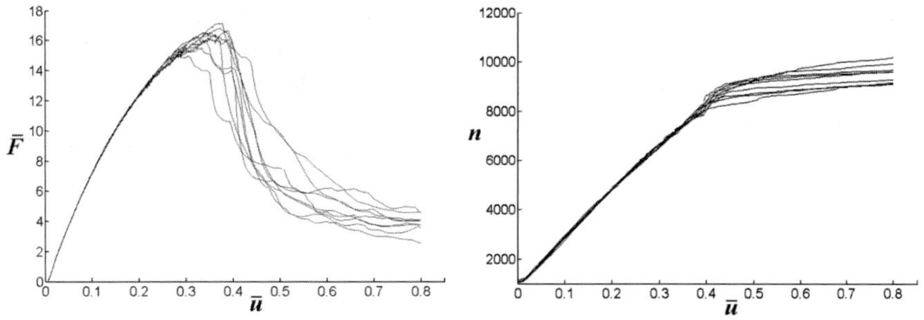

Figure 12. $\overline{F}$-$\overline{u}$ and n-$\overline{u}$ curves for 10 replicates of $N = 192$ (force and displacement values have no physical units here).

Table 1. Number of runs per lattice size from MD simulations

Size N	24	48	72	96	120	192	288
#Replicates	100	34	30	25	20	13	5
#Nodes	576	2304	5184	9216	14400	38416	82944
#Springs	1644	6744	15300	27312	42780	114562	247824

The comparison between responses of different lattice sizes reveals another fundamental tract of the damage process: the sample-size dependence. This can be readily assessed on the $\overline{\sigma}$-$\overline{\varepsilon}$ curves[28] obtained from the uniaxial force responses in the i-th direction via the transformations

$$\overline{\varepsilon} = \overline{\varepsilon}_{ii} = \frac{\overline{u}_i}{L_j} \quad \text{and} \quad \overline{\sigma} = \overline{\sigma}_{ii} = \frac{\overline{F}_i}{L_j}, \quad (i,j=x,y \text{ with } i \neq j) \tag{26}$$

(*e.g.* in the x-axis $\overline{\varepsilon} = \overline{\varepsilon}_{xx} = \overline{u}_x / L_y$ and $\overline{\sigma} = \overline{\sigma}_{xx} = \overline{F}_x / L_y$). The average $\overline{\sigma}$-$\overline{\varepsilon}$ curves from mean data are plotted in Figure 13 for $N = \{24,48,96,192\}$ and demonstrate that the stress response is independent of lattice-size before the peak (and, hence, all four lattice sizes before the peak would render good RVEs) but become size dependent in the softening regime and at failure, in agreement with the

[28] Note that "stresses" in this discussion have no units since the spring stiffness k and length l do not have a metrical correspondence but represent arbitrary scalar values chosen for convenience and regardless of any actual calibration. In real applications mechanical calibration is necessary and is discussed by Rinaldi et al (2008).

phenomenology of damage in real material systems experimentally observed by several authors[29]. Hence, softening phase and failure of quasi-brittle structural and material engineering systems do not depend only on the applied stress and the material properties but also on the specimen size and shape. In mathematical terms, such lattice-size dependence implies that $\bar{D}$ ought to be a simple (random) function of strain $\bar{D}(\bar{\varepsilon}_{kl})$ in the hardening phase and a function $\bar{D}(\bar{\varepsilon}_{kl}, L)$ in the softening phase. In summary the modeling problem in the tensile test would involve two dimensions of variability: *(i)* a "*within-sample*" variability between lattices of same N and *(ii)* a "*between-sample*" variability between lattices of different N, corresponding to the size effect. These two types of variability are often treated as two distinct modeling objectives and, as such, addressed differently and independently in the softening.

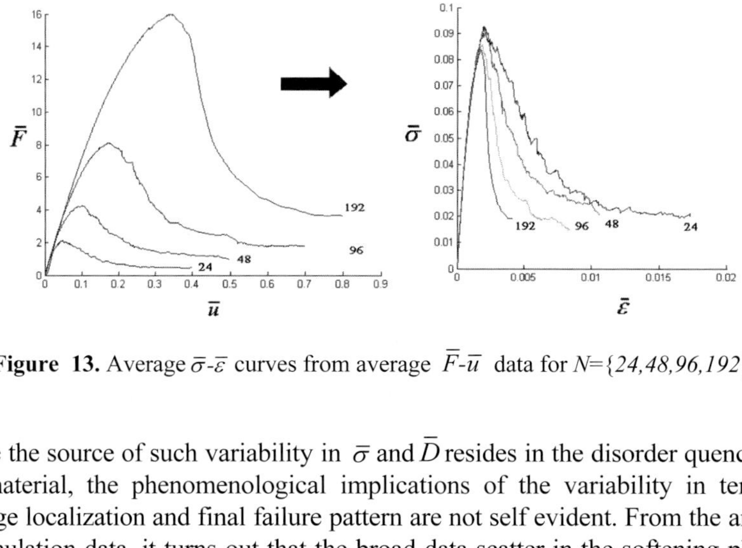

Figure 13. Average $\bar{\sigma}$-$\bar{\varepsilon}$ curves from average $\bar{F}$-$\bar{u}$ data for N=*{24,48,96,192}*.

While the source of such variability in $\bar{\sigma}$ and $\bar{D}$ resides in the disorder quenched in the material, the phenomenological implications of the variability in terms of damage localization and final failure pattern are not self evident. From the analysis of simulation data, it turns out that the broad data scatter in the softening phase is correlated with the large variety of macro-crack patterns resulting from damage localization. Each macro-crack is a random cluster of interacting (not necessarily adjacent) microcracks observable at the threshold of failure and represents the minimum resistance pattern to damage growth. In most cases the crack is observed

[29] Hudson et al. (1971) and Simsch (1993), amongst others, demonstrated that both size and shape of the concrete specimens affect the peak location and the softening phase of the stress-strain curve. Weibull (1939) had already concluded, from experimental data on brittle materials, that the failure threshold depends on specimen size and shape in a manner well described by the homonymous distribution. But apparently the scientific debate about this size effect started much earlier, at the time of Leonardo and Galileo (Bazant (2000)).

to be (statistically) straight and orthogonal to the direction of the applied tensile load (Figure 14) but this sometimes is not the case. "Irregular" patterns like in Figure 15 contain a higher number of broken links and the amount of extra links quantifies the difficulty of microcracks to propagate in the presence of local energy "barriers", like high-strength links at the crack tips or shielding effects.

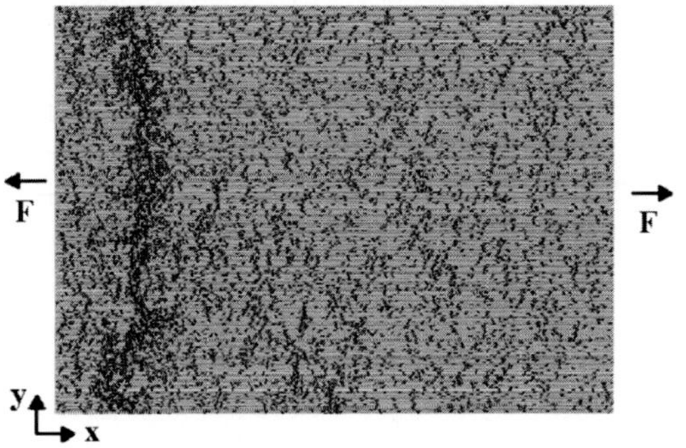

Figure 14. Damage distribution at the threshold of failure for *N=192*.

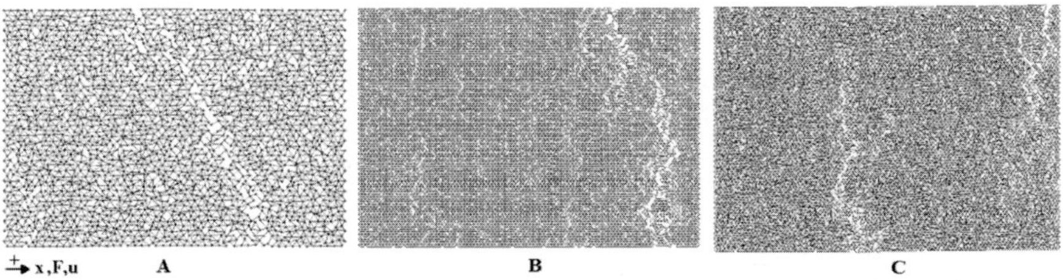

Figure 15. Patterns that are neither straight nor orthogonal to the load direction. (A) *N=48* and (B-C) *N=96*.

For example, the direct comparison between the damage patterns of two lattices of equal size in Figure 16 (top row) indicates that microcracks in case (B) must circumvent many more barriers to propagate and eventually develop into the observed "fat", irregular macro-crack. The density, distribution, and energy level of such barriers determines the damage tolerance of the material. Presumably, more energy is needed in case (B) compared to the well-defined narrower macro-crack in (A). Such an expectation is confirmed by the corresponding test data (bottom row), highlighting larger strain energy dissipation in the $\overline{F}-\overline{u}$ data (from the subtended area) and more cumulated microcracks $n-\overline{u}$ in case (B).

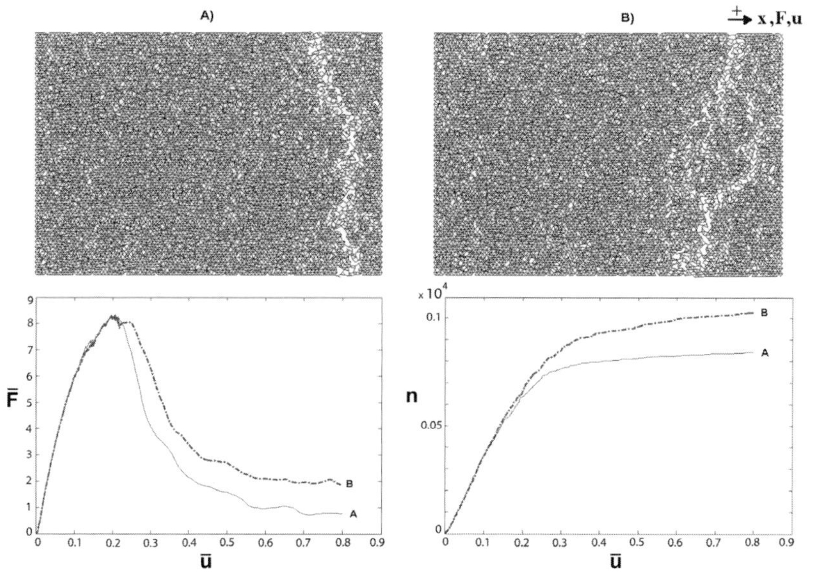

Figure 16. Damage distributions (top) and macro-responses (bottom) for 2 lattices of $N = 96$.

The damage localization coincides with the breakdown of the thermodynamics of equilibrium since complete knowledge of the microstructure is necessary to estimate the macroscopic behavior of the lattice, which is in mechanical equilibrium but not in thermodynamic equilibrium. Large fluctuations in the microscale stress field characterize the softening regime and the large stress density around the macro-crack testifies the ongoing cooperative process. Figure 17 compares the micro-stress distribution (in absolute value) at four states of the damage process. While micro-stress distribution "*1*" is statistically "invariant" for all replicates with same size *N* and corresponds to random damage *nucleation*, distributions "*2, 3,* and *4*" depend on the disorder of the microstructure and correspond to localized damage *propagation*. The $n - \bar{u}$ curve overlaid (out of scale) onto the $\bar{F} - \bar{u}$ curve demonstrates the *knee* marks the occurrence of the transition from homogeneity to heterogeneity (equivalently from nucleation to propagation of damage or from hardening to softening). Thus, damage nucleation and propagation are not simultaneous damage modes. A better perspective comes from plotting the rupture locations *vs.* the time point of the damage process ($\bar{u}$ being the simulation time) in Figure 18. It shows in an exemplificative manner that damage nucleation is seemingly a random diffuse process with higher rupturing rate, whereas damage propagation is a spatially highly correlated with lower rupturing rate.

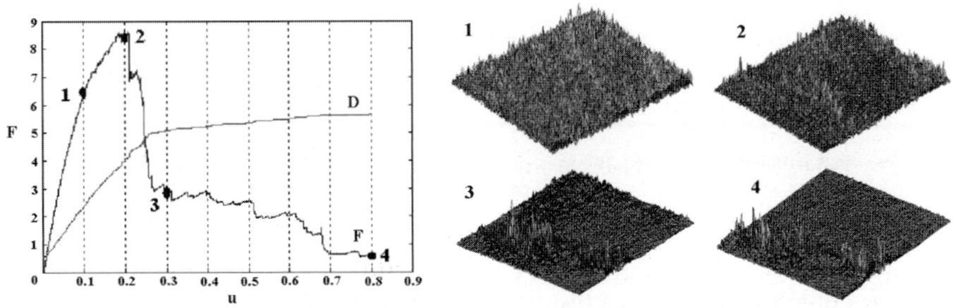

Figure 17. Micro-stress distribution at 4 points of the damage process. Stress (strain) localization is observed at "*2*" from the loss of statistical homogeneity and clusters formation.

Both damage localization and spring-rupturing rate reduction happen suddenly at once, demonstrating a rapid nucleation-propagation transition and preluding to failure. For this reason, a spring-rupturing rate reduction in the $n - \overline{u}$ curve must not be mistaken as a reduction in damage rate, which vice versa increases as demonstrated later in Section 3.3.

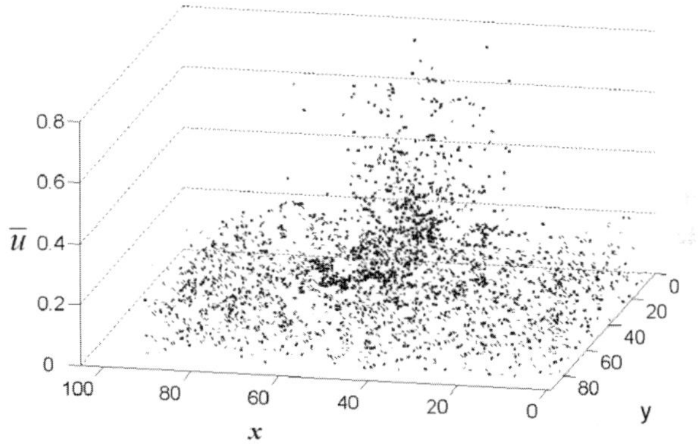

Figure 18. Plot of ruptures location *vs.* simulation time $\overline{u}$ for $N=96$, with evident localization and rate reduction of ruptures.

These observations provide the phenomenological and physical ground for the next detailed analysis leading to the definition of $\overline{D}$ for this system. The damage problem, though, can be cast in terms of a scalar parameter only for the tensile case or for another uniaxial test. The scenario is significantly more complicated when multiaxial loads are imparted. A selection of biaxial stress states for the 2D lattice is examined for illustrative purposes, just to appreciate the influence of the *loading scheme* on damage localization. The following nine cases are simulated aside from those in Table 1:

1. tensile uniaxial test in x ($\bar{\varepsilon}_{xx} > 0, \bar{\sigma}_{yy} = 0, \bar{\sigma}_{xy} = 0$),
2. tensile uniaxial test in y ($\bar{\varepsilon}_{yy} > 0, \bar{\sigma}_{xx} = 0, \bar{\sigma}_{xy} = 0$),
3. biaxial hydrostatic tensile test ($\bar{\varepsilon}_{xx} > 0, \bar{\varepsilon}_{yy} = \bar{\varepsilon}_{xx}, \bar{\sigma}_{xy} = 0$),
4. compressive uniaxial test in x ($\bar{\varepsilon}_{xx} < 0, \bar{\sigma}_{yy} = 0, \bar{\sigma}_{xy} = 0$),
5. compressive uniaxial test in y ($\bar{\varepsilon}_{yy} < 0, \bar{\sigma}_{xx} = 0, \bar{\sigma}_{xy} = 0$),
6. balanced biaxial test with tension in x and compression in y ($\bar{\varepsilon}_{xx} > 0, \bar{\varepsilon}_{yy} = -\bar{\varepsilon}_{xx}, \bar{\sigma}_{xy} = 0$),
7. balanced biaxial test with compression in x and tension in y ($\bar{\varepsilon}_{xx} < 0, \bar{\varepsilon}_{yy} = -\bar{\varepsilon}_{xx}, \bar{\sigma}_{xy} = 0$),
8. biaxial hydrostatic compressive test ($\bar{\varepsilon}_{xx} < 0, \bar{\varepsilon}_{yy} = \bar{\varepsilon}_{xx}, \bar{\sigma}_{xy} = 0$),
9. pure shear ($\bar{\sigma}_{xx} = 0, \bar{\sigma}_{yy} = 0, \bar{\sigma}_{xy} > 0$).

A graphical mapping of each case is provided in Figure 19. All simulations, but Case *9*, are run in displacement controlled fashion through the application of displacement BCs. In particular, in biaxial Cases # *{3,6,7,8}* proportional loading is imposed along x and y such that $|\bar{\varepsilon}_{xx}| = |\bar{\varepsilon}_{yy}|$. Instead, simulations of pure shear load in Case *9* are done in force controlled mode, *i.e.* via traction BCs, since a shear force is more easily applied than a shear deformation. An example of deformed lattice from pure shear is shown in Figure 20.

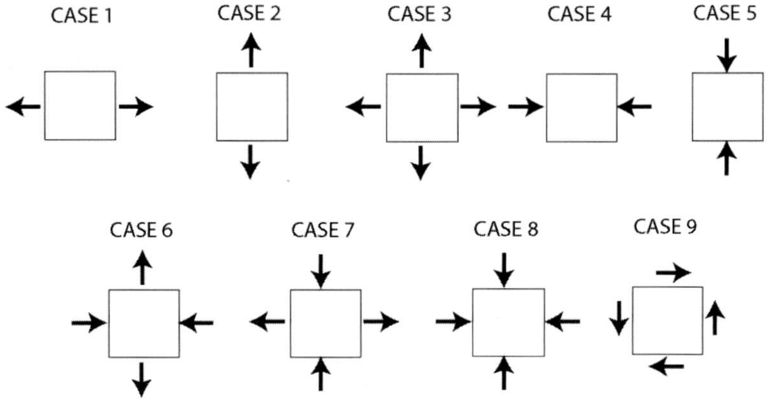

Figure 19. Biaxial loading schemes adopted for the damage simulations.

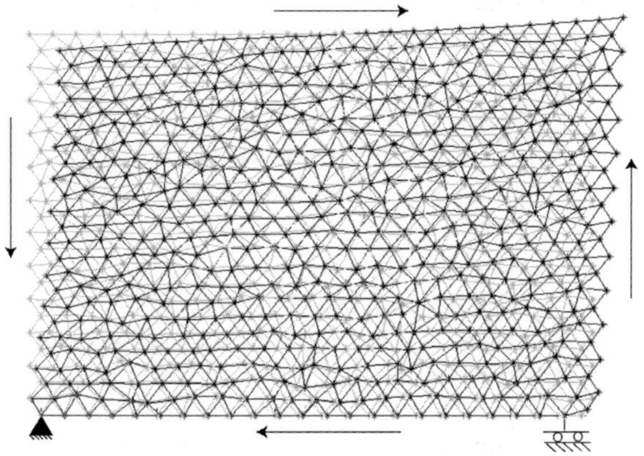

Figure 20. Shear deformation from pure shear obtained by applying traction BCs (Case 9).

The results from these simulations exemplify one distinctive aspect of the multiaxial problem: the shift from one failure mode to another when the loading scheme changes. The failure patterns in Figures 21 to 23 refer to one disordered lattice per size $N = \{24,48,96\}$ and underline the different localization modes from 8 out of the 9 loading cases. The missing Case *8* is unreported for no failure occurred in the lattice under in-plane hydrostatic compression, because all springs are compressed and have indefinitely elastic compressive behavior[30]. The patterns are increasingly sharper as N gets larger and are best appreciated in Figure 23 for $N=96$. The familiar macro-cracks orthogonal to the pulling direction are recognizable in the tensile tests, *i.e.* Cases *1* and *2*. A similar failure mode is also encountered when lateral compression is added as in Cases 6 and 7, although the response becomes more "ductile" in the sense of damage tolerance. The comparison between Case *1* *vs.* Case *7* in Figure 24(A) for the 96×96 lattice indicates that the extra lateral compression reduces the maximum strength but increases ductility, while the pure tensile test leads to a very brittle failure without softening. By a physical standpoint, the transverse compression facilitates incipient damage localization at multiple sites in the lattice, which paradoxically seems to result in gentler transition owing to the presence of several competing clusters of microcracks. The failure pattern from Case *7* in Figure 23 appears indeed less definite than in Case *1*, confirming the difference in the damage process. Accordingly, the number of broken springs at failure is higher for Case *7* in Figure 24(B).

[30] Such a fictitious behavior is an artifact and a limitation of the chosen lattice model. However, buckling or other compressive strength-limiting mechanism can be added into an augmented, refined lattice.

Finally, also the biaxial tensile test (Case *3*) differs from the uniaxial tests in Cases *1* and *2*. It exhibits mixed tendencies between the latter two and the corresponding failure pattern frequently changes direction or branches, alternating between horizontal and vertical alignment. Eventually, though, the damage evolution in one direction prevails randomly and leads the failure.

The types of localization encountered in the remaining compressive and shear loads are, instead, deeply different from what discussed so far. With reference to the schematic depiction in Figure 25, the uniaxial compression tests in Cases *4* and *5*, as well as pure shear in Case *9*, produce *(i)* splitting and *(ii)* shear bands in the lattice, either alone or in combination. Splitting is characterized by macro-cracks aligned in the direction of the applied compression, whereas shear bands are macro-cracks at an angle. In the past, these failure modes have been studied experimentally in many real materials (*e.g.* on concrete, ref. Neville (1981) and Gerstle et al. (1980)) but also in similar lattice models from Krajcinovic and Vujosevic (1998). The latter work contains, in fact, an appealing parametric study showing the transition from splitting to shear banding in compression experiments when lateral confinement is increased from zero (pure uniaxial compression) to a maximum value (biaxial compression). It is interesting, then, that despite no lateral confinement is imposed here in the uniaxial simulations from Case *4* and *5*, splitting (usually in combination with shear bands) happens only in Case *4*, when the external load acts on x. On the contrary, Case *5* is markedly dominated by neat shear bands.

The change between these two failure modes by switching the compression axis in Cases *4* and *5* is intimately connected to the microstructural topology of the lattice and is due to the different role of the Poisson's effect in the two cases. In fact, the selected topology and geometry are determinant to the occurrence of splitting in x. Even in the disordered configurations, the triangular lattice in Case *4* retains a large number of springs nearly parallel to this direction, which essentially contribute to bear the compressive load and are excluded from the damage process. Instead, links more transversal to x (say at $\pm 60^o$ like in a perfect lattice) get pulled by Poisson's effect and are free to break. The Poisson's effect is the driving force of the damage in these latter springs, much alike to a tensile load applied along y (Case *2*). As a proof simulations of lattices with perfect geometry rendered clear splitting failures, resemblant of the one in Case *2* (data not shown). Thus, the spurious mode, mixing splitting and shear banding in Figures 21-23, is attributed to geometrical disorder. Another situation happens in Case *5*. In a perfect triangular lattice, no spring would be closely aligned in the compressive direction y and damage would involve just the springs at $\pm 60^o$ to x. In this regard, the formation of a very sharp zigzag pattern consisting of shear bands segments in Figure 23 is a demonstration of the substantial cross interaction amongst such two orientations also in the disordered lattice. The same can be said for the shear load in Case *9*, which exhibit a similar failure pattern but with a different orientation.

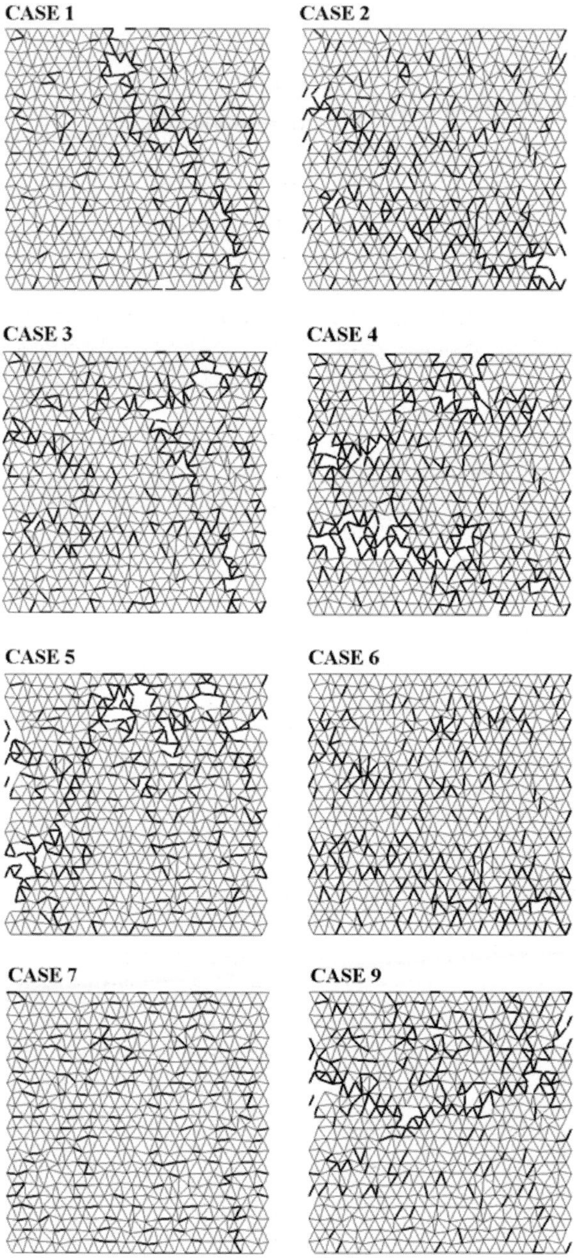

Figure 21. Eight failure patterns from the biaxial cases for one replicate of N=24.

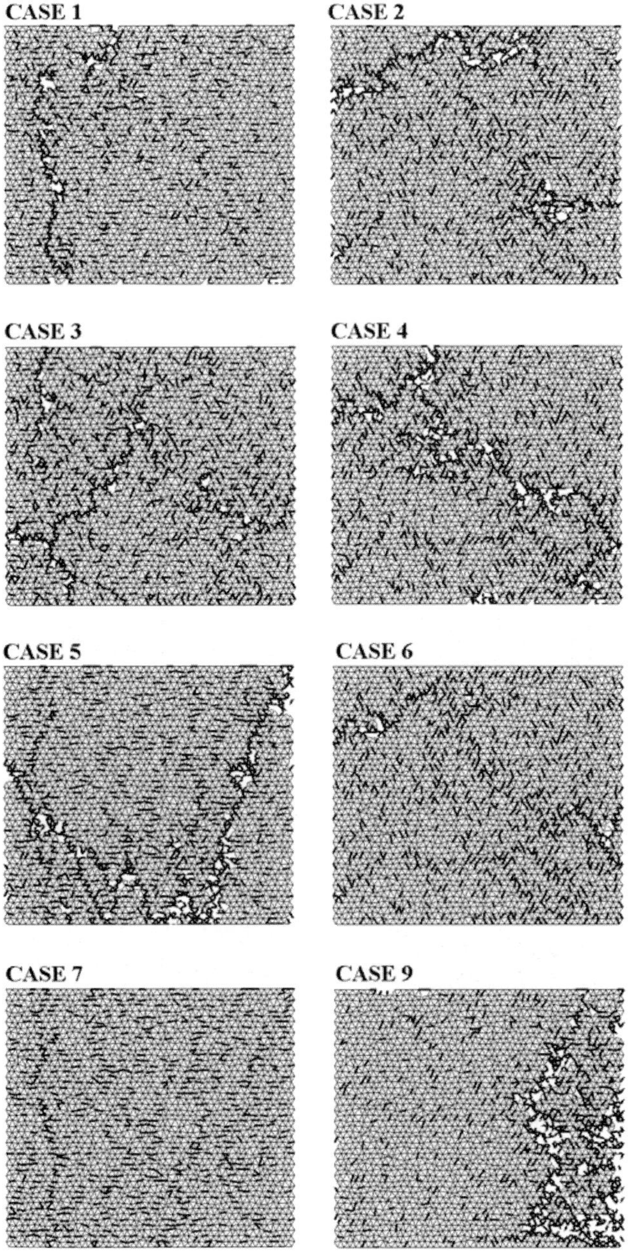

Figure 22. Eight failure patterns from the biaxial cases for one replicate of N=48.

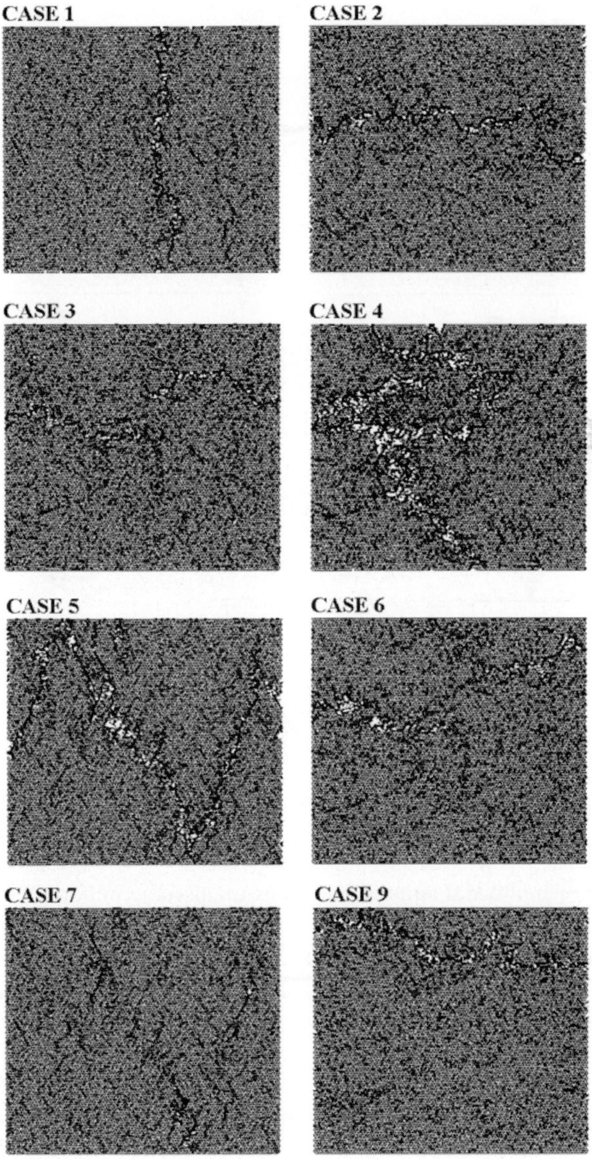

Figure 23. Eight failure patterns from the biaxial cases for one replicate of N=96.

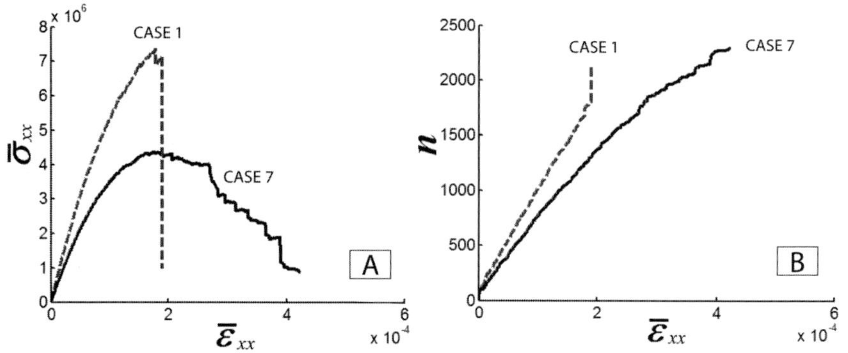

Figure 24. Comparison between Cases *1* and *7* for *N=96*. (A) Stress response; (B) cumulative number of microcracks tracked in the simulated tests.

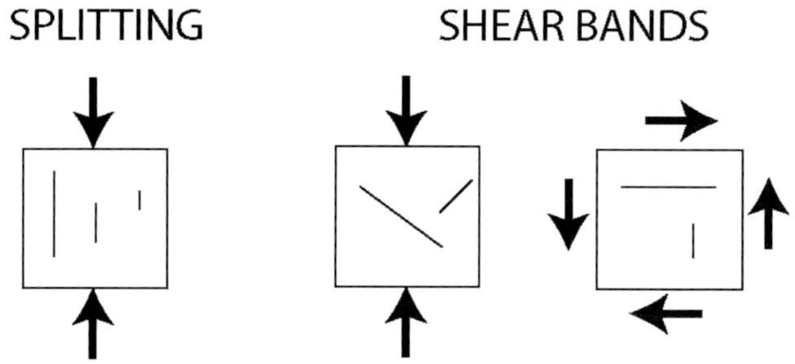

Figure 25. Failure patterns from splitting and shear banding.

The study of multiaxial simulation brings up also the important issue of detecting the localization onset. That is not a trivial task. The localization associated to tensile loads (both uniaxial and biaxial ones) is usually marked by a sudden drop in the lattice stiffness (or in its elastic constants $\bar{C}_{ijkl}$ ($i, j, k, l = 1, 2$) from Eq.(1)), which can be monitored to this purpose. Splitting and shear bands are harder to detect and sometimes do not visibly affect $\bar{C}_{ijkl}$ in spite of large avalanches[31] of micro-ruptures. For example, in Figure 26(A) the damage localization of the 48×48 lattice in Case *4* does not correspond to the highest compressive peak in the stress response. In that case, the transition is instead identified by the first large avalanche (in comparison with the lattice size) highlighted in the curve of cumulative ruptures

[31] An avalanche is a series of simultaneous ruptures that is defined mathematically and examined at length in Section 3.3.

in Figure 26(B). The presence of multiple candidate points, where large avalanches happen, is a significant challenge to the implementation of a robust search algorithm to perform this identification. Compressive failures are much more sudden than in the tensile cases, and the softening phase is rarely observed in these simulations, as indicated by the sample compressive responses in Figure 27. The five replicates (*N=48*) displayed in there exhibit a very brittle behavior in Case *4*, with a sharp localization followed by the uprising rupturing rate in the $n-\bar{\varepsilon}$ curves (note the positive concavity), representing an evident sign of unstable failure.

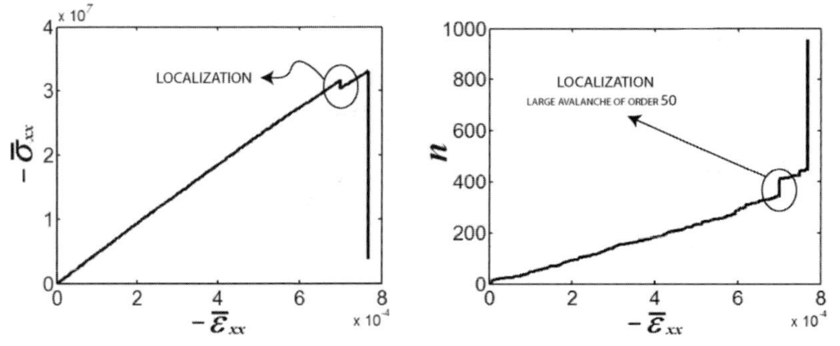

Figure 26. Hard-to-detect localization in Case 4 for a replicate with *N=48*.

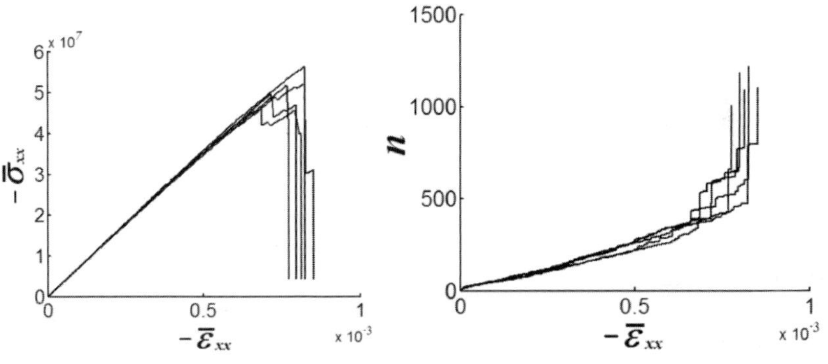

Figure 27. Macro responses for Cases 4 of five replicates of *N=48*.

The rich variety of stress responses and localization patterns reflects the distinctive development of damage-induced elastic anisotropy as a function of the loading scheme. Damage-induced anisotropy is a key aspect in the phenomenology of the damage process. The experiments by Hayhurst (1972), Chow and Wang (1987), Audoin and Baste (1994), and Christopher et al. (2003), amongst others, demonstrated the anisotropic nature of damage even in materials that are isotropic in pristine state. Chow and Wang (1987) put in evidence that isotropic damage models tend to underestimate the effect of damage in comparison to anisotropic damage and that the nature of damage is strongly dependent on the loading path close to failure. Therefore, damage-induced anisotropy needs to be included in the damage analysis to yield robust and reliable results. Later Litewka and Debinski (2003) modeled rock-like materials with an isotropic model in pristine conditions and discussed the elastic anisotropy induced by damage[32]. The load applied to brittle rock causes indeed the formation of oriented microcracks orthogonal to the direction of the maximum principal stress. In this view the loading scheme determines the evolution of anisotropy in the material system, and the corresponding localization pattern is just the end effect of such process. As discussed in the earlier chapters, damage anisotropy has proven hard to model in micromechanics[33] but is seamlessly captured by SDM discrete models such as this one. As proven by the abundance of scalar, vectorial and tensorial damage parameters found in the literature, modeling anisotropic damage evolution is hard and requires not only consideration of number, shape, and size of microcracks but also more detailed knowledge of the loading scheme and of the related damage mechanism active at the microscale.

This concept is elucidated by the data in Table 2 about the numbers of microcracks necessary to cause the failures in the three lattices in Figures 21-23. According to the data, dissimilar damage processes, producing dissimilar localization patterns, may have similar numbers of microcracks (*e.g.* Cases *{1,2,4,7}* or Cases *{3,5,6}* for *N=96*). The same comparison between Cases *1* and *7* featured in Figure 24 returns total numbers of microcracks comparable in both cases, despite differences in $\bar{\sigma}-\bar{\varepsilon}$ and $n-\bar{\varepsilon}$ data. Based on that, the sole consideration of n may not suffice to formulate analytical constitutive relations for the multiaxial damage process. Establishing what "type of knowledge and detailed information" are necessary for this purpose is the objective of Section 3.3.

---

[32] This is an interesting investigation strategy for isotropy represents a neutral initial state, ideal to study the rise of anisotropy as an effect of the sole damage process with regard to a certain loading scheme. This idea is endorsed in Sections 3.3-4 to study the damage-induced anisotropy in the lattice.

[33] While scalar damage parameters were used to model isotropic damage (*e.g.* Krajcinovic and Fonseka (1981a,b), Lemaitre (1985a,b), Tvergaard (1990)), modeling anisotropic damage requires second and higher order tensors (*e.g.* Cordebois and Sidorff (1979), Betten (1983), Krajcinovic (1983), Horii and Nemat-Nasser (1983), Chow and Wang (1987), Chaboche (1988a,b), Murakami (1988), Ju (1990), Voyiadjis and Kattan (1992), Brunig (2004)).

Table 2. Average number of microcracks at the onset of localization per Case and Size

Size *N*	*1*	*2*	*3*	*4*	*5*	*6*	*7*	*8*
N=24	124	140	200	115	141	122	92	/
N=48	440	507	710	450	474	472	350	/
N=96	1670	1530	2410	1750	2300	2000	1600	/

A graphical representation of the strong anisotropy in the simulation data is obtained by plotting all the localization thresholds together in the 3D stress space $\{\bar{\sigma}_{xx}, \bar{\sigma}_{yy}, \bar{\sigma}_{xy}\}$[34]. In particular all stresses at the onset of damage localization of each simulation for the first 8 Cases (ref. Figure 19) plot in the stress plane $\{\bar{\sigma}_{xx}, \bar{\sigma}_{yy}, 0\}$ depicted in Figure 28 for *N={24,48}*. Note that Case *8* is schematically represented as a point at infinity and that data for Case *9* is off such plane and lays on the line $\{0, 0, \bar{\sigma}_{xy}\}$ (not shown). The envelope of the stress points for a given *N* in Figure 28 (drawn with a dotted line) represents an approximate, strongly anisotropic contour correlated with the damage tolerance from different loading schemes when enforced monotonically[35]. Expectedly the lattice exhibits higher

[34] Under the assumption that $\{\bar{\sigma}_{xx}, \bar{\sigma}_{yy}, \bar{\sigma}_{xy}\}$ and $\{\bar{\varepsilon}_{xx}, \bar{\varepsilon}_{yy}, \bar{\varepsilon}_{xy} = \bar{\gamma}_{xy}/2\}$ are the only dynamical and kinematical macro-descriptors of the 2D lattice (which is assimilated to an infinitesimal Cauchy element as an RVE), any biaxial case is represented by a point in such macro-stress and macro-strain conjugate 3D spaces.

[35] Ideally, if all infinite biaxial states were simulated, the stresses corresponding to the onset of damage localization would render a localization surface in the stress space characteristic of the biaxial problem, which can be referred to as pseudo-damage surface for convenience sake. The few stress points in Figure 28 define an envelope (dotted lines) that just approximates the actual $\bar{\sigma}_{xx}$ - $\bar{\sigma}_{yy}$ section of this pseudo-damage surface per a given *N*. This representation is a *de facto* popular way to report results from multiaxial tests on brittle materials (*e.g.* concrete and ceramics) and, as such, deserves mention and consideration here. But, the prefix "*pseudo*" is added to warn the reader about its real physical meaning. That surface only conveys partial information related to the localization threshold associated to a given load scheme enforced monotonically. More complicated, non-monotonic loading paths would have no direct correlation with the pseudo-damage surface so defined. In fact, although a parallelism with the yield surface of plasticity is tempting, this is not correct as the structure of the two problems, although both path-dependent, is very different and the two surfaces are very different objects. The greater difficulty of damage evolution is ultimately and necessarily rooted in the formation of damage-induced elastic anisotropy in the system. While in plasticity, as long as the loading path does not hit the yield surface, the deformation is elastic and the plastic deformation does not change, the points inside the damage localization surface are by no means elastic and the damage process is anything but inhibited. Similarly, there is no

strength and damage tolerance for compressive states, along with a sample-size dependent behavior. In fact, the shrinkage of the envelope between $N=\{24,48\}$ reveals that the ductility reduces with N, thus generalizing for multiaxial loads the sample-size trend highlighted for the uniaxial case[36].

Collectively taken, the numerical data presented so far demonstrate that the lattice model reproduces fairly accurately the phenomenology of the real damage process and captures key physical aspects, such as damage anisotropy and sample-size effects. After gathering a substantial amount of reliable simulation data, the modeling effort shifts on the pursuit of simple, physically based constitutive relations (1) that describe the phenomenon and are suitable for engineering applications. A crucial aspect in this task is the identification of the proper material symmetry group of the subject SDM model, which heavily shapes $\bar{C}^0_{ijkl}$ and, eventually, $\bar{D}_{ijkl}$. The same numerical results are often a source of precious hints in this respect. This lattice, for instance, appears endowed with an orthotropic symmetry because normal strains and strain components are decoupled from the shear ones to a good degree. Consequently, as a first approximation, the tensor constitutive relations for the 2D lattice would specialize as

$$\begin{aligned}
\bar{\sigma}_{xx} &= \bar{C}_{xxxx}\left(\bar{D}_{xxxx}\right)\bar{\varepsilon}_{xx} + \bar{C}_{xxyy}\left(\bar{D}_{xxyy}\right)\bar{\varepsilon}_{yy}, \\
\bar{\sigma}_{yy} &= \bar{C}_{yyxx}\left(\bar{D}_{yyxx}\right)\bar{\varepsilon}_{xx} + \bar{C}_{yyyy}\left(\bar{D}_{yyyy}\right)\bar{\varepsilon}_{yy}, \\
\bar{\sigma}_{xy} &= \bar{C}_{xyxy}\left(\bar{D}_{xyxy}\right)\bar{\varepsilon}_{xy}.
\end{aligned} \tag{27}$$

Nonetheless, the determination of $\bar{D}_{ijkl}$ in the full multiaxial context is a formidable goal and falls out the narrower scope of the present overview[37]. The remaining part of the chapter just deals with strategies and advances limited to the scalar case, *i.e.* the tensile test or, at most, another uniaxial loading.

associated flow rule governing the localization surface evolution when damage increases. In this respect, the pseudo-damage curve is a mathematical entity of limited usefulness but renders meaningful information about the anisotropy and the approximate range of the hardening phase in the various loading cases. For modeling purposes the attention should shift onto the definition of an appropriate stress surface capturing *incipient* damage (similar to a yielding surface) and its anisotropic evolution rules (similar to plastic flow rules).

[36] The statement is reasonable even though the damage process associated to any mixed modes $\{\bar{\sigma}_{xx}, \bar{\sigma}_{yy}, \bar{\sigma}_{xy}\}$ combining non-zero normal and shear stresses (strains) is not contemplated in the simulations.

[37] The lattice results included in this overview are part of a comprehensive, systematic, ongoing study of multiaxial damage, aimed at determining Eq.(27) through the statistical characterization of $\bar{C}_{ijkl}$ and $\bar{D}_{ijkl}$, which require the analysis of many replicates, many sizes, and many load schemes.

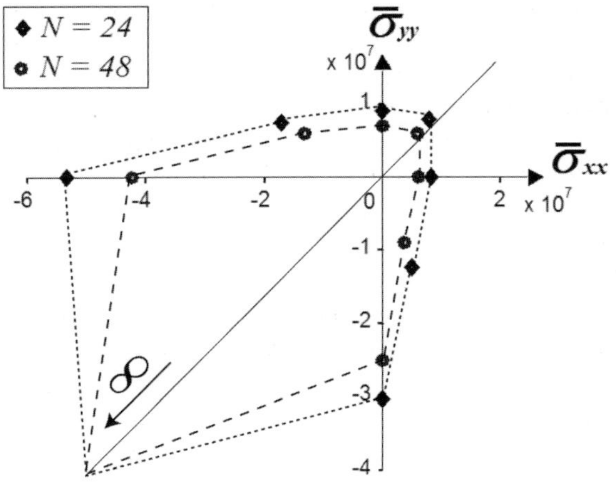

Figure 28. $\bar{\sigma}_{xx} - \bar{\sigma}_{yy}$ sections of localization envelope obtained from the stresses marking the onset of localization in Cases *1* to *8*. Both envelopes for *N=24,48* underline the anisotropy of damage strength and confirm the sample-size dependence behavior of the lattice.

3.3. Physical Foundations of $\bar{D}$

The anisotropic damage evolution illustrated above is expression of microscale processes lying in the microstructure of the lattice. Hence, the definition of $\bar{D}$ should stem from the characterization of the microcracks interactions that, in relation to lattice texture and applied load, shape the damage process and trigger the cooperative phenomenon at localization. This section breaks down the lattice problem into fundamental (microscale) elements and order them to land physical definitions of $\bar{D}$. While closed-form models exist for the one dimensional case, no universally accepted analytical model has emerged yet for the 2D and 3D cases that can capture the complex defects interaction and the non-local effects in the lattice. The following results begin bridging this gap. A preliminary investigation of "thermodynamics" and "energetics" of the lattice will precede and propel the discussion about the damage parameter.

3.3.1. A Perspective from Thermodynamics

The thermodynamic description of the damage process of quasi-brittle materials bears close resemblance with the Griffith criterion and has been broadly discussed in the contest of the continuum damage models of self-consistent type (*e.g.* Rice

(1978), Krajcinovic et. al. (1993))[38]. An elegant treatment is also available for the previous PBS, the discrete counterpart of the self-consistent continuum model (Krajcinovic (1996)). In a self-consistent approach, the load redistribution owed to damage happens always at a global level in the PBS, without a privileged local interaction between a defect and the immediate surrounding material. Limited to the hardening phase a similar description can also be considered for the lattice[39], which in principle resembles real systems better than traditional effective continuum models, at least as far as the local load redistribution is concerned.

By endorsing the notation proposed by Rice (1978) and dropping the bar sign for the effective thermodynamic quantities (just for this section), the internal energy $E(u,S,D)$ of the lattice during a tensile test is defined in the incremental form from the first principle

$$\dot{E} = T\dot{S} + F\dot{u} \quad , \tag{28}$$

where the temperature T and the entropy S are the conjugate variables associated to heat transfer. Instead, D is an internal variable describing the (damage) history. The second law of thermodynamics states that for irreversible transformations

$$T\dot{S} \geq F\dot{u} - \dot{E} \quad . \tag{29}$$

Because working with the Helmholtz free energy $\Phi = \Phi(u,T,D)$ is more convenient, in lieu of the Legendre transform $\dot{\Phi} = \dot{E} - T\dot{S} + \dot{T}S$, Eqs.(28) and (29) yield

$$\dot{\Phi} \leq F\dot{u} - S\dot{T} \quad . \tag{30}$$

For an isothermal process[40] there are two components to $\dot{\Phi}$: one reversible and another irreversible. With reference to the saw-toothed lattice response schematically depicted in Figure 29 (magnified from Figure 11), the reversible part $\dot{\Phi}_{REV}(u)$ is associated to the elastic loading-unloading (*e.g.* B-P) while the irreversible component $\dot{\Phi}_{IRR}(D,\cancel{T})$ is associated to damage transformation (*e.g.* A-B). The areas BPKH and OAB are graphical representations of the gain in strain energy from state B to state P and of the loss from A to B respectively. In agreement with Rice, the reversible part cancels out the right hand side of Eq.(30)

$$\cancel{\dot{\Phi}_{REV}} + \dot{\Phi}_{IRR} \leq \cancel{F\dot{u}} - S\cancel{\dot{T}} \, . \tag{31}$$

[38] Such a parallel is controversial and has also raised legitimate criticism. The critical aspects, though, are pointed out in the discussion.

[39] The scope is restricted since the lattice is not in thermodynamic equilibrium at and after damage localization. Hence, the application of classical thermodynamics and Griffith theory after the transition is cumbersome and perhaps non-sensical during softening.

[40] *i.e.* the lattice can be thought of as connected to a thermal reservoir enforcing $\dot{T} = 0$.

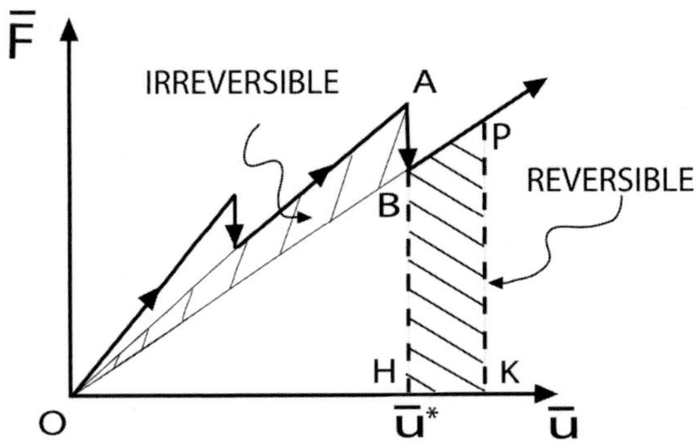

Figure 29. Force response depicted as alternation of two disjoint processes, *i.e.* linear elastic loading and damage.

and

$$\dot{\Phi}_{REV} = \left[\Phi(u+\partial u, T, D) - \Phi(u, T, D)\right] / \partial t = F\dot{u} \quad , \tag{32a}$$

$$\dot{\Phi}_{IRR} = \left[\Phi(u, T, D+\partial D) - \Phi(u, T, D)\right] / \partial t < 0 \quad . \tag{32b}$$

Hence, discrete models underline that elastic loading and damage are two disjoint processes and that the damage parameter D can in principle vary independently from u due other external sources (*e.g.* corrosion, fatigue, ions, radiations, etc.)[41]. Certainly, in the present case, D is intrinsically linked to the maximum displacement $u_{\max}$ reached in the static tensile test, which reflects the finite strength of the microstructure. According to the classical approach by Griffith and neglecting any plastic effect, the Helmholtz energy comprises a bulk term (the strain energy) and a surface term:

$$\Phi = \underbrace{U(u, T, D)}_{BULK} + \underbrace{E_\gamma(D)}_{SURFACES} \quad , \tag{33}$$

and by differentiation

$$\dot{\Phi} = \frac{\partial U}{\partial u}\dot{u} + \frac{\partial U}{\partial D}\dot{D} + \frac{\partial U}{\partial T}\dot{T} + \frac{\partial E_\gamma}{\partial D}\dot{D} \leq F\dot{u} \quad . \tag{34}$$

[41] Therefore, even when D is treated for engineering purposes as a continuous variable (especially in continuum models), it represents an intermittent process (ref. Figure 29) and is a discrete random variable at a microscale resolution.

From Eq.(34), because of the non-contemporaneity of reversible loading ($\partial u > 0$, $\partial D = 0$) and damage process ($\partial u = 0$, $\partial D > 0$), Eqs.(32) specialize as

$$\left(\frac{\partial U}{\partial u} - F\right)\dot{u} = 0 \qquad \partial u > 0\,,\ \partial D = 0\,, \tag{35a}$$

$$\left(\frac{\partial U}{\partial D} + \frac{\partial E_\gamma}{\partial D}\right)\dot{D} \le 0 \qquad \partial u = 0\,,\ \partial D > 0\,. \tag{35b}$$

Equation (35b) can be re-cast in the Griffith formalism by defining the energy release rate G, the resistance R, and the affinity (for damage) $\mathcal{A}$ as

$$\mathcal{A}\,\dot{D} = (G - R)\,\dot{D} \ge 0 \quad , \tag{36a}$$

$$G = -\left.\frac{\partial U}{\partial D}\right|_u \quad , \tag{36b}$$

$$R = \left.\frac{\partial E_\gamma}{\partial D}\right|_u \quad . \tag{36c}$$

$\mathcal{A}$ is the thermodynamic variable conjugate of the damage D and controls the damage process. Damage forms if $\mathcal{A}$=0, otherwise $\mathcal{A}$ is negative and there is no damage increase. In the lattice, then, the microscale counterparts of these macroscopic quantities can be defined for each of the Q pristine spring[42] as

$$\mathcal{A}_p = G_p - R_p\,, \quad \text{with } p = 1..Q \tag{37a}$$

$$G_p = \frac{1}{2} k_p u_p^2 \quad , \tag{37b}$$

$$R_p = \frac{1}{2} k_p u_p^{*2} \quad , \tag{37c}$$

where G_p and R_p are independent random variables associated each to the kinematics and to the strength distribution in the microstructure. It is evident that, unlike other effective parameters, such as the thermodynamic intensive variables and D,

[42] Note that no D_p is defined in Eqs.(37) because the springs in this lattice have a perfectly brittle behavior and cannot accumulate damage (*i.e.* D_p leaps discontinuously from *0* to *1* at rupture when all stored energy is released). An example where a microdamage parameter D_p needs to be accounted for is given by Rinaldi et al. (2006a), featuring a similar lattice adapted to study fatigue.

the macro-parameters G, R, and $\mathcal{A}$ cannot be computed by simply averaging over the corresponding micro-quantities, namely

$$\mathcal{A} \neq < \mathcal{A}_p >, \quad G \neq < G_p > , \quad R \neq < R_p > \ . \tag{38}$$

In fact, the macro-resistance of the lattice is determined by the failure of the link with maximum affinity amongst the Q survivals

$$\mathcal{A} = max\{ \mathcal{A}_p \} \quad , \tag{39}$$

which in general is neither the weakest nor the most elongated one, such that

$$G \neq \ max\{ G_p \} \quad , \tag{40a}$$

$$R \neq \ min\{ R_p \} \quad . \tag{40b}$$

It is not the average properties that matter but rather the "extremes properties" of the random microfields. R reflects an intrinsic material/structural property that depends on the selected loading scheme. Therefore, the operative determination of the macro-parameters G , R , and $\mathcal{A}$ in the lattice is very complex. In general, this challenge arises in any heterogeneous system subject to multi-site cracking, which deviates sensibly from the Griffith configuration consisting of a homogeneous solid medium with just one evolving defect and allowing for clear-cut interpretation and determination of G and R. An exception in that respect is offered by the PBS, for it represents a simplified scenario where all springs have same elongation $u_p = u$. Consequently R can be estimated analytically (Krajcinovic et al. (1993)) from the quenched distribution of critical elongations u_p^* and it is

$$\mathcal{A} = max\{ \mathcal{A}_p \} = min\{ R_p \} \ , \tag{41a}$$

$$G_p = \frac{1}{2} k_p \bar{u}^2 \ . \tag{41b}$$

On the contrary, despite the similarity between lattice and PBS in the hardening phase, the relation between R and quenched distribution (microstructure) is subtle in the lattice, and is affected by local interactions. Furthermore G_p and u_p are very complex random variables depending on boundary effects and on the pre-existing damage in the lattice. These considerations highlight the limitations of thermodynamics in the study of the damage process in the lattice and reiterate the need to approach the problem from below, *i.e.* from the microscale. After all, thermodynamics is an empirically based science that requires no knowledge of microscopic interaction and, as such, cannot suffice alone to determine D. For that, it is necessary to go one step deeper in the analysis and clarify the nature of u_p and of all other relevant microscale variables, similarly to statistical mechanics "that establishes a connection between the microscopic and thermodynamic description of a system" (He and Thorpe (1985)). The next section is dedicated to that, relying on numerical simulations to produce on-demand the microscale information needed to estimate the macro-quantities[43].

3.3.2. Energetics of the Lattice in Uniaxial Tests

Damage is a dissipative process. As said, the formation and propagation of any microcrack entails strain energy consumption. Starting from such a fact, this section develops a novel energetic formulation of the lattice problem to characterize the dissipative process both at the macro- and microscale and set ground for a physically based definition of the $\bar{D}$ properly bridging between the two scales[44]. With reference to Figure 29, the visible and abrupt stiffness drops separating linear elastic regions in the force response reveal the ongoing damage process and imply a strain energy dissipation. For the tensile test the strain energy U stored in the lattice can be expressed in three equivalent forms as

$$U = \underbrace{\frac{1}{2}\bar{K}\cdot\bar{u}^2}_{macro-form} = \underbrace{\frac{1}{2}\mathbf{u}_\mathrm{L}^\mathrm{T}\mathbf{K}_\mathrm{L}\mathbf{u}_\mathrm{L}}_{FE-form} = \underbrace{\frac{1}{2}\sum_{p=1}^{Q} k_p u_p^2}_{micro-form} \quad , \tag{42}$$

each one expressing different viewpoints. The macro-form is a practical definition that requires only the knowledge of macroscopic quantities (*i.e.* $\bar{K}$ and $\bar{u}$) and can be measured graphically as the subtended triangular area of the force-displacement response, such as *OPH* in Figure 30(A). This macro-form holds only for uniaxial tests since it is a special case of $U = 1/2\cdot\sum_m^2\sum_n^2(\bar{K}_{mnpq}\,\bar{u}_{pq})\bar{u}_{mn}$ (with $\bar{u}_{mn} = \bar{\varepsilon}_{mn}\cdot L$) valid for a generic multiaxial loading scheme in two dimensions. It carries no knowledge of the strain distribution within the lattice. Conversely, the FE-form and the micro-form convey the whole knowledge of the field of micro-displacements (and thus micro-strains) but state equivalent information just in terms of different primary variables. The former uses the nodal vectors $\mathbf{u}_\mathrm{L}$ of the N_0 nodes and the latter uses the elongations $u_{(ij)}$ of the Q active springs[45]

[43] Establishing the connection between micro- and macro-variables in an SDM model is the pre-requisite to build damage resistance *R*-curve (analogous to fracture mechanics) of technological relevance for damage tolerant design. Such curves though are only useful in the region of stable structural response (*i.e.* hardening) when the load is controlled.

[44] The reader interested in the definition of the damage parameter can skip these mathematical derivations in the first reading and move directly to Section 3.3.4.

[45] In the discussion a generic spring element can be referred to both as *ij*-th link if the focus is on the end nodes *i* and *j*, or as *p*-th link if the focus is on the element itself (*e.g.* its random critical elongation, the stiffness, etc.). Hence double notation, $\varepsilon_{(ij)}$ and ε_p , is used depending on the circumstances. In the former case, parentheses are used for the subscripts

(see also notation in Appendix 1). All these forms will be used here at different moments[46]. Let the discussion be organized into five main points, marked a) to e).

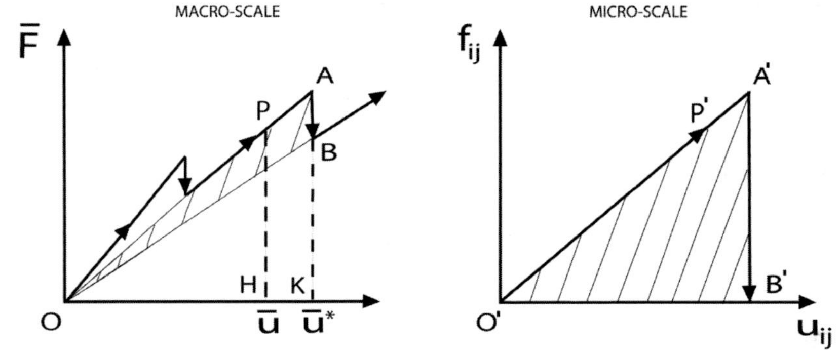

Figure 30. Energy dissipation due to a single rupture (*ij*-th link) as seen at the macro- and micro-scale.

a) Individual Rupture

Consider now the rupture of the p-th link that happens at $\bar{u} = \bar{u}^*$ in Figure 30 when $\bar{u}$ is controlled. By neglecting dynamical effects, friction, and inelastic behaviors other than brittle damage, the passage from state A to state B is an instantaneous vertical load drop. The change in strain energy from Eq.(42) is

$$\Delta U = U^A - U^B = \frac{1}{2}(\bar{K}_A - \bar{K}_B)\cdot \bar{u}^{*2} = \frac{1}{2}\Delta\bar{K}\cdot \bar{u}^{*2} \tag{43}$$

and is equivalent to the marked area OAB. By a microscopic standpoint, instead, one finds that

$$\Delta U = U^A - U^B = \underbrace{\frac{1}{2}k_p u_p^{*2}}_{\Delta U_1} + \underbrace{\frac{1}{2}\sum_{q\neq p}^{Q-1} k_q \Delta\left(u_q^2\right)}_{\Delta U_2} = \Delta U_1 + \Delta U_2 \tag{44}$$

where u_p^* is the critical elongation of the p-th spring. Equation (44) highlights the

to avoid confusion with the index notation ε_{ij} customarily reserved in solid mechanics for second order tensors.

[46] For an imposed $\bar{u}$ at a generic state P, the lattice configuration can be equally specified by the spring-elongation microfield $u_{(ij)}$ and the nodal displacement vector $\mathbf{u}_\mathrm{L}$.

existence of two contributions. The first one, ΔU_1, is the energy released in the brittle rupturing of the p-th spring and is numerically equal to the area $O'A'B'$ in Figure 30(B). This information is not usually available from a real test but it is easily obtained in a computer simulation from the *a priori* complete knowledge of the virtual microstructure. The second term ΔU_2 is the variation due to the change in the microfield of elongations u_q $(q \neq p)$ when the p-th link is removed. The lattice configuration in A is no longer equilibrated and local adjustments in the nodal displacements occur to recreate the equilibrium condition of the neighboring links, which is reached at state B. From the FE-form it is possible to rewrite (44) as

$$\Delta U = U^A - U^B = \frac{1}{2}\left[\left(\mathbf{u}_{\mathbf{L}}^{\mathbf{T}}\right)_A \mathbf{K}_{\mathbf{L}}^A \left(\mathbf{u}_{\mathbf{L}}\right)_A - \left(\mathbf{u}_{\mathbf{L}}^{\mathbf{T}}\right)_B \mathbf{K}_{\mathbf{L}}^B \left(\mathbf{u}_{\mathbf{L}}\right)_B\right], \tag{45}$$

and then to calculate

$$\Delta U_1 = \frac{1}{2} k_p u_p^{*2} = \frac{1}{2}\left(\mathbf{u}_{\mathbf{L}}^{\mathbf{T}}\right)_A \left(\mathbf{K}_{\mathbf{L}}^A - \mathbf{K}_{\mathbf{L}}^B\right)\left(\mathbf{u}_{\mathbf{L}}\right)_A \quad , \tag{46a}$$

$$\Delta U_2 = \frac{1}{2}\sum_{q \neq p}^{Q-1} k_q \Delta\left(u_q^2\right) = \left(\mathbf{u}_{\mathbf{L}}^{\mathbf{T}}\right)_A \mathbf{K}_{\mathbf{L}}^B \Delta\mathbf{u}_{\mathbf{L}} - \frac{1}{2}\Delta\mathbf{u}_{\mathbf{L}}^T \mathbf{K}_{\mathbf{L}}^B \Delta\mathbf{u}_{\mathbf{L}} \quad , \tag{46b}$$

where $\Delta\mathbf{u}_{\mathrm{L}} = (\mathbf{u}_{\mathrm{L}}^A - \mathbf{u}_{\mathrm{L}}^B)$ is the perturbation vector due to damage. Expressions (46) reveal that the redistribution term ΔU_2 does not have a pre-determined sign while ΔU_1 is a quadratic positive definite form and $\Delta U_1 > 0,\ \forall \mathbf{u}_{\mathrm{L}}^A \neq \mathbf{0}$. The strain energy loss ΔU can be computed if both stiffness matrices $\mathbf{K}_{\mathrm{L}}^A$ and $\mathbf{K}_{\mathrm{L}}^B$ are known.

Variational considerations enrich the energetic characterization of the damage process. With reference to the irreversible transformation A to B in Figure 30(A), the discrete functional (since $\mathbf{u}_{\mathrm{L}}$ is not defined on a continuous domain but on a discrete set of N^2 nodes)

$$\Pi\left(\mathbf{u}_{\mathbf{L}}\right) = U - W = \frac{1}{2}\mathbf{u}_{\mathbf{L}}^{\mathbf{T}}\mathbf{K}_{\mathbf{L}}\mathbf{u}_{\mathbf{L}} - \bar{F}\bar{u} \tag{47}$$

is the potential energy of the lattice at any elastic state (say P on OA). The work potential $W = \bar{F} \cdot \bar{u}$ (ref. Gurtin (1975)) is zero for a displacement-controlled test. The field $\mathbf{u}_{\mathrm{L}}$ can be any arbitrary kinematically admissible configuration complying with the given displacement $\bar{u}$. The minimization of Π leads to

$$\left(\partial\Pi / \partial\mathbf{u}_{\mathbf{L}}\right) = \left(\partial U / \partial\mathbf{u}_{\mathbf{L}}\right) = \mathbf{0}, \tag{48}$$

which allows the computation of the only statically admissible (*i.e.* equilibrated) configuration $\mathbf{u}_{\mathrm{L}}$.

b) The Positive Sign Of ΔU

The positive of the strain energy loss $\Delta U > 0$ regardless of the sign of ΔU_2 can be established mathematically. The difference in potential energy between A and B is

$$\Pi^A - \Pi^B = U^A - U^B = \Delta U \,, \tag{49}$$

where ΔU can be expressed as (44) in the FE-form. If none of the passive springs becomes active during the rupture of the p-th spring at A, it follows

$$\left(\mathbf{K}_{\mathbf{L}}^{A}\right)_{rs} = \left(\sum_{q \neq p}^{Q-1} \hat{\mathbf{K}}_{q}^{\mathbf{EL}} + \hat{\mathbf{K}}_{p}^{\mathbf{EL}}\right)_{rs} = \left(\mathbf{K}_{\mathbf{L}}^{B}\right)_{rs} + \left(\hat{\mathbf{K}}_{p}^{\mathbf{EL}}\right)_{rs} \tag{50}$$

for any generic rs-component of the matrices $\mathbf{K}_{\mathrm{L}}^{A}, \mathbf{K}_{\mathrm{L}}^{B}$ and $\hat{\mathbf{K}}_{p}^{\mathbf{EL}}$. The substitution of (50) into (47) leads to the inequality

$$\Pi^A = U^A = \frac{1}{2}\left[\left(\mathbf{u}_{\mathbf{L}}^{\mathbf{T}}\right)_A \mathbf{K}_{p}^{\mathbf{EL}}\left(\mathbf{u}_{\mathbf{L}}\right)_A + \left(\mathbf{u}_{\mathbf{L}}^{\mathbf{T}}\right)_A \mathbf{K}_{\mathbf{L}}^{B}\left(\mathbf{u}_{\mathbf{L}}\right)_A\right] > \frac{1}{2}\left(\mathbf{u}_{\mathbf{L}}^{\mathbf{T}}\right)_A \mathbf{K}_{\mathbf{L}}^{B}\left(\mathbf{u}_{\mathbf{L}}\right)_A \,, \tag{51}$$

as the terms in square brackets are all positive definite. If the lattice configurations $\mathbf{u}_{\mathrm{L}}^{A}$ and $\mathbf{u}_{\mathrm{L}}^{B}$ are the solutions of (48) for A and B respectively, then it must also be

$$\Pi^B = U^B = \frac{1}{2}\left[\left(\mathbf{u}_{\mathbf{L}}^{\mathbf{T}}\right)_B \mathbf{K}_{\mathbf{L}}^{B}\left(\mathbf{u}_{\mathbf{L}}\right)_B\right] < \frac{1}{2}\left(\mathbf{u}_{\mathbf{L}}^{\mathbf{T}}\right)_A \mathbf{K}_{\mathbf{L}}^{B}\left(\mathbf{u}_{\mathbf{L}}\right)_A \,, \tag{52}$$

because $\mathbf{u}_{\mathrm{L}}^{A}$ is kinematically admissible but not equilibrated and the minimum of the functional $\Pi\left(\mathbf{u}_{\mathrm{L}}\right) = 1/2\mathbf{u}_{\mathrm{L}}^{\mathrm{T}}\mathbf{K}_{\mathrm{L}}^{B}\mathbf{u}_{\mathrm{L}}$ corresponds to $\mathbf{u}_{\mathrm{L}}^{B}$. Therefore, the following inequality stems from Eqs.(51) and (52)

$$\Pi^A > \frac{1}{2}\left(\mathbf{u}_{\mathbf{L}}^{\mathbf{T}}\right)_A \mathbf{K}_{\mathbf{L}}^{B}\left(\mathbf{u}_{\mathbf{L}}\right)_A > \Pi^B \,, \tag{53}$$

which is the rigorous proof that $\Delta U > 0$ in Eq.(49) and $\Delta K > 0$ in Eq.(43).

c) Maximum Dissipation Principle for $\Delta\Pi$

The perturbation field $\Delta\mathbf{u}_{\mathrm{L}}$ is constrained by a "maximum" variational principle. By assuming that $\mathbf{u}_{\mathrm{L}}^{B}$ is unknown unlike $\{\mathbf{u}_{\mathrm{L}}^{A}, \mathbf{K}_{\mathrm{L}}^{A}, \mathbf{K}_{\mathrm{L}}^{B}\}$, then Π^A is also known. The field $\mathbf{u}_{\mathrm{L}}^{B}$ that minimizes Π^B is the solution that also maximizes the functional difference $\Delta\Pi(\mathbf{u}_{\mathrm{L}}^{B}) = (\Pi^A - \Pi^B)$, i.e.

$$\frac{\partial\left(\Pi^A - \Pi^B\right)}{\partial \mathbf{u}_{\mathbf{L}}^{B}} = \frac{\partial\left(\Delta U\right)}{\partial \mathbf{u}_{\mathbf{L}}^{B}} = 0 \,. \tag{54}$$

By substituting $\mathbf{u}_{\mathrm{L}}^{B} = \mathbf{u}_{\mathrm{L}}^{A} - \Delta\mathbf{u}_{\mathrm{L}}$ as in Eq.(46b) and using the chain rule, the derivative Eq.(54) can be expressed directly in terms of the perturbation field as

$$\frac{\partial(\Delta U)}{\partial(\Delta\mathbf{u}_{\mathbf{L}})} = \frac{\partial(\Delta U_2)}{\partial(\Delta\mathbf{u}_{\mathbf{L}})} = 0 \ , \tag{55}$$

and finally as

$$\mathbf{K}_{\mathbf{L}}^{B}\Delta\mathbf{u}_{\mathbf{L}} = \mathbf{K}_{\mathbf{L}}^{B}\mathbf{u}_{\mathbf{L}}^{A} \ . \tag{56}$$

The right hand side of Eq.(56) represents the disequilibrium (force) caused by the rupture and constitutes the driving force of the perturbation $\Delta\mathbf{u}_{\mathrm{L}}$. From Eqs.(48) and (50) it is deduced that all the elements of the vector $\mathbf{K}_{\mathrm{L}}^{B}\mathbf{u}_{\mathrm{L}}^{A}$ are the same as $\mathbf{K}_{\mathrm{L}}^{A}\mathbf{u}_{\mathrm{L}}^{A}$ besides the four *dof* of the broken *p*-link. Since $\mathbf{K}_{\mathrm{L}}^{A}\mathbf{u}_{\mathrm{L}}^{A}$ represents the equilibrated state A of the lattice with external forces applied only at the boundary nodes, all the components of $\mathbf{K}_{\mathrm{L}}^{B}\mathbf{u}_{\mathrm{L}}^{A}$ are zeros except for the *dof* of the *p*-link (disequilibrium) and to the outer nodes (external support reactions in state A). Therefore a local perturbation associated to the disequilibrium of four *dof* governs the transformation from *A* to *B*. In conclusion, the relation between the energy dissipation and the elastic perturbation associated to damage may be expressed by the variational principle stated below.

VARIATIONAL PRINCIPLE OF MAXIMUM DISSIPATION:
the elastic perturbation field (i.e. $\Delta\mathbf{u}_{\mathrm{L}}$ *) due to the introduction of new damage in the lattice maximizes the drop of potential energy* $\Delta\Pi$ *from state A to state B.*

Since the result was inferred from the FE-form (which is loading-scheme independent), it is valid for any type of loading scheme, both uniaxial and multiaxial ones. The variational formulation may become useful to implement an incremental approach when the new equilibrated state *B* needs to be computed from state *A* rather than from *O* in Figure 30(A). A similar situation is encountered in presence of damage-plasticity or when an approximate solution is sought by updating the previous undamaged configuration in a sub-domain where the perturbation field $\Delta\mathbf{u}_{\mathrm{L}}$ exceeds a threshold value, *e.g.* in very large lattices and during damage propagation. The perturbation, in purely elastic networks like this one, practically spans the whole lattice during hardening but is confined in a small area when damage localizes (*e.g.* ref. Figure 49).

d) Load Redistribution Effect ΔU_2

The meaning of the redistribution term ΔU_2, as defined in Eqs.(44) or (46b), deserves further attention. This quantity accounts for the interaction effect between a single structural element (*i.e.* a spring) and the surrounding and is characteristic of higher dimensional, highly interconnected systems and real microstructures. For instance, the simplest one-dimensional statistical model, the PBS, lacks a ΔU_2 term and it is just $\Delta U = \Delta U_1$. Two components contribute to ΔU_2 in Eq.(46b).

The first term, $(\mathbf{u}_L^T)_A \mathbf{K}_L^B \Delta\mathbf{u}_L$, is the work of the non-equilibrium forces introduced in the system by the deletion of the p-th link. The $\mathbf{K}_L^B \mathbf{u}_L^A$ vector was already identified in Eq.(56) as the source of perturbation. More precisely, the disequilibrium corresponds to the force pair $\vec{F}_i$ and $\vec{F}_j$ shown in Figure 31, which pull the node i and j away from each other. When the lattice configuration is $\mathbf{u}_L^A$ and the p-th link is removed, only these two forces perform work on the corresponding nodal perturbation vectors $\Delta\mathbf{u}_i = [\ \Delta u_i \ \ \Delta v_i\ ]^T$ and $\Delta\mathbf{u}_j = [\ \Delta u_j \ \ \Delta v_j\ ]^T$. Consequently $(\mathbf{u}_L^T)_A \mathbf{K}_L^B \Delta\mathbf{u}_L > 0$. The intensity of $\vec{F}_i$ and $\vec{F}_j$ is not constant in reality but the transition to the new equilibrated state B is instantaneous in this virtual model.

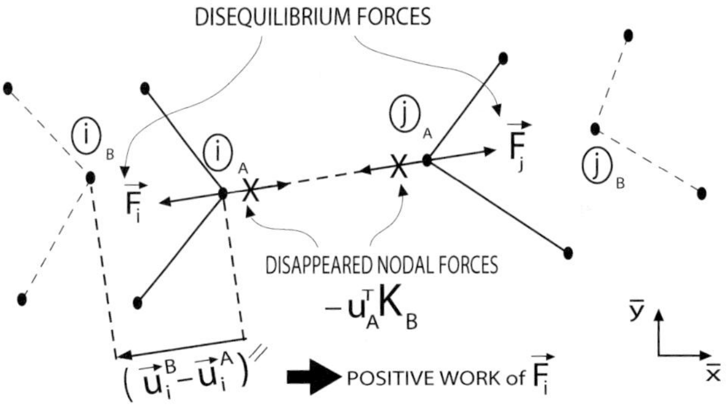

Figure 31. Disequilibrium generated by deletion of *ij*-th spring in the configuration A (full line) and new stable configuration B (dashed line).

On the other hand, the second component of ΔU_2, *i.e.* $-1/2 \cdot \Delta\mathbf{u}_L^T \mathbf{K}_L^B \Delta\mathbf{u}_L$, is always negative because $\Delta\mathbf{u}_L^T \mathbf{K}_L^B \Delta\mathbf{u}_L$ is a positive definite form. Therefore, while the energy $(\mathbf{u}_L^T)_A \mathbf{K}_L^B \Delta\mathbf{u}_L$ associated to the disequilibrium is spent to reach the new configuration, part of it gets stored into the neighboring springs. Then, even if some links relax and other carry higher load, there is always such partial gain/recovery of strain energy. The local details of the disordered microstructure determine the redistribution capability. In conclusion ΔU_2 is a trade-off between these two counter actions and the sign ΔU_2 cannot be predicted *a priori*. Simulations indicate that the area OAB is most of the time greater than $O'A'B'$ in Figure 30 but not always, *i.e.* $\Delta U > \Delta U_1$ most frequently in this lattice. Figure 32 displays, for example, a panel of data plots of a random lattice with $N=48$. Besides the stress response (plot A, full line) and the curve of cumulative ruptures (plot B), it shows also the cumulative curves of ΔU (full line) and ΔU_1 (dashed line) (plot C). The divergence between the latter two curves becomes more visible at the transition and in the softening phase, where ΔU_2 plays an increasingly dominant role. The effect of the transition is also appreciable from the steeper rate of descent of the lattice stiffness $\bar{K}$ (plot D) following the stress peak.

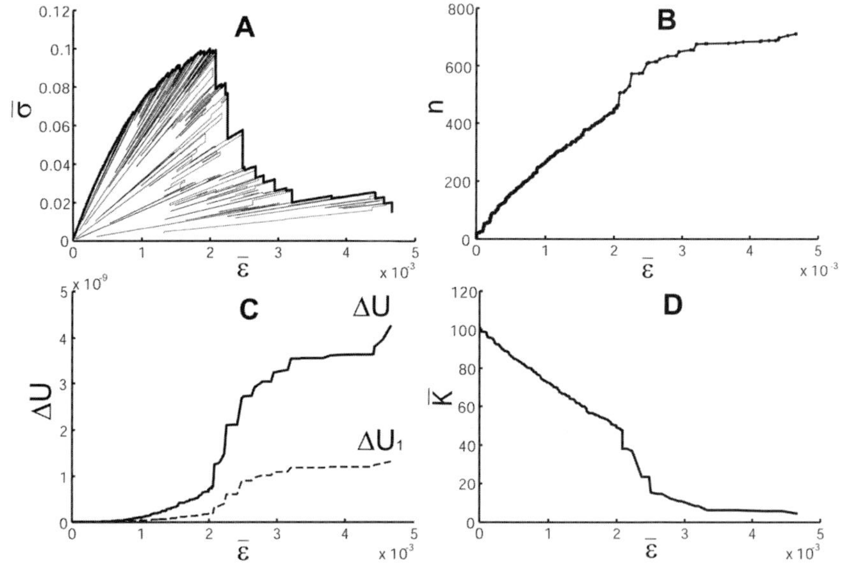

Figure 32. Results from FE simulation of a lattice replicate with $N = 48$. (A) $\bar{\sigma} - \bar{\varepsilon}$ response in damage (dotted) controlled, displacement controlled (full) line; (B) total broken curve $n - \bar{\varepsilon}$; (C) cumulative curve of dissipated strain energy ΔU and ΔU_1 ; (D) secant stiffness curve $\bar{K} - \bar{\varepsilon}$.

e) Cascade of Ruptures: Avalanche and Snapback

In order to complete the energetic analysis, at last it is necessary to extend the results for the individual rupture to the general case of multiple ruptures by introducing the concept of "avalanche". An avalanche consists of a cascade of ruptures of more links in correspondence of the same value $\bar{u}$ in the displacement controlled process. The order of the avalanche is the number of ruptures. One example of order 3 avalanche is shown schematically in Figure 43. The avalanche is an instability phenomenon occurring when the new equilibrated lattice configuration $\mathbf{u}_{\mathrm{L}}^{B}$ in Eq.(45), reached after the first rupture, is kinematically admissible but incompatible with the lattice strength. In fact the static perturbation field associated to the load redistribution can drive at least one more spring to exceed its critical strain under the same imposed displacement $\overline{u}$. A strong perturbation effect can also involve links of superior strength in the avalanche. Avalanches do not occur in a damaged-controlled situation (see Appendix 1), when $\overline{u}$ is a dependent variable and the lattice is allowed to "snapback" to the equilibrated state determined by the element of highest affinity as explained before. Snapbacks due to avalanches are visible in the damage-controlled response shown in Figure 32(A) with dotted lines and denote massive energy redistribution. The gap between ΔU and ΔU_1 (Figure 32(C)) increases particularly in correspondence of the large avalanches in the softening phase.

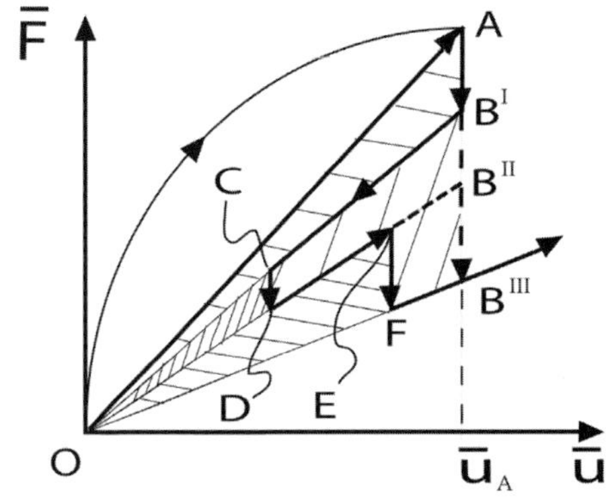

Figure 33. Avalanche of order 3 and snapback effect between states A and B (*i.e.* B^{III}).

This state of affairs can be translated in mathematical terms. From an energetic standpoint, the balance Eq.(44) needs to be generalized by the addition of an extra term ΔU_3 for each spring in the avalanche to account for the extra dissipation associated to snapback. The total drop of strain energy for an avalanche of order q between state A and B is simply the sum of the q contributions

$$\Delta U = U_A - U_B = \sum_{i=1}^{q} \left(\Delta U_1 + \Delta U_2 + \Delta U_3 \right)_i \ , \tag{57a}$$

with $$\Delta U^{(i)} = \Delta U_1^{(i)} + \Delta U_2^{(i)} + \Delta U_3^{(i)} \ . \tag{57b}$$

Under the assumption of instantaneous load redistribution and no dynamic and frictional effects, the lattice would tend to follow the "pattern of minimum energy" $A \to B \to C \to D \to E \to F \to B^{III}$ from A in Figure 33 but it is constrained to stay at $\bar{u}_A$ and it is forced instead into the non-equilibrium pattern $A \to B^{I} \to B^{II} \to B^{III}$. The snapback effect is approximately represented by the area enclosed by the closed loop $B^{I} \to C \to D \to E \to F \to B^{III} \to B^{I}$. For the second rupture, for example, ΔU_3 coincides with the area of the polygon $B^{I} \to C \to D \to B^{II} \to B^{I}$ and can be computed as

$$\Delta U_3^{(2)} = \frac{1}{2}\left[\bar{u}_A^2 \Delta \bar{K}_{B^I B^{II}} - \bar{u}_C^2 \Delta \bar{K}_{CD}\right] . \tag{58}$$

The geometrical similarity between triangles OCD and $OB^{I}B^{II}$ introduces the proportionality factor

$$\lambda^{(2)} = \frac{\Delta \bar{K}_{B^I B^{II}}}{\Delta \bar{K}_{CD}} = \frac{\bar{u}_A}{\bar{u}_C} = \frac{\bar{u}_1}{\bar{u}_2} . \tag{59}$$

Since $1/2 \cdot \bar{u}_C^2 \Delta \bar{K}_{CD}$ is the effect of $\Delta U_1^{(2)}$ and $\Delta U_2^{(2)}$ related to the second rupture, it follows from Eqs.(43) and (44) that

$$\Delta \bar{K}_{CD} = \frac{2\left(\Delta U_1^{(2)} + \Delta U_2^{(2)}\right)}{\bar{u}_C^2} . \tag{60}$$

Finally, for any *i*-th rupture in the avalanche (with $i = 1,..,q$), the substitution of Eqs.(59) and (60) into Eq.(58) delivers

$$\Delta U_3^{(i)} = \left(\left(\frac{\bar{u}_1}{\bar{u}_i}\right)^2 - 1\right)(\Delta U_1^{(i)} + \Delta U_2^{(i)}) = \left(\lambda_i^2 - 1\right)(\Delta U_1^{(i)} + \Delta U_2^{(i)}) , \tag{61}$$

where $\bar{u}_1 = \bar{u}_A$ and $\bar{u}_2 = \bar{u}_C$ for $i = 2$ (Figure 33). The parameter $\lambda_i = \bar{u}_1 / \bar{u}_i$ conveys the effect of the snapback for any *i*-th rupture in the avalanche. Since $\bar{u}_1 \geq \bar{u}_i$, it is always $\lambda_i > 1$ and $\Delta U_3^{(i)} = 0$ if there is no snapback (*e.g.* in the first rupture). However, the quota $\Delta U_3^{(i)}$ is proportional to λ_i^2 and is potentially much larger than $(\Delta U_1^{(i)} + \Delta U_2^{(i)})$ if the snapback is pronounced, like at the peak. The observed energy loss $\Delta U^{(i)}$ during the load-drop $B^{i-1} \to B^{i}$ is from (61)

$$\Delta U^{(i)} = \Delta U_1^{(i)} + \Delta U_2^{(i)} + \Delta U_3^{(i)} = \lambda_i^2 (\Delta U_1^{(i)} + \Delta U_2^{(i)}) > 0 . \tag{62}$$

Therefore, the snapback simply causes an *amplification* of the energy dissipated in a single rupture. Finally, the total observed energy drop AB and the total snapback contributions are respectively

$$\Delta U = U_A - U_B = \sum_{i=1}^{p} \Delta U^{(i)} = \sum_{i=1}^{p} \left[\lambda_i^2 (\Delta U_1^{(i)} + \Delta U_2^{(i)}) \right] > 0 \quad , \qquad (63)$$

and $$\Delta U_3^{TOT} = \sum_{i=1}^{p} \Delta U_3^{(i)} = \sum_{i=1}^{p} \left[\left(\lambda_i^2 - 1\right)(\Delta U_1^{(i)} + \Delta U_2^{(i)}) \right] \geq 0 \quad . \qquad (64)$$

Unfortunately λ_i , $\Delta U_1^{(i)}$ and $\Delta U_2^{(i)}$ are random quantities and are "hidden" on the macro-scale, where only ΔU is usually observable. The experimental assessment of the cumulative snapback effect ΔU_3^{TOT} for an avalanche is indeed a challenging task. Reliable numerical simulations are likely the best tool for this kind of investigation. In conclusion the non-equilibrium pattern $A \to B^{I} \to B^{II} \to B^{III}$ is the pattern of minimum energy dissipation compatible with the microstructure and with the imposed kinematical constraint $\overline{u}_A$. Simulations show that the snapback pattern is a random variable dependent on the microstructural disorder and requires a statistical characterization from many random realizations. The extra energy term ΔU_3^{TOT} is associated to the non-equilibrium pattern $A \to B^{I} \to B^{II} \to B^{III}$ at $\overline{u}$. The energy drop ΔU in Eq.(63) is always positive and is maximized by the perturbation field associated to the avalanche.

Evidently the plot of the avalanche in Figure 34(B) that corresponds to the stress response in Figure 32(A) shows the emergence of large avalanches near the critical point (*i.e.* the stress peak where the homogeneous-heterogeneous transition occurs) and in correspondence with large load drops in the response. The order of the avalanche is an indicator of the range of interaction between defects and Figure 34 demonstrates that the correlation length[47] at the onset of the transition is, indeed, of the order of the lattice size L. This is confirmed by the other two examples in Figure 34 for $N = \{24, 192\}$ (A and C respectively). In particular, for $N = 192$ the magnitude of the avalanches jumps from few units to about 400 and 1200 in correspondence to the critical point. After the sharp transition, this lattice cannot reach a configuration compatible with the imposed displacement and fails. In cases such as this one no softening phase exists and catastrophic failure occurs suddenly at the stress peak. Normally the failure happens either stably by the isolated rupture of a single last spring or unstably during an avalanche.

[47] A diverging correlation length at the force/stress peak is consistent with the general phenomenology of phase transitions, where certain thermodynamic parameters and correlation lengths diverge at the critical points.

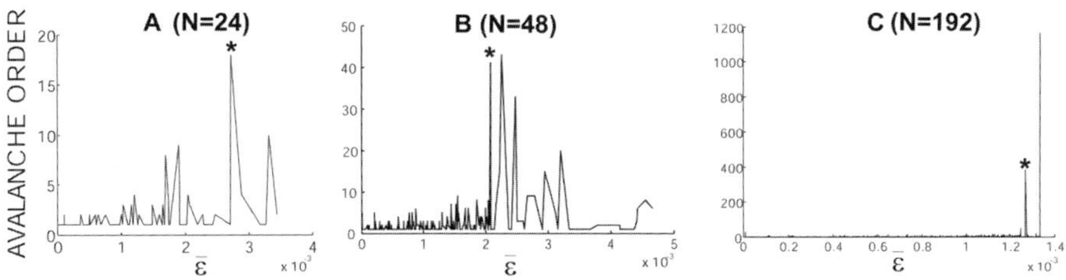

Figure 34. Evolution of order of avalanche for three replicates $N = \{24, 48, 192\}$. Diverging avalanches are observed at the force peak transition (marked with an asterisk).

3.3.3. Rational Damage Model for Uniaxial Tests

On the ground of the mathematical concepts established above, this section turns to the definition of the damage parameter $\bar{D}$ rendering the proper constitutive relation

$$\bar{\sigma}(\bar{\varepsilon}, L) = \bar{K}_0\left[1 - \bar{D}(\bar{\varepsilon}, L)\right]\bar{\varepsilon}\ , \tag{65}$$

that captures the behavior of the lattice subject to uniaxial loading, either tensile or compressive. The emphasis is placed on the statistical foundations of the damage parameter and the derivation proceeds, in a bottom-up fashion, from the results of the energetic analysis. The first precious hint is provided by Eq.(43) stating that each damage event causes the dissipation of a quota of strain energy ΔU that, in turn, is linearly related to a reduction $\Delta\bar{K}$ of the secant stiffness. Owing to the scalar nature of $\Delta\bar{K}$, the sum of all $\Delta\bar{K}$ renders a measure of the accumulated damage. Moreover, because the damage process ends with a failure when $\bar{K}(\bar{\varepsilon}, L) = 0$ and $\bar{K}_0$ is entirely depleted, such a sum is physically constrained to range from nil at zero damage to $\bar{K}_0$ at failure. Thus, by requiring $\bar{D} = 1$ at failure, the damage parameter in Eq.(65) can be defined by normalizing the cumulative stiffness loss to $\bar{K}_0$ as

$$\bar{D} = \frac{\sum_{p=1}^{n(\bar{\varepsilon})} \Delta\bar{K}_p}{\bar{K}_0}\ . \tag{66}$$

The above damage parameter is the cumulative effect of the $n(\bar{\varepsilon})$ ruptures at $\bar{\varepsilon}$ and expresses a general macroscale perspective. In fact, this equation provides the definition of $\bar{D}$ valid not just for the lattice but for any brittle system. To obtain a lattice specific definition, it is necessary to conjugate it to the specific microstructure, as seen in the simpler PBS.

The theory is developed here for honeycomb lattices with random mechanical properties only. An example is given in Figure 35. This responds, in first place, to a "blocking" strategy used in statistics to eliminate the bias otherwise due to geometrical disorder, and to make mechanical disorder the only source of variability in the lattice response. Secondly, the exact triangular lattice (with equal spring constants) in pristine state maps into an isotropic continuum element, as shown by Monette and Anderson (1994), and is convenient to study the development of anisotropic damage. Thirdly, the perfect geometry introduces the important simplification of just three spring orientations $\theta = \{0^o, 60^o, -60^o\}$ with respect to the loading axis x. This configuration is sufficient to develop clear arguments that can be generalized at a later time for a disordered texture. Due to the symmetry of the diagonal elements, the N_0 springs in Eq.(25) are partitioned into two groups:

- Group 1: the set of N_1 horizontal links (*i.e.* $\theta = 0^o$),
- Group 2: the set of N_2 diagonal links (*i.e.* $\theta = \pm 60^o$).

For the lattice in Figure 35, a counting exercise demonstrates that $N_1 = N \cdot (N-1) - N/2$ and $N_2 = 2 \cdot N \cdot (N-1)$, the sum of which is equal to N_0. For $N \to \infty$, N_1 and N_2 tend to the asymptotic values

$$N_1 = 1/3 \cdot N_0 , \tag{67a}$$

$$N_2 = 2/3 \cdot N_0 . \tag{67b}$$

This approximation is acceptable already at $N=24$ and is adopted henceforth.

A rational model for $\overline{D}(\overline{\varepsilon}, L)$ is derived for the entire damage process, accounting for both individual ruptures and avalanches.

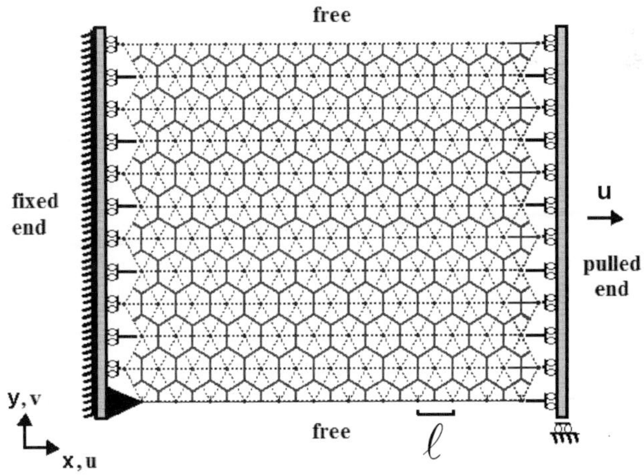

Figure 35. Disordered lattice with perfect geometry ($N = 12$) and BCs for tensile test in x.

The energy loss ΔU (measurable macroscopically) from the failure of the p-th spring is given by Eq.(57b) as the sum between known $\Delta U_1 = k_p u_p^{*2}$ (estimable *a priori* from knowledge of k_p and u_p^*) and the unknowns ΔU_2 and ΔU_3 . The "redistribution factor" η_p is introduced to express ΔU_2 and ΔU_3 in terms of ΔU_1

$$\eta_p = \frac{\Delta U - \Delta U_1}{\Delta U_1} = \frac{\Delta U_2 + \Delta U_3}{\Delta U_1} \quad , \tag{68}$$

such that ΔU can be rewritten as

$$\Delta U = \left(1+\eta_p\right)\Delta U_1 = \frac{1}{2}\left(1+\eta_p\right)k_p u_p^{*2} \quad , \tag{69}$$

where $\eta_p \geq -1$ is the only prescription on η_p stemming from $\Delta U > 0$. Since the macro-form (43) holds for uniaxial loading, the loss of secant stiffness $\Delta\bar{K}$ is

$$\Delta\bar{K} = \frac{\left(1+\eta_p\right)k_p u_p^{*2}}{\bar{u}^{*2}} . \tag{70}$$

Because all springs have same length $\ell_{(ij)} = \ell_0$ (with $\ell_0 = 1$ in this model) and stiffness $k_{(ij)} = k$, the lattice and link strains are respectively

$$\bar{\varepsilon}^* = \bar{u}^* / L \qquad \text{and} \qquad \varepsilon^*_{(ij)} = u^*_{(ij)} / \ell_0 , \tag{71}$$

the substitution of which in Eq.(70) renders

$$\Delta\bar{K} = \left(1+\eta_p\right)k\left(\frac{\ell_0}{L}\right)^2\left(\frac{\varepsilon_p^*}{\bar{\varepsilon}^*}\right)^2 . \tag{72}$$

This result provides the fundamental linkage to endow definition (66) with the lattice-specific microscale meaning, namely

$$\bar{D}(\bar{\varepsilon}, L) = \frac{\sum_{p=1}^{n(\bar{\varepsilon})} \Delta\bar{K}_p}{\bar{K}_0} = \frac{k}{\bar{K}_0}\left(\frac{\ell_0}{L}\right)^2 \sum_{p=1}^{n(\bar{\varepsilon})}\left(1+\eta_p\right)\left(\frac{\varepsilon_p^*}{\bar{\varepsilon}^*}\right)^2 . \tag{73}$$

The above equation represents the full rational model for any uniaxial loading, both in the hardening and in the softening phases, and renders the cumulative effect of the $n(\overline{\varepsilon})$ ruptures occurred at $\overline{\varepsilon}$. The model is stochastic and requires following three random fields as inputs:

1. the critical strain $\varepsilon_p^*(\overline{\varepsilon};L)$ of the spring failing at $\overline{\varepsilon}$, linking the kinematics on the macroscale and on the microscale;
2. the redistribution parameter $\eta_p(\overline{\varepsilon};L)$, regarding the load redistribution capability of the microstructure and the snapback effect;
3. the number of broken bonds $n_p(\overline{\varepsilon};L)$, related to the accumulated damage[48].

To prove that Eq.(73) is the rational model, exact inputs { $\varepsilon_p^*, \eta_p, n_p$ }, as tracked during the simulation, are fed into it to verify the perfect overlap between the model prediction and the output of the native simulation. Figure 36 (left column) display plots of the predicted stress response from the rational model (73) for the same *48×48* lattice used as validation example under Case *1* (tensile test in *x*), Case *2* (tensile test in *y*), Case *4* (compression test in *x*), and Case *5* (compression test in *y*) as defined in Figure 19. The corresponding model output from Eqs.(65) and (73) is plotted therein too. As expected simulation data and model data are indistinguishable. The model response marked by "*" in the plots in the left column of Figure 36 overlaps, indeed, perfectly with the $\overline{\sigma}-\overline{\varepsilon}$ data from simulation drawn underneath with solid bold line. Such a lack of model error legitimates the term "rational". The damage curves $\overline{D}-\overline{\varepsilon}$ from Eq.(73) are plotted on the right column (solid bold line), along with the three components $\overline{D}_i(\overline{\varepsilon})$ (solid line with markers) corresponding to the following spectral representation of the damage parameter based on spring orientation

$$\overline{D}(\overline{\varepsilon};L) = \overline{D}_1(\overline{\varepsilon};L) + \overline{D}_2(\overline{\varepsilon};L) + \overline{D}_3(\overline{\varepsilon};L) = \tag{74}$$

$$= \frac{k}{\overline{K}_0}\left(\frac{\ell_0}{L}\right)^2 \left[\sum_{p=1}^{n1(\overline{\varepsilon})} (1+\eta_p)\left(\frac{\varepsilon_p^*}{\overline{\varepsilon}^*}\right)^2 + \sum_{p=1}^{n2(\overline{\varepsilon})} (1+\eta_p)\left(\frac{\varepsilon_p^*}{\overline{\varepsilon}^*}\right)^2 + \sum_{p=1}^{n3(\overline{\varepsilon})} (1+\eta_p)\left(\frac{\varepsilon_p^*}{\overline{\varepsilon}^*}\right)^2 \right]$$

with " $n(\overline{\varepsilon}) = n_1(\overline{\varepsilon}) + n_2(\overline{\varepsilon}) + n_3(\overline{\varepsilon})$ " and subscripts {*1,2,3*} standing for orientations { $0^o, 60^o, -60^o$ }. The consistent usage of data markers { $*, O, \nabla$ } respectively for θ = { $0^o, 60^o, -60^o$ } in all plots let appreciate the different role of each spring group in each loading scheme. That allows – for instance – grasping the symmetry between $\overline{D}_2$ and $\overline{D}_3$ in every plot. Also the component $\overline{D}_1$ from the horizontal springs appears always predominant but in Case *2* of tension along *y*. As pointed out already, in the latter case the horizontal springs do not break easily since they are orthogonal to the tensile axis and compressed in the pristine state. The damage

[48] Because of the lattice-size effect (Section 3.4), the micro-parameters ε_p^* , η_p , n_p are functions of $\overline{\varepsilon}$ but implicitly depend on *L*, which is dropped hereafter for readability.

patterns[49] of all four cases are shown in Figure 37 and demonstrate different failure modes, *i.e.* transverse cracking for both the tensile cases, "splitting" from Poisson's effect for Case *4* and shear bending in Case *5*. Please note the following: *(i)* the open cracks drawn in Figure 37 with dashed line are just schematics marking the location and the orientation of the main macrocracks observed at failure, and *(ii)* in Case *5* the failure pattern closely resembles a Mode II sliding crack that sprouts winged secondary cracks near the loading edges. The lattice is much more resistant when compressed along x (Case *3*) due to preferentially oriented springs. In such case no damage "seems" to occur in the hardening phase – which is actually a false impression – and the stress response is seemingly linear up to the peak, with the magnitude of the maximum sustained stress being two orders of magnitude greater than in tension and one order of magnitude more than in y-compression. Since the extent of softening is much narrower for tensile load in y than in x, this network is actually more damage tolerant along the x direction[50].

Input analysis

A deeper comprehension of the damage evolution descends from exploring the correlation between the three types of inputs in Eq.(73) and the associated damage curve in Figure 36. Such input data $\{ \varepsilon_p^*, \eta_p, n_p \}$ are shown for the four cases respectively in Figures 38-41. All plots feature the spectral representation along with the agreed markers $\{ *, \mathrm{O}, \nabla \}$, which is an informative way to interrogate, collect, and present the input data. For argument sake, let us first look at a smaller lattice with $N=24$ and examine the corresponding $\varepsilon_p^*(\bar{\varepsilon})$ curve in Figure 42 (bottom) for Case *1*, which is more "diluted" and readable. The plot of the critical strain on the left side reports data in aggregate form, while the spectral form is given on the right for $\theta = \{ 0^o, \pm 60^o \}$, with the diagonal ruptures being pooled together for simplicity. The aggregate form clearly reveals the existence of two regimes corresponding to the hardening and softening phases of the force response (Figure 42, top). Recalling that ε_p^* are sampled in the interval $[0,10^{-2}]$, the plot suggests that the hardening phase is characterized by the progressive and ordered rupturing of weaker links, according to the relation $\varepsilon_p^* \le \bar{\varepsilon}$ between the spring thresholds and the applied macro-strain. Hence the maximum ε_p^* is related linearly to $\bar{\varepsilon}$ (a fact to be exploited later in the derivation of approximate closed-form solutions).

[49] Ref. Rinaldi and Lai (2007) for color figures, where broken springs are colored red to distinguish them from unstable ones colored green (dandling links or mechanisms).

[50] The shape of the damage curves further supports this statement because the upward concavity in the y-cases (2 and 4) hints to an unstable process, as opposed to the x-cases (1 and 3) where flat plateaus can be regarded as typical signs of a more gradual, curbed damage evolution. Consequently, damage patterns for the y-cases in Figure 37 are neater.

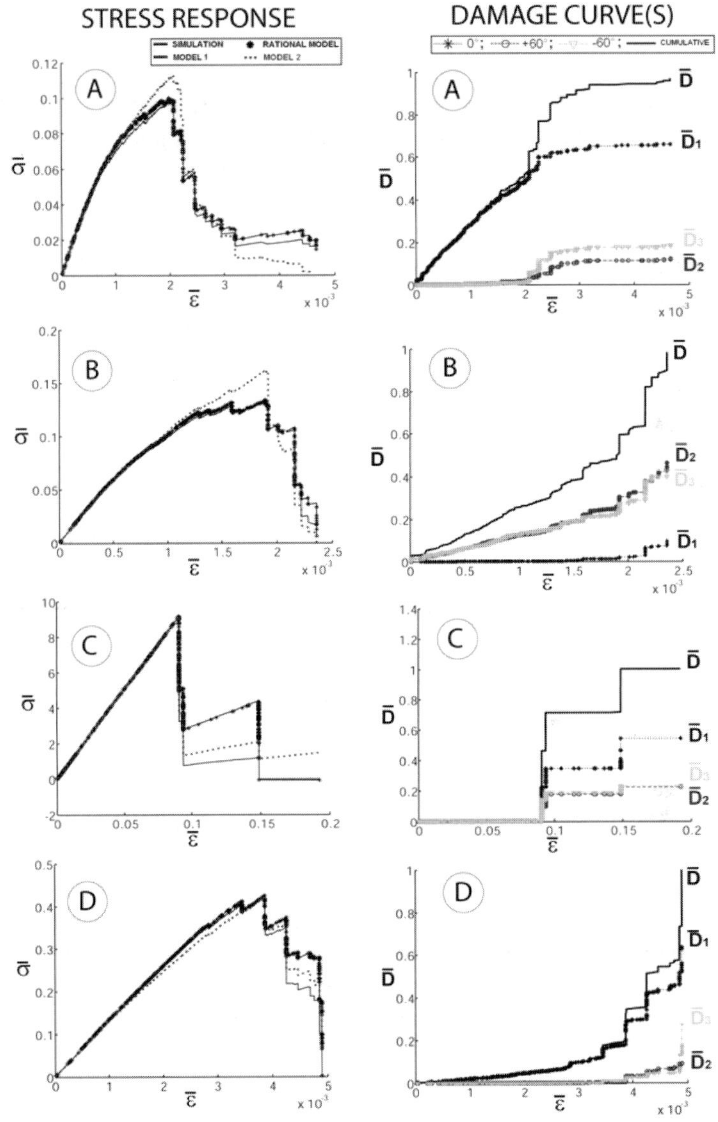

Figure 36: Outputs of rational model for the four Cases $\{1,2,4,5\}$ of x-tension (A), y-tension (B), x-compression (C), and y-compression (D). The stress responses are shown in the left plots along with simulation data and two simplified models. The damage curves and the components from the rational model are shown on the right.

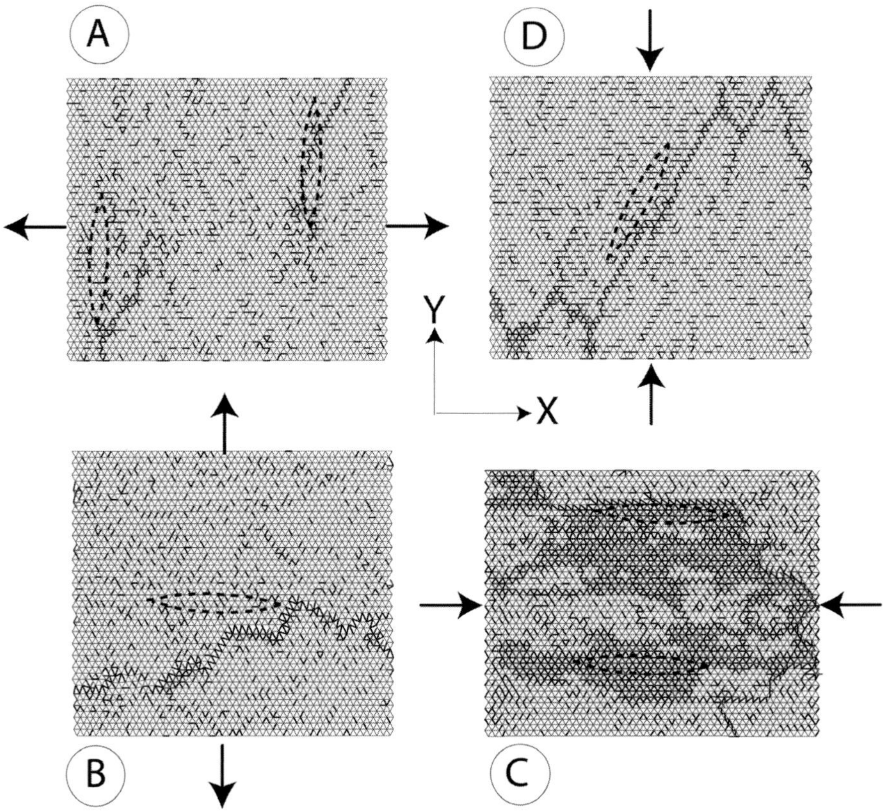

Figure 37. Patterns for the four cases in Figure 36 outlining different failure modes.

Such "linearity" breaks down in correspondence of the stress peak at $\overline{\varepsilon}_{peak} \simeq 2.7 \cdot 10^{-3}$, when rupturing becomes disordered, high-strength springs get involved in the damage process, a large avalanche occurs (see force response), and softening begins. Therefore the transition[51] is caused by microcracks interactions able to generate increasing perturbation in the various microfields (*e.g.* stress, strains, deformation energy, affinity, etc.), the local fluctuations of which become much larger than the mean-field value and drive the damage localization randomly

[51] The change in the "structure" of the series $\varepsilon_p^*(\overline{\varepsilon};L)$ sheds new light on the transition and renders with extreme clarity how the material microstructure passes from a status of statistical homogeneity to one of heterogeneity. As damage localizes, the randomness of the micro-texture assumes control of the macro-response.

on the weaker cross-section of the lattice. Plotting the critical strains ε_p^* *vs.* $\overline{\varepsilon}$ is an effective way of tracking the magnitude of the microstructural fluctuations. Plotting the same data in the spectral form is even more enlightening and meaningful because portrays a finer image of the damage process. So, dismembering $\varepsilon_p^*(\overline{\varepsilon})$ into the two components in Figure 42(B) shows that most ruptures in the hardening occur along $\theta = 0^o$, while the weaker diagonal spring breaks at much lower rate. This distinction enables stating that not all microcracks are "equal" and that in this case the horizontal springs play a dominant role in the damage process. For that, the latter ones deserve the appellative "primary", whereas the diagonal springs behave as "secondary" or "slave" (the sense of this terminology will become even more apparent later on). This scenario is in agreement with experimental observations by Christofer et al. (2003), seconding that grain boundaries orthogonal to the loading direction have a much higher probability of cracking. Of course which spring (or grain boundary) orientation plays what role depends on the loading scheme and needs to be established case by case, as deduced from data in Figures 38-41. Furthermore, the example in Figure 42(B) shows that the role of a group of springs may change during in the damage process. In fact, the spectral representation unveils a "role-switching" between the master and slave groups happening at the transition, where the latter one becomes dominant and its damage rate increases abruptly. The large avalanche, marking the crossover point, encompasses ruptures of diagonal links with inferior strength as well as high-strength ones, like a diagonal spring with $\varepsilon_p^* \simeq 8\cdot10^{-3} >> \overline{\varepsilon}_{peak}$. Identifying the roles of different set of links (or grain boundaries in a polycrystalline microstructure) at different stages of the damage process is crucial to the study of damage localization by a physics perspective, as well as for the derivation of simplified models as seen later.

The considerations about the phase transition and the spectral representation illustrated for $\varepsilon_p^*(\overline{\varepsilon})$ are also valid for the other two inputs $\{\eta_p, n_p\}$. At a glance, the redistribution factor η_p appears to be the most "chaotic" of the three inputs. It varies in a wide range of values and just a limited range in shown in Figures 38-41. The η_p values are reported in the plot at the location where the data series exceeded the chosen pictorial window. For example, in Figure 38 η_p is lower than 40 for the horizontal ruptures but reaches maximum values of 700 and 11000 for ruptures at $+60^o$ and -60° respectively. Such bursts in η_p typically occur during avalanches with large ΔU_2 and ΔU_3 (*e.g.* a very weak link with large ΔU_2 breaking at large $\overline{\varepsilon}$). The plot reports the minimum theoretical value of the redistribution factor $\eta_p = -1$, corresponding to full local redistribution of the released energy ΔU_1 (*i.e.* $\Delta U = 0$ for $\Delta U_1 = -\Delta U_2$). Mean values and ranges of η_p appear markedly different along the three orientations, yet the behavior of the springs at $+60^o$ and -60° can be guessed to be statistically symmetric from the plots.

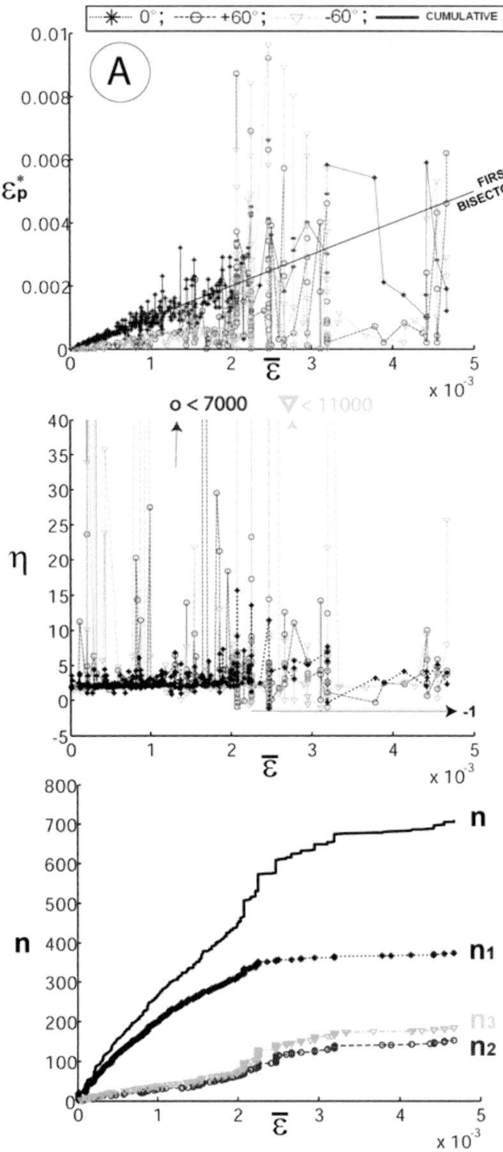

Figure 38. Inputs { $\varepsilon_p^*, \eta_p, n_p$ } of rational model for Case *1* (tension along x).

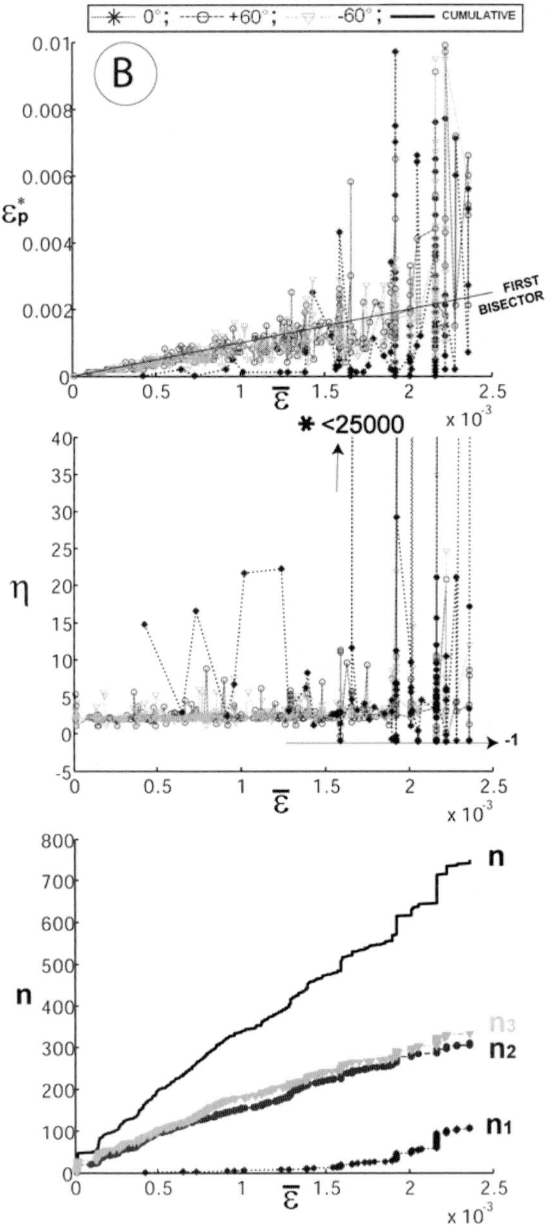

Figure 39. Inputs $\{ \varepsilon_p^*, \eta_p, n_p \}$ of the rational model for Case *2* (tension along y).

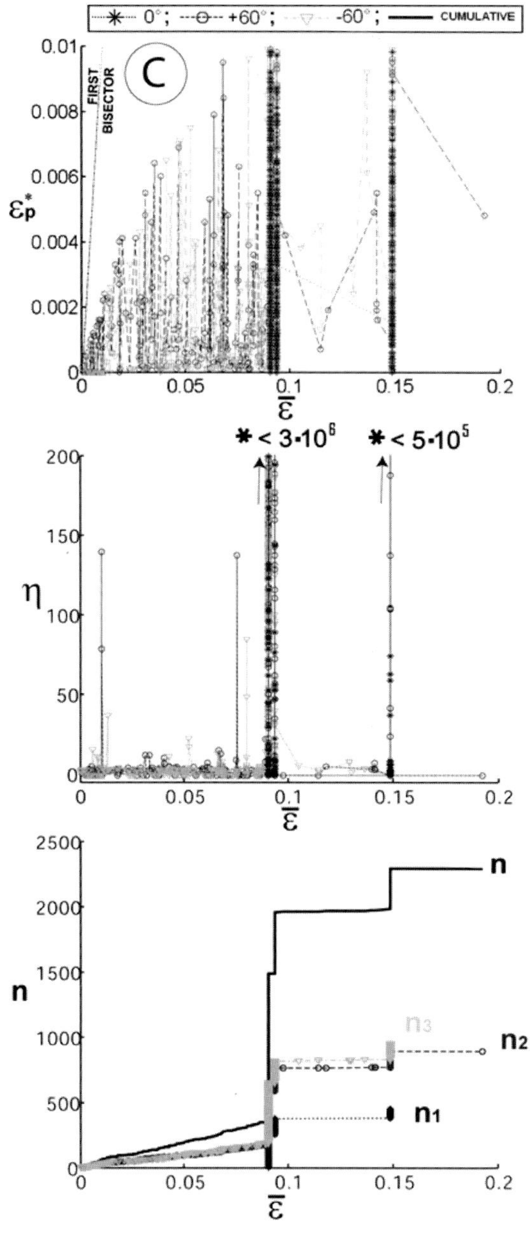

Figure 40. Inputs { $\varepsilon_p^*, \eta_p, n_p$ } of the rational model for Case *4* (compression along x).

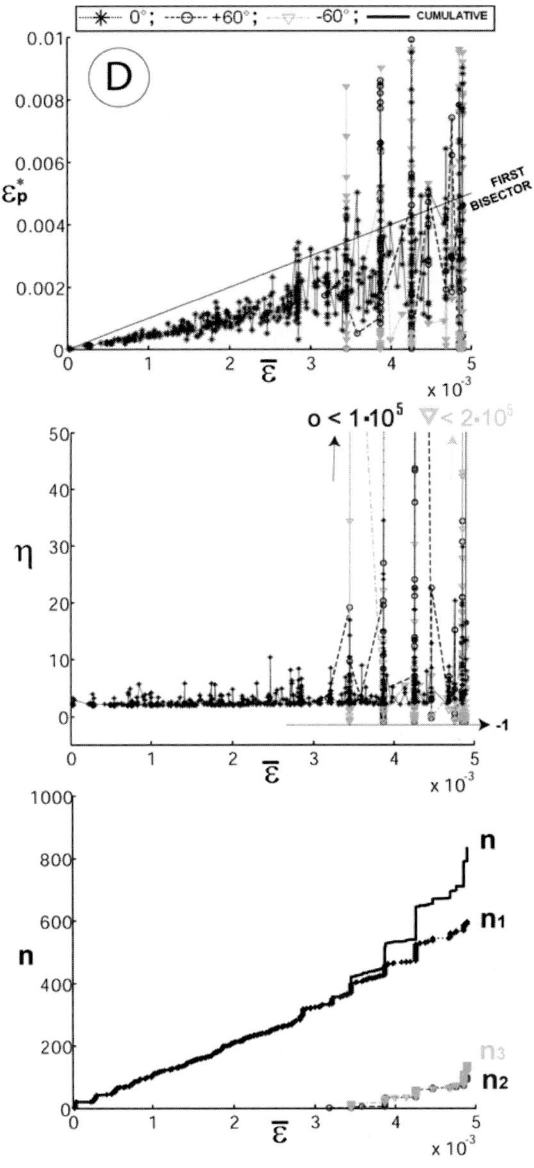

Figure 41. Inputs $\{ \varepsilon_p^*, \eta_p, n_p \}$ of the rational model for Case *5* (compression along y).

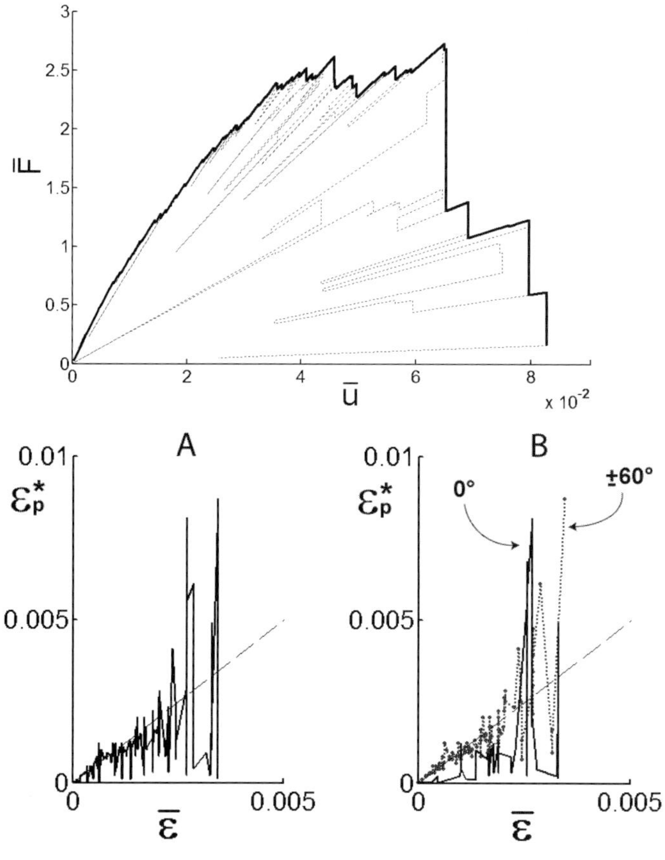

Figure 42. Force response (top) and critical strains of broken springs $\overset{*}{\varepsilon}_p$ *vs.* $\bar{\varepsilon}$ (bottom) in aggregate form (A) and in spectral representation (B) for a lattice with $N=24$ in Case *1*. The avalanche of diagonal ruptures at $\bar{\varepsilon} = 2.7\ 10^{-3}$ is the signature of damage localization.

The number of the ruptures $n(\bar{\varepsilon})$ is the last input of the model. Being a monotonically increasing random variable, this parameter and its spectral components are easier to represent than $\{\overset{*}{\varepsilon}_p, \eta_p\}$. The $n(\bar{\varepsilon})$ data in Figures 38-41 bear similarity with the corresponding damage curves in Figure 36 but demonstrate that $\overline{D}(\bar{\varepsilon}) \neq n(\bar{\varepsilon})$ and that their difference is not a matter of affine transformation. Focusing on Case *4* for example, the damage measured with Eq.(66) remains zero for most of the hardening phase (see Figure 36(C)) despite diagonal ruptures take place at sustained rate as shown by $n_2(\bar{\varepsilon})$ and $n_3(\bar{\varepsilon})$ in Figure 40. Of course, the "latent damage" due to such "unseen" ruptures would become manifest in case of load reversal or change in the loading axis. Thus, the fact that the rupturing (and hence the damage) is ongoing but is not accrued in $\bar{D}$ is ultimately a consequence of having restricted the scope to the consideration of just one direction at the time. That is an inherent limitation of the scalar formulation Eq.(65), which yields correct results but renders only part of the "story". Such drawback would disappear only with the recourse to the full multiaxial formulation Eqs.(27). In general, ignoring damage effects on directions other than the loading axis may be a deliberate and legitimate operation to obtain a scalar model that fits the purpose in many practical circumstances, but this should be done with the awareness that damage is intrinsically a multiaxial problem affecting the full stiffness tensor. Knowledge of the limits and overheads of the scalar model (65) are essential for avoiding model misuse and for proper interpretation of the output. These observations, in conjunction with Eq.(73), point out that the relation between $\bar{D}(\bar{\varepsilon})$ and $n(\bar{\varepsilon})$ is not an affine transformation but rather a weighted sum. For a unit increment Δn due to the p-th rupture, the damage increment is expressed by

$$\Delta \bar{D}_p(\bar{\varepsilon}) = w_p \Delta n = w_p \tag{75}$$

and, hence, $\bar{D}(\bar{\varepsilon}) = \Sigma^{n(\bar{\varepsilon})} w_p$, where the weight w_p is defined by

$$w_p = c\left(1+\eta_p\right)\left(\frac{\varepsilon_p^*}{\bar{\varepsilon}^*}\right)^2 \tag{76}$$

and $c = k / \bar{K}_0 \cdot (\ell_0 / L)^2$. If either "$(1+\eta_p) = 0$" or "$\varepsilon_p^* / \bar{\varepsilon}^* = 0$", then $w_p = 0$ and $\Delta \bar{D}_p = 0$. This is what occurs before the stress peak during the x-compression (Figure 40), where only weak diagonal links break ($\varepsilon_p^* << \bar{\varepsilon}^*$) and full energy redistribution takes place ($\eta_p \simeq -1$). In Case *4*, $\bar{D}$ increases from zero only when horizontal springs start breaking nearby the peak, accompanied by a large release of strain energy. The general lesson learned from this example is that the measure of $\bar{D}$ is sensitive only to the ruptures of those springs (read grain boundaries) where the strain energy gets actually stored and that affect the stiffness in the loading direction. Those are the "master" springs invoked earlier.

3.3.4. Simplified Closed-Form Models of $\bar{D}$

The importance of the rational model (73) is twofold. By a design perspective, its predictive power can effectively aid in devising structural networks with controlled properties. By a modeling perspective - which is the main concern here - it allows *(i)* exposing the "mathematical structure" of the damage problem, *(ii)* identifying the fundamental parameters, and *(iii)* setting a reference for the creation of simpler practical models. A rational model effectively shifts the task from modeling $\bar{D}$ to modeling the input parameters $\{\overset{*}{\varepsilon}_p, \eta_p, n_p\}$. Given the random nature of these parameters, this task should be carried out by means of statistical methods.

The main goal of this section is to seek ways to simplify the rational definition Eq.(73). Proper simplifying assumptions may descend from the analysis of the "structure" of each input variable. For example, given the presence of both hardening and softening regimes where numerical data exhibit two distinct behaviours, it is unlikely that approximations and statistical estimates suitable for the hardening would hold for the softening and vice versa. A distinction can be drawn between two modeling philosophies. On one side there "direct modeling", which pursues closed-form analytical models of $\overline{D}$ through a bottom-up strategy that processes and combines the individual inputs following a rigorous statistical reasoning. Alternatively, there exist a variety of "indirect methods" that, in a top-down fashion, start out from the simulation outputs to obtain approximations for the inputs based on statistical regression criteria, such as "ordinary least squares" (OLS), maximum likelihood principle, non-parametric statistics, and generalized linear models. An exhaustive treatment of each option would be lengthy and beyond the present scope. Instead, explicative examples are discussed limited to the tensile test in Case *1*. Two mechanistic models (*i.e.* MODELS 1 & 2) from the "direct approach" are illustrated for the hardening phase and two regression models (*i.e.* MODELS 3 & 4) from the "indirect approach" are given for the softening phase.

a) Direct approach: mean-field approximation (MODEL 1)

The damage parameter defined by Eq.(73) can be computed for each lattice only if η_p, $\overset{*}{\varepsilon}_p$ and $\overline{\varepsilon}^*$ are known for all $n(\overline{\varepsilon})$ springs. In practice it is necessary to establish a user-friendly average model $<\overline{D}>$ of broader statistical valence, representative of the entire class of lattices with same statistical disorder and same size. This is done by introducing approximations for each random variable that capture in a simple manner the kinematics and the mechanical disorder at the microscale. MODEL 1 represents a relatively simple approximate model of Eq.(73) during hardening based on the following set of assumptions:

1. the lattice is much larger than the spring length, *i.e.* $L \gg \ell_0$, and is treated as an infinitesimal Cauchy element of isotropic material having Young's modulus $E = \overline{K}_0$ and Poisson's ratio ν,
2. the strains associated to the Cauchy element are small,
3. the rupturing process consists of three *independent* rupturing processes along the three orientations $\theta = \{0^o, 60^o, -60^o\}$, *i.e.* $\overline{\varepsilon}^*$ of Groups 1 and 2 are not spatially *correlated* (ref. Appendix 2),
4. the rupturing process on each orientation is not spatially *autocorrelated* (ref. Appendix 2),

5. the threshold ε_p^* for each orientation are sampled independently of each other from the same quenched distribution.
6. the redistribution factor is assumed $\eta_p = 2$ in the whole hardening phase.

The primary, glaring effect of these assumptions is the elimination of the random variable η_p that is replaced with a constant value. Sample simulation data in Figure 43 for *N* = *{24,48,192}* suggests that η_p varies sensibly during the damage process but is well approximated by its average value[52] $\eta_p \simeq 2.2$ at the beginning of the hardening phase. To simplify mathematics even further, assumption # 6 takes $\eta_p = 2$ regardless of spring orientations. Accordingly Eq.(73) modifies to

$$<\bar{D}> = \frac{3k}{\bar{K}_0}\left(\frac{\ell_0}{L}\right)^2 \sum_{p=1}^{n(\bar{\varepsilon})}\left(\frac{\varepsilon_p^*}{\bar{\varepsilon}^*}\right)^2 . \qquad (77)$$

The other assumptions focus on the lattice kinematics and help relating ε_p^* to $\bar{\varepsilon}^*$. Their usage is better explained with reference to the spectral form

$$<\bar{D}> = \frac{3k}{\bar{K}_0}\left(\frac{\ell_0}{L}\right)^2 \left[\sum_{p=1}^{n_1(\bar{\varepsilon})}\left(\frac{\varepsilon_p^*}{\bar{\varepsilon}^*}\right)^2 + \sum_{p=1}^{n_2(\bar{\varepsilon})}\left(\frac{\varepsilon_p^*}{\bar{\varepsilon}^*}\right)^2 + \sum_{p=1}^{n_3(\bar{\varepsilon})}\left(\frac{\varepsilon_p^*}{\bar{\varepsilon}^*}\right)^2\right] . \qquad (78)$$

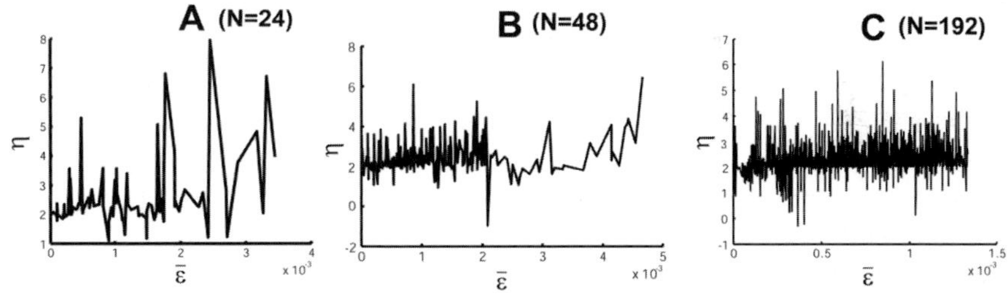

Figure 43. Redistribution factor η for replicates of *N* = *{24,48,192}*. The assumption $\eta = 2$ in the hardening phase is more accurate as the lattice size increases. Note that replicate *N=192* does not have a softening phase.

[52] The robustness of the results to this approximation has been verified against numerical data. Yet, this is a rough approximation that neglects the correlation between η_p and the spring orientation θ_p , which does exist as pointed out in the spectral representations in Figures 38-41. The η_p is regarded as insensitive to orientation and just one value (actually representative only of the primary springs) is assigned to any spring. A better choice would be considering the spectral representation Eq.(74) and at least assign one appropriate constant to each orientation, which will be done in MODELS 3 and 4 (ref. Table 5).

The kinematical variables $\overset{*}{\varepsilon}_p$ and $\bar{\varepsilon}^*$ are not independent. The microscale random variable $\overset{*}{\varepsilon}_p$ is linked to the chosen quenched distribution, while $\bar{\varepsilon}^*$ is a macroscopic random variable that depends in a non-trivial manner on $\overset{*}{\varepsilon}_p$ and on the range of microcracks interaction. The first two assumptions imply that for any applied $\bar{\varepsilon}$ the average strain field in the lattice is expressed in global coordinates (see Figure 35) by the small-strains tensor

$$\mathbf{E_{xy}} = \begin{vmatrix} \bar{\varepsilon}_{xx} & \bar{\varepsilon}_{xy} \\ \bar{\varepsilon}_{xy} & \bar{\varepsilon}_{yy} \end{vmatrix} = \begin{vmatrix} \bar{\varepsilon} & 0 \\ 0 & -\nu \cdot \bar{\varepsilon} \end{vmatrix} . \tag{79}$$

The strain field can be represented in any other local frame x'-y' via the transformation $\mathbf{E}_{x'y'}(\phi) = \mathbf{R}^{\mathbf{T}}(\phi)\mathbf{E}_{xy}\mathbf{R}(\phi)$, where ϕ is the inclination of x' in x-y and $\mathbf{R}(\phi)$ is the rotation matrix around the z-axis

$$\mathbf{R}(\phi) = \begin{vmatrix} \cos\phi & \sin\phi \\ -\sin\phi & \cos\phi \end{vmatrix} . \tag{80}$$

Therefore, for $\phi = 60^0$ (or $\phi = -60^0$) it follows

$$\mathbf{E_{x'y'}}(\phi = 60^0) = \frac{\bar{\varepsilon}}{4} \begin{vmatrix} 1-3\nu & \sqrt{3}(1+\nu) \\ \sqrt{3}(1+\nu) & 3-\nu \end{vmatrix} . \tag{81}$$

In a mean-field sense, the components $\bar{\varepsilon}_{xx}$ in (79) and $\bar{\varepsilon}_{x'x'}$ in Eq.(81) are the estimates of the strain levels in the horizontal and diagonal links respectively. A simplified relation between the kinematics on the macroscale and on the microscale is obtained by assuming

$$\varepsilon_{(ij)}(\theta = 0^0) = \left\langle \varepsilon_{(ij)}(\theta = 0^0) \right\rangle = \bar{\varepsilon}_{xx} = \bar{\varepsilon} , \tag{82a}$$

$$\varepsilon_{(ij)}(\theta = \pm 60^0) = \left\langle \varepsilon_{(ij)}(\theta = \pm 60^0) \right\rangle = \bar{\varepsilon}_{x'x'} = \frac{1-3\nu}{4}\bar{\varepsilon} , \tag{82b}$$

where the fluctuations in the $\varepsilon_{(ij)}$ microfield due to finite lattice size and microcracks are neglected. This is consistent with the third and fourth assumptions of non-correlation between macro-thresholds $\bar{\varepsilon}^*$. The Eqs.(82) introduce a perfect (linear) correlation between the random pair $\overset{*}{\varepsilon}_p$ and $\bar{\varepsilon}^*$ in Eq.(78) for all three orientations, when a link reaches the rupturing condition $\varepsilon_{(ij)} = \varepsilon^*_{(ij)}$. The most important consequence is that the variable $\varepsilon^*_{(ij)} / \bar{\varepsilon}^*$ becomes deterministic and Eq.(78) can be rewritten as

$$\overline{D} \cong \frac{3k}{\overline{K}_0}\left(\frac{\ell_0}{L}\right)^2 \left[\sum_{p=1}^{n_1(\overline{\varepsilon})}(1)_p^2 + \sum_{p=1}^{n_2(\overline{\varepsilon})}\left(\frac{1-3\nu}{4}\right)_p^2 + \sum_{p=1}^{n_3(\overline{\varepsilon})}\left(\frac{1-3\nu}{4}\right)_p^2\right], \tag{83a}$$

i.e.
$$\overline{D} \cong \frac{3k}{\overline{K}_0}\left(\frac{\ell_0}{L}\right)^2 \left[n_1(\overline{\varepsilon}) + \left(\frac{1-3\nu}{4}\right)^2 \left(n_2(\overline{\varepsilon}) + n_3(\overline{\varepsilon})\right)\right]. \tag{83b}$$

Despite the effort, such an approximate $\overline{D}$ still depends on the three random variables n_1 , n_2 and n_3 . Knowledge of the quenched distribution $\varepsilon^*_{(ij)}$ comes to rescue in the guise of assumption # 5. Since, the values ε^*_p are sampled from the uniform distribution with the probability density function $p(\varepsilon_p) = 1/\varepsilon^{MAX}$ shown in Figure 44, the rupturing probability of the p-th link at $\overline{\varepsilon}$ is

$$P\left(\varepsilon^*_p < \varepsilon_p\left(\overline{\varepsilon},\theta_p\right)\right) = P\left(\overline{\varepsilon},\theta_p\right) = \int_0^{\varepsilon_p(\overline{\varepsilon})} p(\varepsilon)\,d\varepsilon = \frac{\varepsilon_p\left(\overline{\varepsilon},\theta_p\right)}{\varepsilon^{MAX}}, \tag{84}$$

with ε_p given by Eqs. (82) for Groups 1 and 2 respectively. Since the links at $+60^o$ and -60° have same rupturing probability and the Poisson's ratio measured from simulations is roughly $\nu = 0.33$ (accurate for $N>48$), the ratio between the probabilities $P_1 = P(\overline{\varepsilon},\theta_p = 0^0)$ and $P_2 = P(\overline{\varepsilon},\theta_p = \pm 60^0) = 2\cdot P(\overline{\varepsilon},\theta_p = 60^0)$ of these Groups is

$$\frac{P_1}{P_2} = \frac{\varepsilon_p(\overline{\varepsilon},\theta_p = 0^0)}{2\varepsilon_p(\overline{\varepsilon},\theta_p = +60^0)} = \frac{\overline{\varepsilon}_{xx}}{2\overline{\varepsilon}_{x'x'}} = \frac{2}{1-3\nu} = 200\,. \tag{85}$$

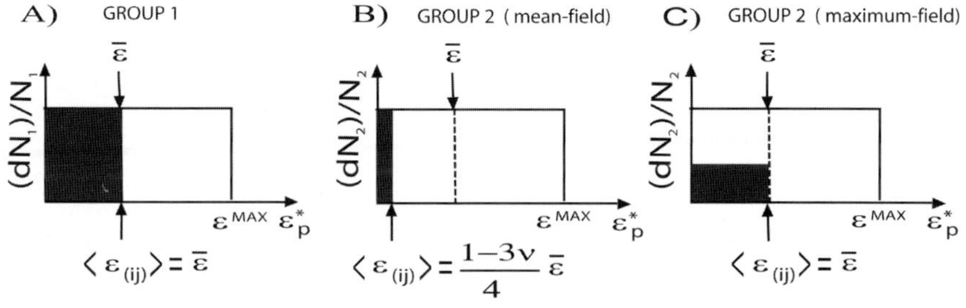

Figure 44. Schematic representations of the probability (shaded area) of sampling a critical value $\varepsilon^*_{(ij)}$ less than the average microstrain $<\varepsilon_{(ij)}>$ for the given $\overline{\varepsilon}$ with respect to the selected quenched distribution. The three plots are a pictorial representation of the kinematic assumptions Eq.(82) and Eq.(98) relating the values $<\varepsilon_{(ij)}>$ and $\overline{\varepsilon}$. While (A) refers to Group 1, the other schemes indicate the failure probability of a diagonal link in Group 2 according to mean-field (B) and maximum-field estimates(C).

Hence a horizontal spring has a rupture probability more than hundred times higher than a diagonal one. In the limit $\nu \to 1/3$, no diagonal ruptures are expected and the ratio diverges. From assumption #5 and Eq.(82a) the set of N_1 threshold values ε_p^* of Group 1 forms at any given $\overline{\varepsilon}$ a IID random sample of Bernoulli trials (ref. footnote # 12), where a "success" trial is a "rupture" event $(i.e.\ \varepsilon_p^* < \varepsilon_p(\overline{\varepsilon}, \theta_p))$ and has a probability to occur equal to $P(\overline{\varepsilon}, \theta_p)$ from Eq.(84). Thus, the probability density function of the random variable $n_1(\overline{\varepsilon})$ is the binomial distribution

$$P_{N_1}\left(n_1(\overline{\varepsilon})\right) = \binom{N_1}{n_1} P_1^{n_1} \cdot \left(1 - P_1\right)^{N_1 - n_1} \quad , \tag{86}$$

with the binomial coefficient expressing the number of possible combinations of sampling n_1 ruptures out of N_1 total "trials" in the Bernoulli parlance. The expected value $< n_1(\overline{\varepsilon}) >$ is defined by the weighted average

$$\left\langle n_1(\overline{\varepsilon})\right\rangle = \sum_{n_1=1}^{N_1} \left[P_{N_1}\left(n_1(\overline{\varepsilon})\right) \cdot n_1 \right] \quad , \tag{87}$$

which can be evaluated explicitly from the first derivative of $\Sigma_{n=1}^{N_1} P_{N1}(n) = 1$ with respect to P_1 in Eq.(86), namely

$$\left\langle n_1(\overline{\varepsilon})\right\rangle = P_1 \cdot N_1 = \frac{P_1 \cdot N_0}{3} \quad . \tag{88}$$

Identically, the expected values $< n_2(\overline{\varepsilon}) >$ and $< n_3(\overline{\varepsilon}) >$ are computed from Eq.(88) by permuting P_1 and N_1 with $P_2/2$ and $N_2/2$ respectively[53]

$$\left\langle n_2(\overline{\varepsilon})\right\rangle = \left\langle n_3(\overline{\varepsilon})\right\rangle = \frac{P_2}{2}\frac{N_2}{2} = \frac{P_2 \cdot N_0}{6} = \frac{P_1 \cdot N_0}{1200} = \frac{\left\langle n_1(\overline{\varepsilon})\right\rangle}{400} . \tag{89}$$

Simple algebraic manipulations over Eqs.(82), (88), and (89) produce

$$\left\langle n_1(\overline{\varepsilon})\right\rangle = \frac{\overline{\varepsilon}}{\varepsilon^{MAX}}\frac{N_0}{3} = \frac{2}{3 - 3\nu}\left\langle n(\overline{\varepsilon})\right\rangle , \tag{90a}$$

$$\left\langle n_2(\overline{\varepsilon})\right\rangle = \left\langle n_3(\overline{\varepsilon})\right\rangle = \frac{1 - 3\nu}{4}\left\langle n_1(\overline{\varepsilon})\right\rangle = \frac{1 - 3\nu}{2(3 - 3\nu)}\left\langle n(\overline{\varepsilon})\right\rangle \quad , \tag{90b}$$

$$\left\langle n(\overline{\varepsilon})\right\rangle = \frac{1 - \nu}{2}\frac{\overline{\varepsilon}}{\varepsilon^{MAX}} N_0 \quad . \tag{90c}$$

[53] Note that P_2 is the probability of diagonal ruptures of any orientation from Eq.(85).

Then, applying the expectation operator to (83b) leads to the relation

$$\left\langle \bar{D}(\bar{\varepsilon}) \right\rangle = \frac{3k}{\bar{K}_0}\left(\frac{\ell_0}{L}\right)^2 \left[\left\langle n_1(\bar{\varepsilon}) \right\rangle + \left(\frac{1-3\nu}{4}\right)^2 \left(\left\langle n_2(\bar{\varepsilon}) \right\rangle + \left\langle n_3(\bar{\varepsilon}) \right\rangle \right) \right], \quad (91)$$

which through Eqs.(90) yields MODEL1 as

$$\left\langle \bar{D}(\bar{\varepsilon}) \right\rangle = \frac{2k}{\bar{K}_0(1-\nu)}\left(\frac{\ell_0}{L}\right)^2 \left[1 + 2\left(\frac{1-3\nu}{4}\right)^3 \right] \left\langle n(\bar{\varepsilon}) \right\rangle . \quad (92)$$

Expression (92) finally establishes a connection between the two macro-quantities $< \bar{D}(\bar{\varepsilon}) >$ and $< n(\bar{\varepsilon}) >$ based on the lattice microstructure. It is clear that the damage parameter depends on the ratio ℓ_0 / L between the characteristic lengths of the microscale and macroscale. MODEL1 can also be written in the form

$$\left\langle \bar{D}(\bar{\varepsilon}) \right\rangle = \frac{\left\langle n(\bar{\varepsilon}) \right\rangle}{\alpha_1 \cdot L^2}, \quad (93a)$$

$$\alpha_1 = \frac{\bar{K}_0(1-\nu)}{2k\ell_0^{\ 2}} \left[1 + 2\left(\frac{1-3\nu}{4}\right)^3 \right]^{-1} \simeq 0.36, \quad (93b)$$

where the entire complexity of Eq.(73) is encapsulated into the one constant α_1. Since $\bar{K}_0 = 103$, $k = 100$, and $\ell_0 = 1$ for these simulations (*e.g.* Figure 36(A)), it results $\alpha_1 \cong 0.36$. To validate MODEL 1, a comparison in made with the value $\hat{\alpha}_1 = 0.4$ that best-fits simulation data. Being $\alpha_1 < \hat{\alpha}_1$, MODEL 1 apparently tends to slightly overestimate the actual damage $\bar{D}(\bar{\varepsilon})$ such that

$$\bar{D}(\bar{\varepsilon}) \cong 0.9 \left\langle \bar{D}(\bar{\varepsilon}) \right\rangle . \quad (94)$$

While this performance is not barren, alone it does not warrant sufficient validation either. A second check has to be done about the damage rate $< \dot{n} > = \partial < n(\bar{\varepsilon}) > / \partial \bar{\varepsilon}$, which can be decomposed in the two spectral contributions $< \dot{n}_1 > = \partial < n_1(\bar{\varepsilon}) > / \partial \bar{\varepsilon}$ and $< \dot{n}_{23} > = \partial < n_{23}(\bar{\varepsilon}) > / \partial \bar{\varepsilon}$, with $n_{23}(\bar{\varepsilon}) = n_2(\bar{\varepsilon}) + n_3(\bar{\varepsilon})$. The numerical estimates of these rates obtained by differentiation of Eq.(90c) are reported in the second column of Table 3 (ref. column "MODEL 1") against the corresponding estimates $\dot{n}$, $\dot{n}_1$ and $\dot{n}_{23}$ from the regression analysis of numerical data for *N={24,48,96,192}* (ref. first column "SIMULATION"). Such reference values are computed away from the transition over the strain range $\bar{\varepsilon} \in [0; 7 \cdot 10^{-4}]$, as depicted in Figure 45 for *N=48*. The comparison in Table 3 is crucial to discover that even though $< \dot{n} >$ and $< \dot{n}_1 >$ compare well against $\dot{n}$ and $\dot{n}_1$, there is a big discrepancy between $< \dot{n}_{23} >$ and the actual $\dot{n}_{23}$, which is underestimated by one order of magnitude (OM) in MODEL 1. In that respect, MODEL 1 is not acceptable. In conclusion, despites of the numerous assumptions, MODEL 1 is an extremely simplified statistical model that captures the order of magnitude of both $\bar{D}(\bar{\varepsilon})$ and $\partial n(\bar{\varepsilon})/\partial\bar{\varepsilon}$ in the hardening phase but, at a closer look, falls short to describe the formation of "secondary" microcracks. Never the least, this model stands out as a valuable reference, being the discrete counter part of continuum micromechanical models derived under similar assumptions, *e.g.* the RVE from the cited dilute concentration models. A refinement is discussed next to improve the accuracy of $< \bar{D}(\bar{\varepsilon}) >$ towards damage rates.

b) Direct approach: maximum-field approximation (MODEL 2)

A better statistical model can be formulated from a better understanding of the damage nucleation process. The weaknesses of the previous attempt have to be preliminarily examined and solved. The assumption of total non-correlation (lack of both correlation and autocorrelation) between and within the rupturing processes along each orientation was very convenient for derivation but has a great impact on the accuracy of the model. This assumption can be relaxed to derive a "partially correlated model". A few observations from simulation data help identifying more realistic model assumptions.

Table 3. Prospect of damage rates $\dot{n}$ and components $\dot{n}_1$ and $\dot{n}_{23}$ estimated from simulations data and from the available models for the hardening phase.

Size *N*	*RATES*	*SIMULATIONS*	*MODEL 1*	*MODEL 2*
	$\dot{n}$	63892	55878	74100
24	$\dot{n}_1$	54187	55600	55600
	$\dot{n}_{23}$	8870	278	18500
	$\dot{n}$	277430	227532	301867
48	$\dot{n}_1$	209120	226400	226400
	$\dot{n}_{23}$	68310	1132	75467
	$\dot{n}$	1039100	918168	1218133
96	$\dot{n}_1$	762280	913600	913600
	$\dot{n}_{23}$	276820	4568	304533
	$\dot{n}$	4427500	3688752	4893867
192	$\dot{n}_1$	3383600	3670400	3670400
	$\dot{n}_{23}$	1043900	18352	1223467

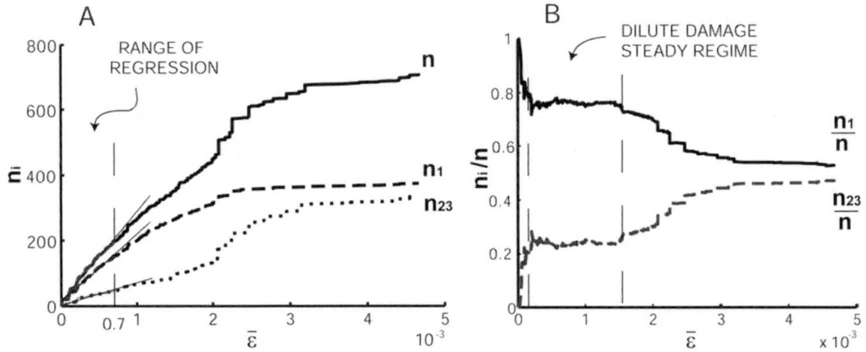

Figure 45. (A) Plot of overall curve of broken bonds $n(\overline{\varepsilon})$ and components $\dot{n}_1(\overline{\varepsilon})$ and $\dot{n}_{23}(\overline{\varepsilon})$ for replicate *N=48*. The range of regression used for Table 3 is marked by dashed vertical line. In correspondence to the transition the $n-\overline{\varepsilon}$ data takes the shape of the $n_{23}-\overline{\varepsilon}$ curve. (B) Plot of fractions of broken bonds n indicating the percentages of Group 1 and Group 2 of accumulated damage. The plot highlights that for most of the hardening phase there is a steady state regime during which the percent contributions stay constant (*i.e.* damage rates are about constant).

First observation. The strain levels in the horizontal and diagonal links are not deterministic, as held in Eqs.(82), but are Gaussian-like distributed random variables since the pristine state, primarily because of the finite lattice size and of the presence of free boundaries. Figure 46 displays the three plots of maximum, mean, and standard deviation of the strain distributions of horizontal (full line) and +60° diagonal (dashed line) over the entire damage process for the *48×48* lattice in Figure 32(A). The vertical drops observable in the plots denounce multi-valued estimates in correspondence to avalanches of order greater than one. In the hardening phase, away from the peak, the plot in Figure 46(A) confirms that the average strain level in the horizontal springs is approximately equal to the macro strain, as stated in Eq.(82a). The large difference between the mean values of Group 1 and Group 2 is also evident and their ratio, as plotted in Figure 47(A) for the same simulation, indicates that the average strain level is approximately two OM greater in the horizontal than in diagonals links over the entire damage process. Such ratio peaks at about 550 in the limit of zero damage. This exceeding difference may induce to think that diagonal links take part to the damage process at rates a hundred times slower than the horizontals. And, in reality, this is exactly what stated by Eq.(85), where the same ratio was estimated in 200. Yet this is wrong. Simulation data assert that diagonals contribute substantially to damage also in the hardening (see Figure 45). The solution to this "apparent" paradox is comprehended from the exam of the maximum strain values in Figure 46(C). The difference between the maximum strains in the two Groups of links is not as large as for the means, and their ratio, as plotted in Figure 47(B), proves these quantities to be of the same OM. Thus, while most diagonal links experience a strain of the same OM than the mean-field estimate Eq.(82b), a smaller subset undergoes strain levels about one OM higher. In conclusion, the kinematical assumption Eq.(82) is accurate for the horizontal links but is too simplistic for the diagonals. In order to pull the damage rate from $<\dot{n}>\simeq 227{,}532$ (see MODEL 1 in Table 3) towards the reference value $277{,}340$ from the simulation, the refined model should embed the behavior of the highly strained diagonal elements, which are the sole ones more actively involved in damage. Figure 48 displays the strain distributions of survival springs for Group 1 and Group 2 after the 30 ruptures (only two of which are diagonals) accumulated at $\bar{\varepsilon} = 0.45 \cdot 10^{-3}$ for the size *N=24*, and indicates that both distributions are peaked and "heavy" tailed. This happens because the ruptures of horizontal springs induce an intense perturbation which "pulls" the right tail of the diagonal distribution (circled in Figure 48) and cause few diagonal springs to experiment higher strain levels, typical of horizontal ones. As damage grows the effect of the perturbation becomes so strong that eventually the heavy tail separates from the "parent" distribution of diagonal links, giving rise to an independent "satellite distribution". Then, it is now evident that the maxima in Figure 46(C) refer to such satellite distribution, while the mean values reflect the strain level of the parent distribution. The ratio maximum-to-mean strains in the diagonals, as plotted in Figure 47(C), confirms the two OM difference between maximum and mean strains. The values of standard deviation of the diagonal links in Figure 47(B) are clearly inflated by this process too, and could be misinterpreted if such partition of diagonals into "parent" and "satellite" subgroups were overlooked and the horizontals compared directly *vs.* (pooled) diagonals.

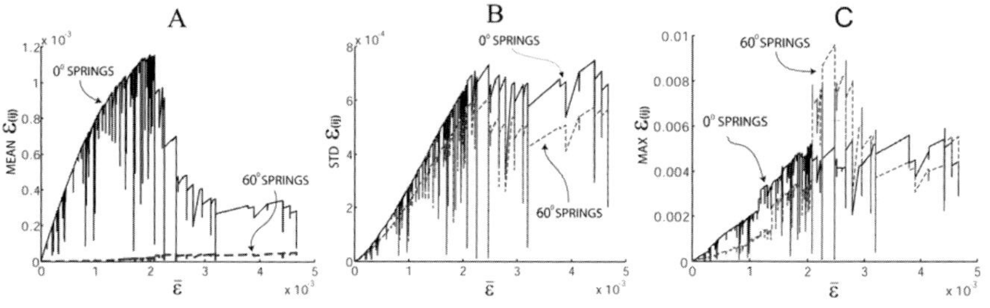

Figure 46. Plots of the mean (A), standard deviation (B) and maximum (C) of the $\varepsilon_{(ij)}$ distribution vs. $\bar{\varepsilon}$ of survival horizontal (solid) and +60° diagonal (dashed) links for the replicate *N=48*. The last plot shows the crossover in correspondence of the transition where diagonal springs experience a higher value of maximum strain.

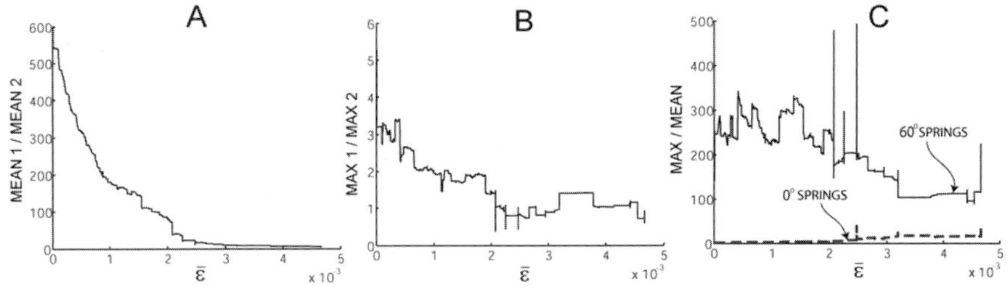

Figure 47. (A) Ratio between the mean of microstrains $\varepsilon_{(ij)}$ of horizontal and diagonal links, (B) ratio between the maximum strains $\varepsilon_{(ij)}$ in horizontal and diagonal links, and (C) ratio max-to-mean per horizontals and +60° diagonals for the replicate *N=48*.

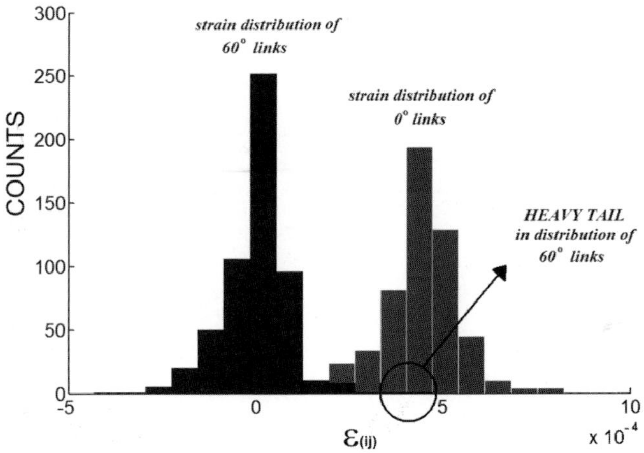

Figure 48. Comparison between strain distributions of 0° and +60° links for *N=24*. The right (heavy) tail for diagonal links is highlighted with a circle.

<u>*Second observation*</u>. A quantitative measure of the influence of the rupturing process on the microscale kinematics is possible from the plots of the percent perturbation fields produced by the first rupture of one horizontal link in Figure 49. The percent perturbation for the *ij*-th unbroken element is computed as

$$\%\ Strain\ Perturbation = \frac{\varepsilon_{(ij)} - \varepsilon_{(ij)}^{REF}}{\varepsilon_{(ij)}^{REF}} \times 100\ , \tag{95}$$

where $\varepsilon_{(ij)}^{REF}$ is the strain of the *ij*-th spring measured under the same macrostrain $\overline{\varepsilon}$

if the lattice were in pristine condition. The magnitude of the perturbation decays away from the damaged location but the maximum tensile perturbation induced on diagonal links is 10^3 % to 10^4 %, against the modest 20% of the horizontal links. Such a remarkable amplification of the strain field of the diagonal links appears certainly capable of triggering the ruptures of the weaker diagonal elements, already after the failure of just one horizontal spring. In practice, as verified in Appendix 2, horizontal springs break randomly in the lattice, producing "primary" microcracks that induce in the surrounding a high elastic perturbation responsible for a fast increase in the cracking probability of unfavorably oriented grain boundaries located nearby. As a result, those primary ruptures drive the formation of "secondary" microcracks (purposely named "slave" in Section 3.3.3) in their vicinity (the strong spatial correlation between the two groups is assessed in Appendix 2). This analysis promotes a refined perspective about damage nucleation, which evidently cannot be considered a random process *tout court*, but limited to the primary microcracks.

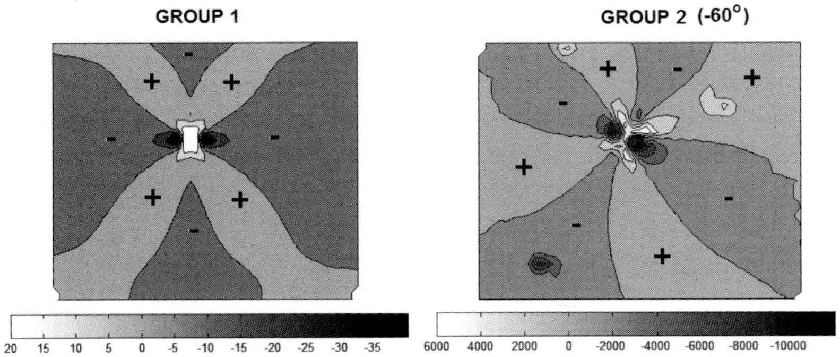

Figure 49. Percent perturbation fields in survivor horizontal (A) and diagonal at -60° links (B) produced by the first (horizontal) rupture for $N = 24$.

<u>Third observation</u>. A final observation concerns the damage rates. If the relative agreement between $\dot{n}_1$ and $<\dot{n}_1>$ in Table 3 on the one hand validates the estimate of $n_1(\bar{\varepsilon})$ from MODEL 1, on the other hand the same renders a faulty prediction of $<\dot{n}_{23}> \simeq 1{,}132$ *vs.* $\dot{n}_{23} = 66{,}469$ for the subject simulation. The refined model should eliminate or reduce the gap between $<\dot{n}_{23}>$ and $\dot{n}_{23}$ for Group 2. In that respect, Table 4 reports the ratios $\dot{n}_1 / \dot{n}_{23}$ from simulation data. The agreement of the estimates for $N \geq 48$ suggests the approximate target value for $< n_{23}(\bar{\varepsilon}) >$ of

$$\langle n_{23}(\bar{\varepsilon})\rangle \simeq \langle n_1(\bar{\varepsilon})\rangle / 3 . \tag{96}$$

Table 4. Dependence of ratio $\dot{n}_1 / \dot{n}_{23}$ on lattice size, as estimated from simulations data in the limit of dilute damage.

RATIO	*N=24*	*N=48*	*N=96*	*N=192*
$\dot{n}_1 / \dot{n}_{23}$	6.1	3.14	3.12	3.13

Consequences. Following these three observations, a modified set of assumptions is offered below to develop a refined "partially correlated" model that reflects a maximum-field perspective:

1. same as for the non-correlated model,
2. same as for the non-correlated model,
3. the rupturing process consists of one *independent* process for Group 1 ($\theta = 0^o$) and of another *dependent* process for Group 2 ($\theta = \{ +60^o, -60^o \}$); the latter is "*one-way correlated*" since this is influenced by the damage process in Group 1 but does not influence it in return (ref. Appendix 2) ,
4. same as for the non-correlated model,
5. same as for the non-correlated model,
6. same as for the non-correlated model.
7. only a subset $N_2^* \leq N_2$ of the diagonal springs is subject to a significant tensile perturbation induced by the horizontal ruptures and participates to the damage process; on an experimental basis the active set is assumed to be

$$N_2^* = N_2 / 6 = N_1 / 3 \ . \tag{97}$$

Because of the similarity with MODEL 1 assumptions, the Eqs.(77,78) hold still. The third assumption introduces a modification and accounts for the "trigger" effect of Group 1 ruptures on Group 2. By neglecting any reciprocity-effect from diagonal ruptures onto the rupturing of Group 1, the kinematical approximation Eq.(82a) and the derivation of $< n_1(\bar{\varepsilon}) >$ hold unchanged as well. The seventh assumption is the other novelty and means that only the ruptures of diagonal links piloted by Group 1 are relevant to the model. The damage process in the other diagonal springs, unexposed to a significant tensile perturbation, is too slow and negligible by a this modeling standpoint. If the strain field in the "active" diagonal springs is assumed to be the same as in Group 1, Eq.(82b) can be replaced by

$$\varepsilon_{(ij)}(\theta = \pm 60^0) = \left\langle \varepsilon_{(ij)}(\theta = 0^0) \right\rangle = \bar{\varepsilon} \ . \tag{98}$$

The Eq.(98) implies that the failure probability of one diagonal spring in the active set is the same for a horizontal spring, *i.e.*

$$P\left(\bar{\varepsilon},\theta_p=60^0\right)=P\left(\bar{\varepsilon},\theta_p=0^0\right)=P_1\left(\bar{\varepsilon}\right), \tag{99a}$$

$$\frac{P_1\left(\bar{\varepsilon},\theta_p=0^0\right)}{P_2\left(\bar{\varepsilon},\theta_p=\pm 60^0\right)}=\frac{1}{2}. \tag{99b}$$

The diagonal ruptures are regarded as an IID random sample such that the probability associated to the $n_2\left(\bar{\varepsilon}\right)$ or $n_3\left(\bar{\varepsilon}\right)$ ruptures is

$$P_{N_3}\left(n_3\right)=P_{N_2}\left(n_2\right)=\begin{pmatrix} N_2^*/2 \\ n_2 \end{pmatrix}\left(P_1\right)^{n_2}\cdot\left(1-P_1\right)^{N_2^*/2-n_2}. \tag{100}$$

Then, this time the expected values $<n_2(\bar{\varepsilon})>$ and $<n_3(\bar{\varepsilon})>$ in Eq.(99) change to

$$\left\langle n_2(\bar{\varepsilon})\right\rangle=\left\langle n_3(\bar{\varepsilon})\right\rangle=\frac{P_2}{2}\cdot\frac{N_2^*}{2}=\frac{P_1\cdot N_1}{6}=\frac{\left\langle n_1(\bar{\varepsilon})\right\rangle}{6}, \tag{101}$$

where the number of the active elements in Group 2 is given by Eq.(97) in assumption # 7. The latter one is an empirical position made to calibrate the damage rate $<\dot{n}_{23}(\bar{\varepsilon})>$ against simulations. The proof of Eq.(101) from rigorous statistical arguments is not yet available and is neither pursued here[54]. Such equation intuitively states that roughly one third of the lattice domain is spanned by a perturbation strong enough to cause diagonal ruptures. Figures 44(B-C) offer conceptual schemes of the different rupturing modes of diagonal springs assumed in MODEL 1 and MODEL 2 with respect to the selected quenched distribution of critical strains. The actual histograms of broken springs at failure for $N=24$ are displayed in Figure 50 providing the numerical confirmation for the damage modes featured in Figures 44(A-C) and endorsed in MODEL 2.

[54] Eq.(97) is one of the heuristic assumptions deduced from simulations and necessary (to date) to achieve analytical models. Such assumptions lack generality and represent weak points or pitfalls in a rational derivation chain. Future investigations should seek to abandon (or limit) progressively the recourse to similar "hidden" numerical shortcuts and rely, instead, on simplifications from physical or statistical arguments only.

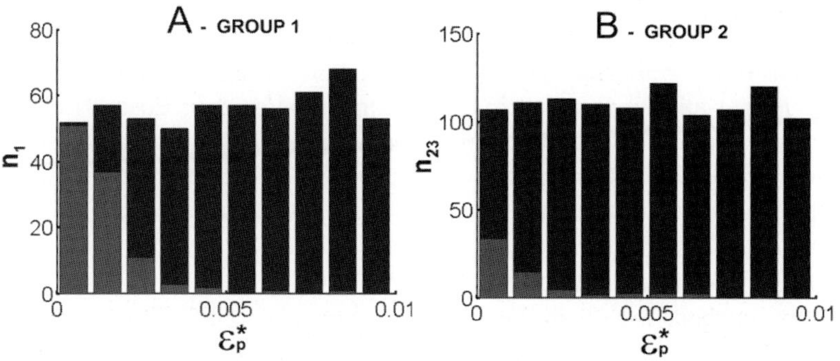

Figure 50. Histograms of critical strengths ε_p^* of the n_1 broken bonds in Group 1 (A) and the n_{23} in Group 2 (B) at the failure (light) overlaid on top of the corresponding histogram of critical strengths quenched in the lattice (dark) for *N=24*. The ordinate axis reports counts rather than frequencies/probabilities, attainable by normalization. The histogram of diagonal links supports the maximum-field hypothesis depicted in Figure 44.

In analogy to previous derivation and substituting Eqs.(98) and (101) where appropriate, the expression $<\bar{D}(\bar{\varepsilon})>$ from the partially-correlated model is finally

$$\left\langle \bar{D}(\bar{\varepsilon}) \right\rangle = \frac{3k}{\bar{K}_0} \left(\frac{\ell_0}{L} \right)^2 \left[1 + 2 \cdot (1)^3 \frac{1}{6} \right] \left\langle n_1(\bar{\varepsilon}) \right\rangle , \tag{102a}$$

and
$$\left\langle \bar{D}(\bar{\varepsilon}) \right\rangle = \frac{3k}{\bar{K}_0} \left(\frac{\ell_0}{L} \right)^2 \left\langle n(\bar{\varepsilon}) \right\rangle . \tag{102b}$$

The new estimate α_1 defined in Eq.(93a) is

$$\alpha_1 = \frac{\bar{K}_0}{3k\ell_0^{\ 2}} = 0.34 , \tag{103}$$

which is close to the aforementioned reference value $\hat{\alpha}_1 = 0.4$, as well as to the value from MODEL 1. Instead, the accuracy of predicted damage rates $<\dot{n}>$, $<\dot{n}_1>$, and $<\dot{n}_{23}>$ is much bettered, as reported in the third column of Table 3 (ref. column "MODEL 2") . The significant improvement in damage rates validates the revised assumptions used in the partially-correlated model. By tying the damage in diagonal springs to the dominant damage process along the most favorably oriented horizontal spring, the refined model succeeds to capture both damage and damage rate(s). In essence, this result stems from replacing the usual mean-field rationale with a maximum-field rationale, which corresponds to a modeling criterion focused on the highly strained springs susceptible to damage.

MODEL 1 and MODEL 2 are closed-form approximations of $\overline{D}(\overline{\varepsilon})$ limited to hardening and demonstrate how immanent disorder (*i.e.* quenched distribution and orientations of the grain-boundaries) and lattice kinematics, at both macroscale and microscale, get all compounded into the damage parameter. Eqs.(93) and (102) make a clear connection between applied load, spring orientations, and distribution of critical strains. The statistical reasoning deployed in the derivation is somewhat involved but is rewarding in terms of physical insight and consistency.

c) Regression Approach (MODEL 3 and MODEL 4)

Regression methods are a valid alternative to pursue simplified expression of $\overline{D}$ from Eq.(74). The concept is demonstrated by means of two simple models valid for the entire damage process, not just for the hardening. The regression effort is limited to $\{ \varepsilon^*_p, \eta_p \}$ but the rationale can be readily extended to n_p. The first model (MODEL 3) is aimed at assessing whether or not η_p can be always approximated by a constant, properly averaged value $<\eta>$ such that each damage component is estimated in a spectral fashion as

$$\hat{D}_i = c\left(1+\langle\eta\rangle\right)\sum_{p=1}^{n_i(\overline{\varepsilon})}\left(\frac{\varepsilon^*_p}{\overline{\varepsilon}^*}\right)^2 \tag{104}$$

Unlike MODELS 1 and 2, an optimal $<\eta>$ is sought for each i-th orientation (i=1..3). To this end, a suitable optimization criterion must be selected first amongst the many possible choices. For simplicity the condition " $\Delta\overline{D}_i - \Delta\hat{D}_i = 0$ " is enforced over a given strain interval [$\overline{\varepsilon}_{START}, \overline{\varepsilon}_{END}$]. Then simple manipulations yield

$$\langle\eta\rangle_i = \frac{\sum_{START}^{END}\left(1+\eta_p\right)\left(\frac{\varepsilon^*_p}{\overline{\varepsilon}^*}\right)^2}{\sum_{START}^{END}\left(\frac{\varepsilon^*_p}{\overline{\varepsilon}^*}\right)^2} - 1, \tag{105}$$

which can be solved piecewise - separately for the hardening and the softening - to obtain $<\eta>$ from simulation data. The corresponding six values $<\eta>$ are listed in Table 5, not just for Case *1* but for all loading cases on display in Figure 36[55]. This is possible because regression models such as Eq.(104) are no-brainer in comparison with *mechanistic* models from the previous direct approach, and can be implemented easily to process large datasets quickly and automatically. The

[55] Note that the tabulated value $<\eta_1> = 0$ for Case *4* (C) means that an estimate could not be obtained due to the said lack of horizontal ruptures before the stress peak.

corresponding stress-strain estimates are shown in Figure 36 (light solid line) and match very closely the simulation data in both regimes. This is a remarkable result when reckoning that six constants so effectively replace the complication of the $\eta - \bar{\varepsilon}$ data series plotted in Figures 38-41. But what would happen if the optimization is carried out over two random inputs $\{ \varepsilon^*_p , \eta_p \}$?

The second regression model (MODEL 4) attempts to answer this question in a simple, modular manner. Following a step-wise logic (typical of regression methods and optimization theory), the optimization of ε^*_p is done alone while holding still the estimates $<\eta>$ from MODEL 3. Hardening and softening are treated separately to exploit the peculiar data structure of the ε^*_p data discussed in Figures 38-41.

Table 5. Fitted parameters for the simplified models. MODEL 3 is based on one fitted parameter $<\eta>$ whereas MODEL 4 receives $<\eta>$ and either $<\varepsilon^*_p / \bar{\varepsilon}>$ or $<\varepsilon^*_p>$, depending on the regimes.

		MODEL 3		*MODEL 4*			
Case	*Spring Angle*	*Hardening* $<\eta>$	*Softening* $<\eta>$	*Hardening* $<\eta>$	$<\varepsilon^*/\bar{\varepsilon}>$	*Softening* $<\eta>$	$<\varepsilon^*>$
	0°	*2.65*	*4.12*	SAME AS MODEL 3	*0.95*	SAME AS MODEL 3	*0.0042*
1	*+60°*	*1.07*	*1.95*		*0.21*		*0.0029*
(x-tension)	*-60°*	*3.06*	*0.93*		*0.25*		*0.0042*
	0°	*1.35*	*0.17*		*0.19*		*0.0030*
2	*+60°*	*2.71*	*2.85*		*0.80*		*0.0049*
(y-tension)	*-60°*	*2.56*	*3.48*		*0.74*		*0.0037*
	0°	*0*	*1825.00*		*0.00*		*0.0038*
4	*+60°*	*2.59*	*394.00*		*0.07*		*0.0038*
(x-compr.)	*-60°*	*2.81*	*356.00*		*0.06*		*0.0039*
	0°	*3.40*	*10.30*		*0.57*		*0.0032*
5	*+60°*	*12.40*	*2.60*		*0.14*		*0.0037*
(y-compr.)	*-60°*	*0.47*	*7.12*		*0.17*		*0.0038*

In the hardening, where $\varepsilon_p^* \leq \bar{\varepsilon}$, the micro-strain seems well approximated by the linear relation $<\varepsilon_p^*> = \alpha \cdot \bar{\varepsilon}$, which implies $<\varepsilon_p^* / \bar{\varepsilon}> = \alpha$ and yields MODEL 4 in the form

$$\hat{D}_i = c\left(1+\langle\eta\rangle_i\right)\left\langle\varepsilon_p^* / \bar{\varepsilon}^*\right\rangle^2 \cdot n_i(\bar{\varepsilon}) \ . \tag{106}$$

Again, optimizing against the condition " $\Delta\bar{D}_i - \Delta\hat{D}_i = 0$ " but solving just for $<\varepsilon_p^* / \bar{\varepsilon}>$ renders one value per orientation, as reported in Table 5 for each case.

In the softening, where ε_p^* varies randomly over the whole (large) sampling strain range [0, 0.01], the rationale is different. At first it makes sense to look simply for a constant value $<\varepsilon_p^*> = \alpha$. Correspondingly, MODEL 4 in the softening is

$$\hat{D}_i = c\left(1+\langle\eta\rangle_i\right)\alpha_i^2 \cdot n_i^{peak} + \Delta\hat{D}_i \ , \tag{107a}$$

$$\Delta\hat{D}_i = c\left(1+\langle\eta\rangle_i\right)\left(\left\langle\varepsilon^*\right\rangle_i / \bar{\varepsilon}^*\right)^2 \cdot \Delta n_i(\Delta\bar{\varepsilon}) \ , \tag{107b}$$

where only $\Delta\hat{D}_i$ has to be determined. This time the optimization is less trivial, though, because guessing the representative value $<\varepsilon_p^*>$ is not self-evident. A worst case estimate would clearly be $<\varepsilon_p^*> = \varepsilon_{MAX}^* = 0.01$ but that would deliver too conservative of an estimate. On the other hand, the usual fitting strategies do not yield satisfactory results owing to the relatively large strain range and to a marked cross-correlation between ε_p^* and η in the softening, which is completely overlooked when $<\varepsilon_p^*>$ and $<\eta_p>$ are fitted separately - as inhere. Therefore, a trial-and-error search is used instead by trying progressive adjustments on $<\varepsilon_p^*>_i$ to maximize the agreement between MODEL 4 and the stress response from simulations (ref. Figure 36). This procedure suggests best values $<\varepsilon_p^*>$ of less than the median of the quenched distribution of critical strains (*i.e.* less than $\varepsilon_{MAX}^* / 2 = 0.005$), which confirms that only a small subset of the total springs, consisting of the weaker ones, participates in the damage process.

The complete list of MODEL 4 parameters $<\varepsilon_p^* / \bar{\varepsilon}>_i$ and $<\varepsilon_p^*>_i$ estimated for all cases is reported in Table 5. The predicted stress response is plotted (dotted line) in Figure 36 for a ready comparison with MODEL 3 and simulation data. As expected from the greater simplification of the original numerical data and from the step-wise fitting procedure, the performance of MODEL 4 is sensibly degraded. In the hardening, though, the agreement with the rational model is still good, which proves that the model assumptions are reasonable in that phase and that $\{\varepsilon_p^*, \eta_p\}$ are likely not correlated[56]. In the softening the divergence is more evident but the

[56] It is worth mentioning that also the estimates from OLS perform similarly well in the hardening. This is a relevant result since the lattice response before the phase transition is

model still captures the original profile reasonably well. Interestingly, the estimate $\hat{D}$ from Eqs.(106-107) compared in Figure 51 against simulation data for Case *1* confirms the goodness of fit of MODEL 4 in regard to the damage parameter, in both hardening and softening and for each spectral component. In conclusion, MODEL 3 and MODEL 4 prove that simplified regression models containing a few constant parameters (*i.e.* 6 and 12 values respectively) are indeed effective and suffice to manage the complexity of the input data $\{\varepsilon_p^*, \eta_p, n_p\}$.

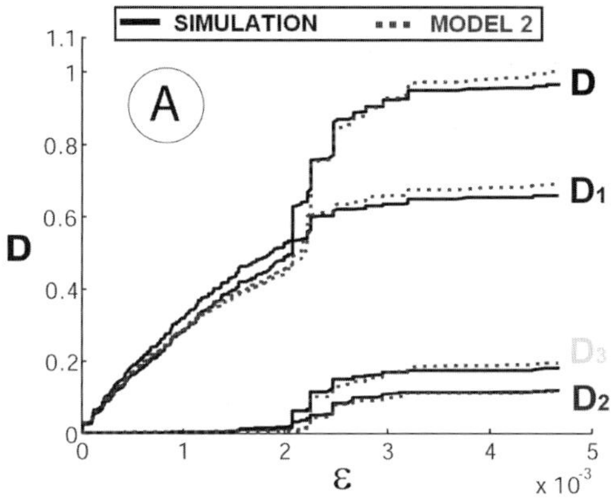

Figure 51. Damage (total and components) from simulation *vs*. MODEL 4 Eqs.(107).

d) Side Notes: hierarchy of microcracks subsets and anisotropic damage

The damage partition featured in the spectral representation was a key idea, exploited in all four models discussed above. It serves to emphasize the different role that grain boundaries have in the damage evolution based on their orientation relative to the applied load. The hierarchical partition of microcracks into "primary" and "secondary" stems from that perspective and is an essential part of direct modeling. It is the primary subset that shapes the development of elastic anisotropy during the damage process and the macrocrack at failure. The hierarchical ranking of different orientations in a microstructure is tightly related to the given loading scheme. To illustrate it with a simple example, a comparison is made between two uniform tensile tests, one in x and another in y, performed on one 96×96 triangular lattice with just one horizontal broken spring in it.

robust to the microstructural details and is representative of any random realization o a lattice class. Furthermore, softening does not sometimes exist, while the hardening phase always does. The model predictions are reasonably close to the true response but not as accurate as in the hardening.

The perturbation fields from Eq.(95), plotted in Figures 52-53 for 0° and –60° respectively, demonstrate that the effect of this microcrack changes drastically between the two tests, both in sign and magnitude. Figure 52 compares the perturbation fields investing the survival horizontal springs (Group 1) in the two cases, highlighting a relatively modest strain amplification that is even milder for the *y* test. Figure 53 does the same check on the diagonal springs at -60° (as representative of Group 2) and registers, instead, a dramatic difference, consistent with Figure 49(B). Thus a large strain effect is observed when tension is applied in *x*, while a rather negligible tensile perturbation irradiates when loading along *y*[57]. Clearly if *y* is the loading axis, the horizontal springs do not trigger any damage acceleration in diagonal links and, hence, are no longer a "primary" group. The damage evolution in the two cases proceeds according to two different patterns, as well as the elastic anisotropy. This is why even a material isotropic in the pristine state (like our triangular lattice) acquires a specific anisotropy during damage nucleation that reflects the faster depletion of the grain-boundaries that are primary for the given loading scheme.

The dependence of the "hierarchy of microcracks subsets" on the loading conditions is of great importance to elaborate comprehensive physically-based constitutive relations in analytical form. In the ambit of multiaxial loading, a lengthy characterization of the whole damage tensor is required case by case and the framework underlying rational model (73) needs to be generalized beyond the uniaxial formulation for a perfect triangular lattice[58]. The ideas presented here may also inspire (and aid) the creation of innovative refined RVEs of continuum micromechanics that would introduce conditional probabilities of microcracking dependant on loading and microstructure, similarly to what done above with MODEL 2. In the long run, this approach may deliver a consistent and mature theoretical framework for damage mechanics that bridges correctly between micro- and macroscale.

---

[57] Note that although in the latter case the chosen horizontal spring is associated to a compressed high angle grain boundary and would not be the first to break spontaneously, the scenario is still plausible because such microcrack could originate in the microstructure during prior loadings or from manufacturing and environment (*e.g.* load reversal, wear, plasticity, fatigue, corrosion, cooling, etc.).

[58] To model anisotropic damage in geometrically disordered microstructures containing "*p*" orientations with the spectral approach, the vector $\mathbf{n}(\bar{\varepsilon}) = [n_1(\bar{\varepsilon}), n_2(\bar{\varepsilon}), ..., n_p(\bar{\varepsilon})]$ should be known and taken into account in place of the usual aggregate scalar $n(\bar{\varepsilon})$, in agreement with the observations from Section 3.2. In the continuum limit, this vector is replaced in 2D by the function $\mathbf{n}(\bar{\varepsilon}, \theta_p)$ defined on a continuous spectrum of orientations in the interval $\theta_p = [0, \pi]$.

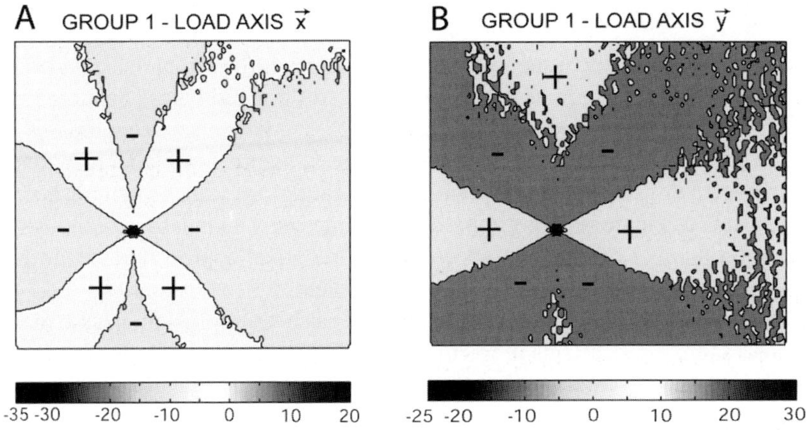

Figure 52. Percent perturbation fields of survivor horizontal links in a lattice $N=96$ with only onc horizontal rupture under uniform tension. The load axis is x in (A) and y in (B). The effect is milder in the latter case.

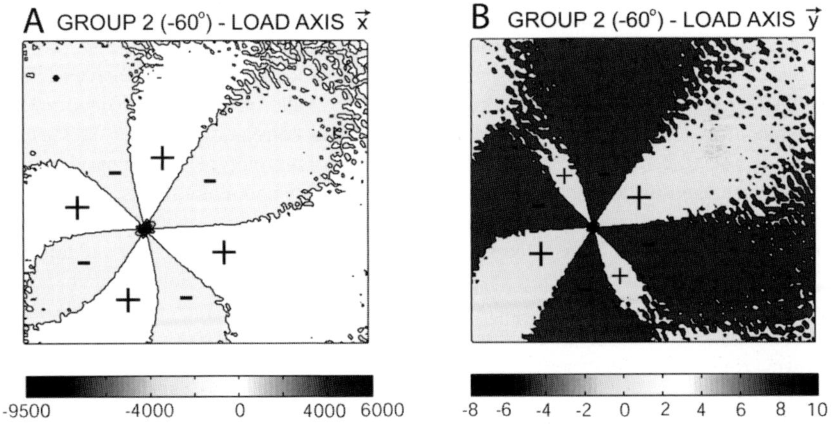

Figure 53. Percent perturbation fields in survivor diagonal links at -60° in a lattice of size $N=96$ with only one horizontal rupture under uniform tension. The load axis is x in (A) and y in (B). The situations are totally different as evident from the scale range.

3.4. Sample-Size Effects

An outstanding modeling issue in damage mechanics is the sample-size dependence of softening phase and failure (ref. Figure 13). As pointed out in Section 3.2, the damage tolerance of the lattice reduces as N (*i.e.* the lattice size L) increases. This makes it hard to deduce the behavior of a large structure from laboratory tests on similarly shaped samples unless a so-called scaling law can be established. The problem has been under the spotlight for sometimes now and many researchers have attempted a number of different approaches. One novel approach based on fractal theory is covered here applied to the lattice problem.

Over the past two decades the fractal theory has been applied in diverse manners to address sample-size effects in fracture and damage mechanics[59]. Scaling laws are based indeed on the scale-invariance of fractal objects[60]. In practice, if the statistics of a process are known on one scale and are governed by some scale-invariant parameters, it becomes possible to infer the statistics of the same process on any other scale. The fractal dimension[61] $\mathcal{D}_0$ of a generic sample-size dependent quantity

[59] The present discussion voices the method developed by Krajcinovic and Rinaldi (Krajcinovic and Rinaldi (2005a,b), Rinaldi et al. (2006b)). Different views are offered, *e.g.*, by Hansen and Roux (2001), Alava et al. (2006) or Carpinteri et al.(2009).

[60] Unlike design equations used by engineers, scaling laws based on fractals are not deterministic and exploit universal trends of a phenomenon. In statistical physics, "the attribute *universal* distinguishes the behavior of a parameter which is weakly influenced by the details of the microstructural disorder, *i.e.* those details that are not replicated with any regularity in different realizations of the same statistics. Universality enables unambiguous identification of the macro properties and the material parameters independent of the specimen size. This is relevant to establish the correlation between the test data measured on specimens and the expected properties of a prototype" (Krajcinovic (1996)). Lattices such as ours are very useful to study universality and scaling laws. Modeling of a solid by simple reticulated assemblages (networks or lattices) of axial springs and/or beams may appear to a large fraction of applied mechanicians (whose time and talents are devoted to very sophisticated and ever more accurate numerical models based on the finite or boundary element methods) as a vestige of the ancient times which is too crude, inexact and simplistic to be considered as being a modern, computer age research and design tool. However, the lattice analyses are blessed with a few important advantages, such as simplicity and computational efficiency and economy, which enables consideration of many realizations of some statistics. This simplicity makes the search for universal trends feasible due to the fact that the analyses are not encumbered by the computational details needed to achieve the accuracy necessary in engineering design.

[61] A generalized fractal dimension $\mathcal{D}_q$, with $q \in (-\infty, \infty)$, can be defined when more fractal sets coexist in *I(L)*. However, so-called multifractal dimensions require the introduction of a "natural density" for *I(L)* and are more difficult to interpret since a spectrum of fractal dimensions exist. Since not strictly necessary to our treatment, to avoid diverting from the main topic, the reader interested in basic ideas and definitions of fractal and multifractal

$y(L)$ is a characteristic parameter and is defined as

$$\mathcal{D}_0 = \lim_{L \to \infty} \frac{\ln y(L)}{\ln (L)} \quad . \tag{108}$$

Proving the existence of fractals significant to the damage process is an essential step in the formulation of scaling relations for damage. In the lattice this search is done separately for hardening and softening - as done in the regression method from the previous section - since the damage mechanisms are different in the two phases. The objective of this innovative fractal formulation is the derivation of scaling relations for the number of broken springs $n(\bar{\varepsilon}, L)$. The damage $\bar{D}(\bar{\varepsilon}, L)$ is obtained eventually by normalizing $n(\bar{\varepsilon}, L)$ by a proper factor, in agreement with the closed-form models Eq.(93a) and Eq.(102b). However, because the fractal approach truly expresses a different and independent modeling perspective compared to Section 3.3, the results previously derived are ignored in the derivation, pretending that nothing is known yet about $\bar{D}$ and starting upright from simulation data.

3.4.1. Hardening Phase

It is not readily clear whether it would make sense or not to seek a connection with fractals and fractal theory in this regime. But actually - as it turns out - fractal theory offers a proper framework to understand and model this regime in an effective and original way. In fact, the hardening phase is affected by the lattice size (see stress response in Figure 13) but the occurrence of the transition marking the end of this phase depends on the lattice size and is governed by a fractal.

The determination of $\bar{D}$ is carried out in a heuristic fashion by guessing inductively the model form starting from phenomenological observations and fractal theory. The first relevant observation is that hardening can be associated to a special case of fractals (Rinaldi and Lai (2007)). At small damage density the average number of broken springs $n(\bar{\varepsilon}, L)$ is indeed a fractal[62] with exponent α=2

$$n(\bar{\varepsilon}, L) = \beta(\bar{\varepsilon}) L^{\alpha} \approx \beta(\bar{\varepsilon}) L^{2}, \tag{109}$$

where β is a function of $\bar{\varepsilon}$. As a visual confirmation, the logarithmic plot of n *vs.* L at an arbitrary $\bar{\varepsilon}$ in Figure 54(A) is a straight line with slope $\simeq 2.003 \pm 0.014$ (95%

theory is referred to the specialized literature on the subject (*e.g.* Mandelbrot (1977), Vicsek (1989), Gouyet (1996), Rinaldi (2004)).

[62] This is actually a "fat" fractal (Ott (2002)), which may liken damage nucleation to the transient chaos phenomena observed in some non-linear dynamical systems governed by a fat fractal.

confidence intervals)[63]. Interestingly, since the material is statistically homogeneous, the function β would take up the geometrical meaning of uniform microcracks density if the lattice were a square of area L^2 . This implies that the coefficient $\beta(\bar{\varepsilon}) = n(\bar{\varepsilon}, L)/L^2$ is scale-invariant according to the substantial overlap of $\beta - \bar{\varepsilon}$ data for any L in Figure 55(B), which happens only if n is normalized by L^2. To obtain $\bar{D}(\bar{\varepsilon})$, it is necessary to normalize n in Eq.(109) by an appropriate constant factor as done in the PBS. However, the estimate of such a proportionality constant is not obvious[64].

The proper constant is inferred here by conjecturing the equivalency between the lattice and the PBS up the stress peak. Because damage nucleation in the lattice is a statistically homogeneous process where (primary) defects do not interact with each other, the damage evolution should not differ much from the PBS until the transition happens when the same strength distribution is used (like inhere). The linearity of the $n - \bar{u}$ data exhibited by both systems prior to the transition seems to support such an expectation. We recall that the damage $\bar{D}$ of the PBS was defined in Eq.(4) by normalizing n to the number of total bars, equaling twice the microcracks number at the stress peak N_P (see Figure 4), as

$$\bar{D} = \frac{n}{2N_p} \quad . \tag{110}$$

We can now verify if the conjecture is correct and this same definition holds for the lattice. The first step is the computation of N_P. It would be erroneous to determine N_P from Eq.(109) since $N_P(L) \neq \beta(\bar{\varepsilon}) \cdot L^2$. The quantity N_P is in fact observed to be a fractal object itself in Figure 54(B) and the regression analysis renders a significantly smaller fractal exponent $\alpha \simeq 1.942 \pm 0.044$ and a more accurate power law $N_P(L) \simeq e^{-1.36} L^{1.94}$. Accordingly, the damage parameter (110) can be rewritten for the lattice in the hardening as

$$\bar{D} \simeq \frac{\beta(\bar{\varepsilon}) L^2}{2e^{-1.36} L^{1.94}} = \frac{\beta(\bar{\varepsilon})}{2e^{-1.36} L^{-0.06}} \, . \tag{111}$$

[63] The coefficient of determination R^2 reported by MINITAB 14© is rounded automatically to 100% in both graphs in Figure 4.54, proving the high significance of the regression analysis (R^2=100% states perfect correlation). The regression renders a 2 digits precision for the fractal exponents. However, since this type of modeling is sensitive to accuracy and robustness of fractal exponents, as much data as possible should be collected over an extended size range for regression purposes and engineering applications.

[64] The proportionality in the hardening between damage parameter and density of broken springs was recognized already by Hansen et al. (1989) but the determination of the correct proportionality coefficient remained controversial.

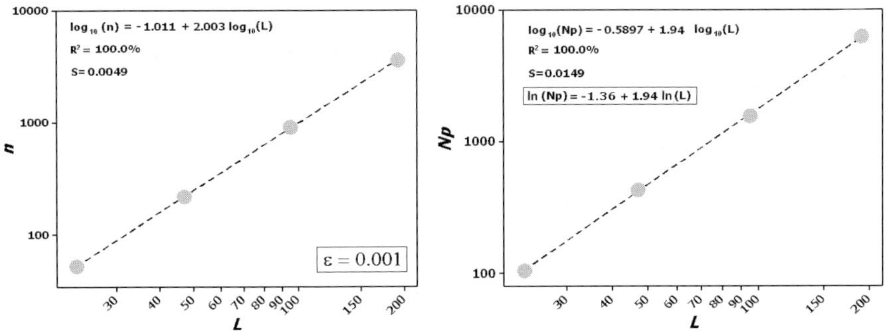

Figure 54. Fractal behavior of $n(\bar{\varepsilon}, L)$ during the hardening phase, both away from the peak (*e.g.* $\bar{\varepsilon} = 0.001$) and at the peak ($N_P$), demonstrated by the linear plot log(n) *vs.* log(L). In the former case the exponent is ~2 whereas at the peak it lowers to ~1.94. For convenience-sake, the power law of N_P is expressed also in base e such that $N_P \sim e^{-1.36} L^{1.94}$.

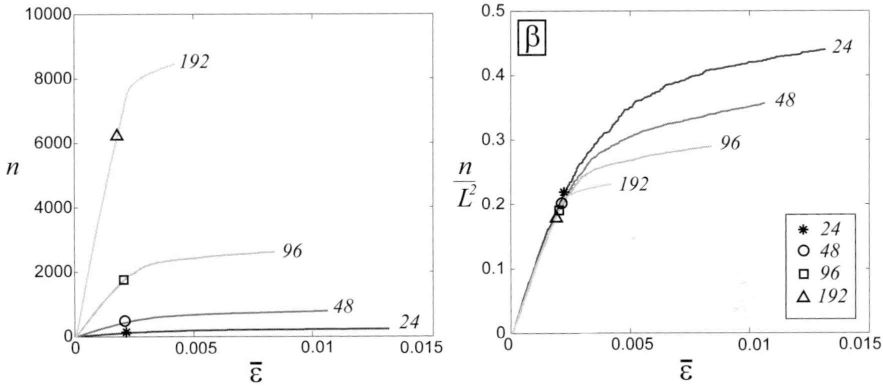

Figure 55. Plots $n - \bar{u}$ and $\beta - \bar{\varepsilon}$ showing how scaling n by L^2 produces a substantial data overlap in the hardening phase regardless of L. Yet, the stress peak location depends on L.

To express it as a function of $\bar{\varepsilon}$, we can linearize the function β as $\beta(\bar{\varepsilon}) \simeq (\partial\beta / \partial\bar{\varepsilon}) \cdot \bar{\varepsilon}$ and introduce the constant $\bar{\varepsilon}_0 = e^{-1.36} / (\partial\beta / \partial\bar{\varepsilon})$ such that

$$\bar{D}(\bar{\varepsilon}, L) \simeq \frac{\bar{\varepsilon}}{2\bar{\varepsilon}_0 L^{-0.06}} = \frac{\bar{\varepsilon}}{2\bar{\varepsilon}_p} . \tag{112}$$

This final form of the damage parameter is linear in $\bar{\varepsilon}$ and implies a parabolic relationship between stress and strain. The good agreement of this model (dotted line) with simulation data (solid line) up to the peak, as shown in Figure 56, provides the validation of such conclusion and of the entire deduction. Even though Eq.(112) is formally identical to the PBS, note that the peak of the parabola in the lattice is located at the transition point $\overline{\varepsilon}_p(L) \simeq \overline{\varepsilon}_0 \cdot L^{-0.06}$, which is a fractal quantity and captures the lattice size dependency of the stress peak[65]. At the peak $\overline{D} = 0.5$, which is reached faster as L increases.

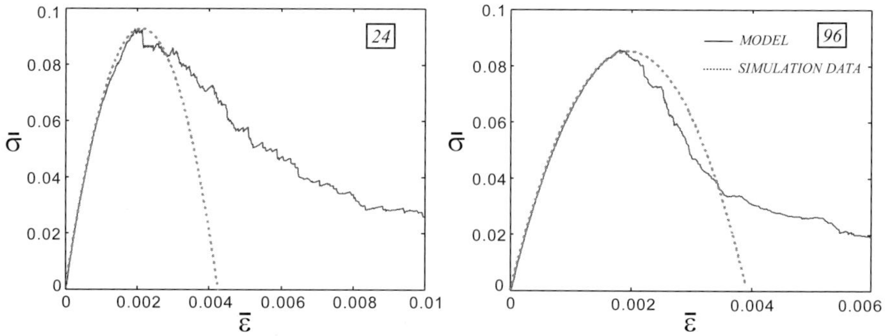

Figure 56. Mean response from simulation data (solid line) plotted *vs.* constitutive model Eq. (65) (dotted line) from the damage definition Eq.(112) for N=24,96. The model appears accurate in the hardening and, expectedly, deviates from simulations after the peak.

3.4.2. Softening Phase

The scenario is much different in the softening phase, which is affected throughout by the lattice size and needs to be approached with another mindset. The normalized number of microcracks n/L^2 in Figure 55(B) is a function $\beta(\overline{\varepsilon},L)$ of L in the softening. Such size dependence cannot be removed by an affine transformation or by scaling by L^{α} like in Eq.(109). From a modeling perspective, the size dependence is a significant challenge that makes it necessary to develop a universal $\overline{D}$ model applicable to softening for any L. By proceeding like in the hardening, such a goal is achieved via a two-steps procedure, which first establishes a scaling law for $n(\overline{\varepsilon},L)$ and then identifies the proper normalization factor for n that yields $\overline{D}$.

The first step is accomplished by means of an empirical scaling procedure borrowed from the study of phase transitions and clusters growth. In fact, the damage problem itself can be regarded as a process consisting of nucleation and growth of a microcracks clusters, which bears some resemblance with, say, the

[65] The fractality of $\overline{\varepsilon}_p$ descends evidently from the fractal exponent of N_p being 1.94 and less than 2. Estimating N_p from Eq.(109) would have yielded a constant peak strain $\overline{\varepsilon}_p = \overline{\varepsilon}_0$, representing a first order, less accurate approximation.

growth of an advancing solidification front at the liquid-solid interface (ref. Family and Vicsek (1985), Vicsek (1990), Barabasi and Stanley (1995), Gouyet (1996)). To explain basic rationale the scaling method, a hypothetical representative scenario is portrayed in Figure 57(A) where the generic quantity $y(x,L)$ is taken to depend not only non-linearly on a variable x but also on the cluster size L. The *knee-shape* depicted therein typically marks the occurrence of a phase transition at the critical point location (x^*, y^*). If a scaling relation exists, then the data $y(x,L)$ would map into one universal scale-invariant curve such that $f(x/L^\beta, \not{X}) = y/L^\alpha$ for any L, as indicated in Figure 57(B), and the scaling relation would take the general form

$$y(x,L) = L^\alpha f\left(\frac{x}{L^\beta}\right) \quad . \tag{113}$$

The scaling function $f(x/L^\beta)$ can be approximated analytically from the scaled data in Figure 57(B) via regression methods. Three conditions must be met for this scaling procedure to be feasible:

1. the y value at the transition must be a fractal such that $y(x^*,L) \propto L^\alpha$,
2. the location of the transition must be a fractal such that $x^* \propto L^\beta$,
3. before the transition the data must follow a power law $y(x,L) \propto x^\gamma$.

Under these conditions only two out of the three exponents $\{\alpha,\beta,\gamma\}$ are independent due to the constraint

$$\gamma = \alpha / \beta \quad . \tag{114}$$

The determination of α and β can be performed either by implementing a search algorithm that automatically maximizes the accuracy of the scaling or by proceeding in a trial-and-error way, which tends to converge fast.

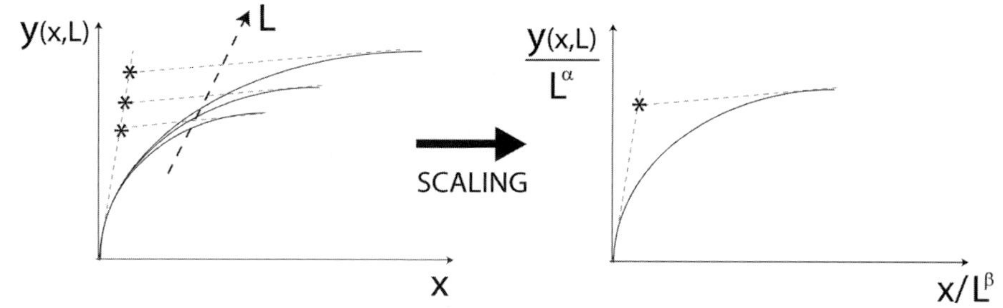

Figure 57. Illustration of the scaling procedure applicable to some non-linear systems that experience a transition governed by a universal law. The response y depends on the controlled variable x but also on the system size L, which controls the occurrence of the transition. A scaling law exists if $y(x,L)$ maps into a scale invariant curve upon normalizing y and x by L^α and L^β respectively.

For our specific lattice problem, this method is applied to the softening data n / L^2 in Figure 58(A), replotted from Figure 55(B) with $\Delta n(\bar{\varepsilon}, L) = n - N_p$ and $\Delta\bar{\varepsilon} = \bar{\varepsilon} - \bar{\varepsilon}_p$. By taking $\Delta n(\bar{\varepsilon}, L)/L^2$ and $\Delta\bar{\varepsilon}$ respectively as y and x, the application of Eq.(113) produces the scaled data on display in Figure 58(B) (dotted lines), with $\alpha = \beta = -0.6$. The equality between α and β is not an arbitrary simplification but a consequence of Eq.(114) expressing a constraint on the exponent γ (*i.e.* the best power law fit to data in the early stage of softening). Since all four datasets $\Delta n(\bar{\varepsilon}, L)/L^2$ are virtually linear near $\Delta n = 0$, the only possible choice that preserves such data agreement after scaling is indeed $\gamma = 1$ and, consequently, the scaling relation emerging from Figure 58(B) is simply

$$\frac{\Delta n(\bar{\varepsilon}, L)}{N_p} = L^{\alpha} f\left(\frac{\Delta\bar{\varepsilon}}{L^{\alpha}}\right) \quad . \tag{115}$$

The scaling function $f(\Delta\bar{\varepsilon}/L^{\alpha})$ is well approximated by the following analytical function[66] (Figure 58(B) - solid line)

$$f(\Delta\bar{\varepsilon}/L^{\alpha}) = a_1\frac{\Delta\bar{\varepsilon}}{L^{\alpha}} + b_1\left[1 - exp\left(-c_1\frac{\Delta\bar{\varepsilon}}{L^{\alpha}}\right)\right], \tag{116}$$

where the three fitting parameters a_1, b_1, c_1 are determined from simulations ($a_1 = 6,\ b_1 = 1.25,\ c_1 = 75.2$). The latter ones are not independent and the two

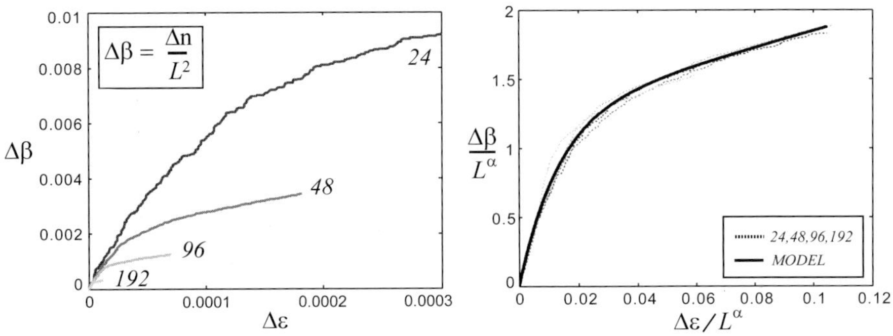

Figure 58. Softening data from Figure 55(B) before and after scaling Eq.(115).

[66] The structure of this seemingly complex function is simply the sum of two contributions: a linear term and an exponential transient, which only contributes before the *knee* in Figure 58 and vanishes afterwards, when the rupturing rate decreases to another steady value.

following conditions can be imposed onto $\partial n / \partial \bar{\varepsilon}$ to maintain a C^1 continuity with the hardening data from Eq.(111) and insure the correct asymptotic failure rate:

$$a_1 + b_1 c_1 = \partial n / \partial \bar{\varepsilon}\big|_{\bar{\varepsilon}_p} \quad , \tag{117a}$$

$$a_1 = \partial n / \partial \bar{\varepsilon}\big|_{\infty} \quad . \tag{117b}$$

where the terms on the right-hand side are measurable from Figure 58(B). Thus, once a_1 is chosen, only one free parameter is left and the pair $\{ b_1, c_1 \}$ is selected to optimize the data fit of Eq.(116). Combining Eqs.(114) and (115) yields at last the scaling relation for Δn

$$\Delta n(\bar{\varepsilon}, L) = 2N_p a_1 \Delta\bar{\varepsilon} + 2N_p L^{\alpha} b_1 \left[1 - exp\left(-c_1 \frac{\Delta\bar{\varepsilon}}{L^{\alpha}} \right) \right] \quad , \tag{118}$$

which concludes the first step. The search of the actual damage parameter for the softening phase is the next step. By invoking the additivity of damage[67], the new damage simply sums up to the amount previously cumulated in the hardening as

$$\bar{D}(\bar{\varepsilon}, L) = \bar{D}_p + \Delta\bar{D}(\bar{\varepsilon}, L) \quad , \tag{119}$$

where $\Delta\bar{D} = \bar{D} - \bar{D}_p$ and $\bar{D}_p = 0.5$ from Eq.(112). In analogy to Eq.(110), a normalization factor for Δn needs to be identified to estimate the damage contribution in the softening as

$$\Delta\bar{D}(\bar{\varepsilon}, L) = \frac{\Delta n}{X(L)} \quad . \tag{120}$$

Unlike the hardening, this time there is no cue prompting for an *a priori* guess of the normalization factor $X(L)$. The only guidance is provided by the physical requirements that the "correct damage parameter" must be $0 \le \bar{D} \le 1$ and should monotonically increase to one at failure (*i.e.* $\bar{K}_{failure} = 0$), either discontinuously or continuously depending on the failure mode. Therefore, as opposed to the inductive reasoning deployed for the hardening, the determination of $X(L)$ is pursued through statistical methods (ref. Section 3.3.4) to reach a "data driven" decision. According to standard regression methods, the stress estimate $\hat{\sigma}_i$ in correspondence to $\bar{\varepsilon}_i$ and Δn_i at the generic *i*-th point of the stress response can be expressed by

$$\hat{\sigma}_i(\bar{\varepsilon}_i, \Delta n_i, X) = \bar{K}_0 \left[\left(1 - N_p / (2e^{-1.3} L^2) \right) - \Delta n_i / X \right] \bar{\varepsilon}_i \quad , \tag{121}$$

[67] As silently assumed in Eq.(65) and throughout Section 3.3.3. This assumption is a reasonable first order approximation in absence of load reversal like in a monotonic loading.

where X is the target parameter to be estimated. Since this constitutive model is linear in $1/X$, the OLS method can be used to compute a minimum unbiased estimator $\widehat{X}$ (Montgomery et al. (2001)). After selecting Q points from the actual simulation, the mean square error function is computed as

$$Err\left(\frac{1}{\widehat{X}}\right)=\sum_{i=1}^{Q}\left(\bar{\sigma}_i-\hat{\sigma}_i\right)^2=\sum_{i=1}^{Q}\left(\bar{\sigma}_i-\bar{K}_0\left[\left(1-D_p\right)-\frac{1}{\widehat{X}}\Delta n_i\right]\bar{\varepsilon}_i\right)^2 , \tag{122}$$

which is the sum of the squared deviations between the simulation data and the predicted values over Q. When the error function is minimized with respect to X by setting $\partial Err/\partial\widehat{X}=0$, the calculations lead to

$$\hat{X}\left(L\right)=\frac{\sum_{i=1}^{Q}\bar{K}_0\bar{\varepsilon}_i^2\Delta n_i^2}{\sum_{i=1}^{Q}\left(\left[\left(1-D_p\right)\bar{K}_0\bar{\varepsilon}_i-\bar{\sigma}_i\right]\Delta n_i\bar{\varepsilon}_i\right)} . \tag{123}$$

The results from Eq.(123) are displayed in Figure 59 with circles (note that extra data for $N=\{72,120\}$ are also used for validation purposes). Such plot discloses that a regression line

$$X(L)=a_2L+b_2 , \tag{124}$$

fits the six $\hat{X}(L)$ values very well (for $a_2=65$ and $b_2=-1038$ it is $R^2=0.997\simeq1$) and can be taken as the normalization function $X(L)$ in Eq.(120). That result fully solves the softening problem, with Eqs.(118) and (120) delivering the desired scaling relations for the average constitutive model for the softening, *i.e.*

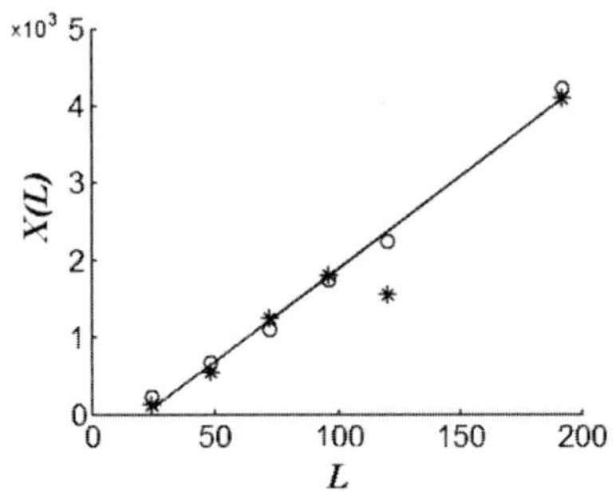

Figure 59. Comparison of *X(L)* estimated from Eq.(123) from average data Δn (circles) *vs.* from 6 random replicates for $N=\{24,48,72,96,120,192\}$ (asterisks).

$$\sigma(\bar{\varepsilon},L)=\bar{K}_0\left[1-\bar{D}_p-\frac{2N_p a_1\Delta\bar{\varepsilon}+2N_p L^{\alpha} b_1\left[1-\exp\left(-c_1\frac{\Delta\bar{\varepsilon}}{L^{\alpha}}\right)\right]}{a_2 L+b_2}\right]\bar{\varepsilon}\,. \tag{125}$$

The actual performance of this fractal-based model is assessed next. However, it is first noticed that for very large lattices, when the intercept becomes negligible (*i.e.* $1038 \ll 65L$), the damage parameter from Eq.(120) can be approximated as

$$\Delta\bar{D}(\bar{\varepsilon},L)=\frac{\Delta n(\bar{\varepsilon},L)}{a_2 L}\quad. \tag{126}$$

In this view, $X(L)$ appears to be a fractal with exponent equal to one[68]. This circumstance is consistent with the phenomenology from Section 3.2, with regard to the high degree of spatial correlation amongst microcracks recorded during propagation, which tend to localize along a narrow strip. Thus, an invariant set with unitary fractal exponent "driving" damage propagation would intuitively make as much sense as a fat fractal for the damage nucleation in the hardening. This interpretation is appealing because it enables endowing the damage parameter in the softening with an explicit physical meaning.

3.4.3. Model Validation and Final Remarks

The final model, obtained by combining the "half" models independently developed for hardening and softening, needs to be validated, especially as far as the scaling law (125). This is done in Figure 60(A) by comparing the model estimates from analytical expression (112) and scaling model (125) against simulation data for *N=96*. The close agreement in the hardening is expected as it was already acknowledged (Figure 56). In the softening, instead, the model captures the data sufficiently well only in the first part[69]. Even if the model is reliable for a strain

[68] Like $2N_P$ in Eq.(111), $X(L)$ amounts to about twice the number of softening microckracks necessary to cause failure from the stress peak.

[69] The Eqs.(111) and (120) imply a discontinuity in the derivative of the damage parameter $\bar{D}$ due to the abrupt change in the denominator from L^2 to L at the hardening-softening transition. Thus, while $n(\bar{\varepsilon})$ is a C^1 continuous function at the transition, the damage rate is just C^0, which agrees with the phenomenology (e.g. see Figure 60(A)). The choice of the Q data points to be used in the evaluation of Eq.(123) is crucial to the accuracy of $\hat{X}(L)$ and depends on what part of the response is of interest. This analysis concentrated on capturing the steepest descent of the early phase of the softening regime and only the corresponding points (about 25% of softening dataset) were selected for the statistical OLS computation.

range satisfactory for many engineering applications, the outlined disagreement at failure is puzzling and solicits further consideration for the purpose of model validation. Two questions arise naturally:

1. does the accuracy of the analytical estimate $\Delta n(\bar{\varepsilon}, L)$ from scaling (118) affect model performance?
2. more importantly, is the selected form of the damage parameter in Eqs.(111) and (120) really appropriate?

The first question addresses the fact that the scaling law (118) provides a convenient *smooth* analytical relation for $\Delta n(\bar{\varepsilon}, L)$ but is clearly a trade-off between the need of having an average estimate of the microcracks number valid for any lattice size and the inevitable loss of "details" characterizing each individual curve (ref. the data scatter in Figure 58(B)). Unfortunately the loss of details is relevant, judging from Figure 60(B), which displays the results from damage definitions (111) and (120) when actual numerical data for n , N_P, and Δn are used directly in place of the analytical expressions (112) and (118) deployed in Figure 60(A). The comparison between the two cases demonstrates that in the latter case the model "shadows" the actual response in the softening phase visibly better than before (the difference in the hardening exists but is much less noticeable). For these simulations the difference between the estimate of Δn from scaling (118) and the real value from simulations never exceeded 10% for any lattice size. Therefore, although the difference between (A) and (B) in Figure 60 is not huge and the predictive power of both models does not differ dramatically, it is reasonable to answer affirmatively to question #1 and conclude that the inaccuracy on Δn is partly responsible for the alleged mismatch between model and simulations in Figure 60(A).

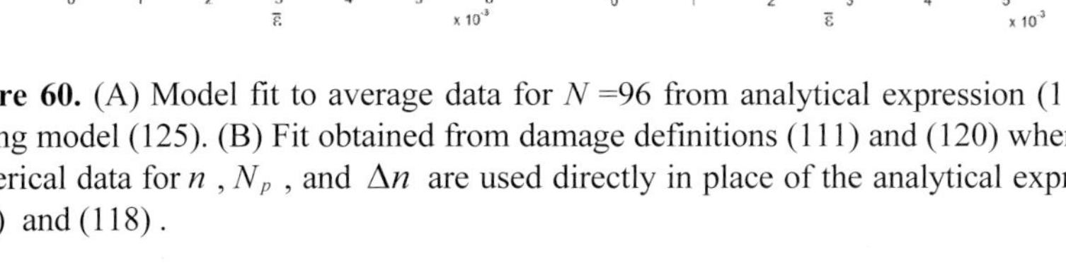

Figure 60. (A) Model fit to average data for N =96 from analytical expression (112) and scaling model (125). (B) Fit obtained from damage definitions (111) and (120) when actual numerical data for n , N_p , and Δn are used directly in place of the analytical expressions (112) and (118) .

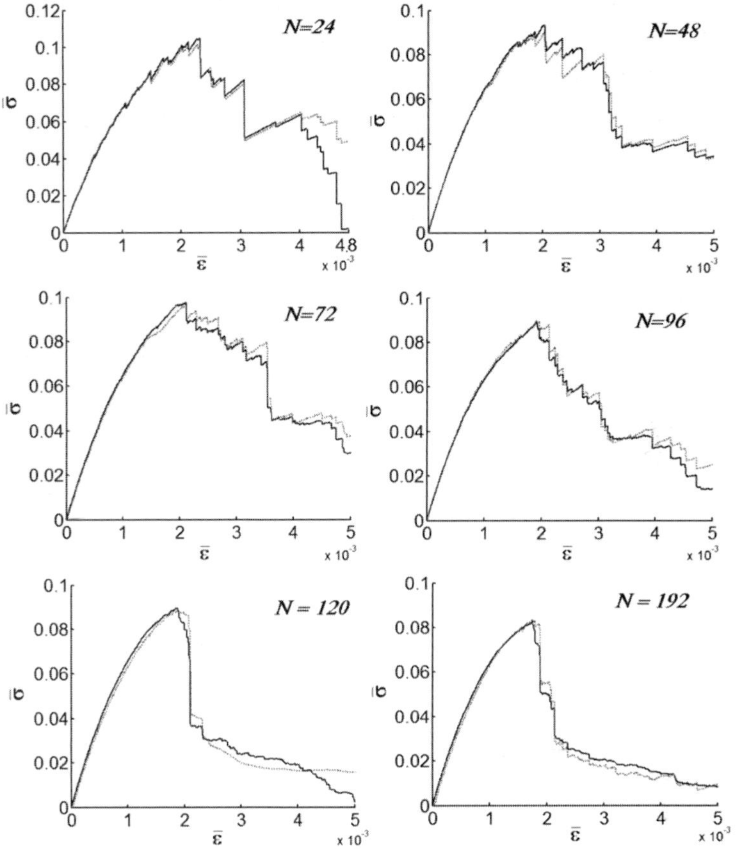

Figure 61. Responses of customized model Eqs.(111) and (120) (full line) *vs.* simulation data (dotted line) for six individual random replicates for $N = \{24, 48, 72, 96, 120, 192\}$.

Question #2, on the other side, is of more substantial nature and disputes the appropriateness of the proposed model form. Stated in other terms, it questions whether or not the founding hypotheses, expressed mathematically by Eqs.(111) and (120), are well taken and, consequently, whether the mismatch in Figure 60(A) can be traced back to this type of fundamental mistake. To answer effectively it is sufficient to check the performance of Eqs.(111) and (120) against an individual lattice simulation when no analytical approximation is used. In practice, besides using direct numerical data for n, N_P, and Δn, also the parameter $X(L)$ is estimated *ad hoc* from each simulation rather than from Eq.(125). The results of this method are valuable. As shown by the six randomly picked replicates of

$N = \{24, 48, 96, 72, 120, 192\}$ in Figure 61, the model is clearly able to reproduce the actual damage response from simulations when no artifact is introduced via analytical approximations. The agreement between models (full line) and numerical data (dotted line) is indeed remarkable when acknowledging the extreme simplicity of definitions (111) and (120). Complex profiles and discontinuities of the softening phase are well-captured in Figure 61. The results therein are achieved by merely re-estimating each time the sole parameter X, the six values of which are reported in Figure 59 with asterisks. Although those data points gather around the regression line Eq.(125) used in Figure 60(B), a small error in X evidently translates into a large difference in the model response. In conclusion, to answer question #2, the model form appears indeed suitable.

From these observations the divergence in Figure 60(A) is essentially imputable to approximations (118) and (125) on Δn and X. In particular, the continuous approximation for Δn is critical to the model accuracy because overlooks the large discontinuities registered in the $n(\overline{\varepsilon}, L)$ plots (*e.g.* ref. Figures 38-41(C) and 45(A)) in correspondence to large avalanches at the onset of damage localization and afterwards. Recall that the saw-toothed shape of the lattice response becomes macroscopically appreciable in the softening owing to large avalanches responsible for the sudden vertical "drops" in the stress response between two elastic "climbs" (see Figure 61). Such a "roller-coaster" profile disappears almost entirely in the average data in Figure 13, introducing instead a "fictitious" smoother descent. That is an intrinsic limitation of this approach, solely aimed at capturing the average behavior of the lattice and focused on the sample-size effect. An accurate model can be fit case by case to closely match an individual test but the specific details need to be properly re-accounted for.

Never the less the analytical approximation (120) is precious - by an engineering perspective - when the goal is to capture the general "trend" and get a first order estimate of the softening behavior for any lattice size. In this view, the scaling model inhere is a significant advance in damage modeling and represents, perhaps, the first successful trial in "bridging the scales" with SDM[70].

4. More Related Work

The selection of results in this overview is by no means comprehensive and is just aimed at emphasizing a few outstanding issues. Extensive literature is available to

[70] The investigation of sample size effects in this 2D spring network began nearly 20 years ago. The work by Hansen et al. (1989) was among the early attempts to formulate a scaling procedure and to apply fractal theory in the contest of damage mechanics. The lattice simulations were then limited to N=24 maximum.

expand on the topic. For instance, the scaling problem for the same lattice is treated differently by Rinaldi et al. (2007b) with an approach based on "Extreme Value Theory" (EVT), which delivers a semi-empirical model that offers a viable design strategy and is as effective as the fractal-based counterpart examined above. While EVT probabilistic models lack the rigor and the physical bearing of some analytical formulations discussed here, for practical purposes they may sometimes offer a better description of data scatter of the lattice properties, delivering scaling laws not just for the mean response but for the entire stress distribution in the strain softening and at failure. The importance of modeling also variability was already stated in Section 3.2 and only if the scaling laws capture the whole statistical distributions then the softening behavior and the failure can be predicted with known precision from discrete models, which is crucial to design and maintenance in structural engineering. Robust models for estimating damage tolerance and survivability of structural components subject to multi-site microcracking could benefit many industrial compartments, such as (but not limited to) aircraft, aerospace, nuclear and micro-electronic industries.

Space limitation forced also to leave out other relevant results beyond the static case. Fatigue and dynamic problems were not covered, for instance, although significant advances have been achieved. An application of the same lattice model to fatigue can be found in Rinaldi et al. (2006b), whereas an extensive treatment of SDM modeling applied to impact and shock loading can be found in the monographic account by Mastilovic (2008).

5. Concluding Remarks

This brief discourse about advances in SDM shows the potential of the methodology. Starting from the lower length scale characteristic of the microstructure, SDM multiscale approach delivers a surprising amount of new insight and makes for a precious companion to micromechanics methods discussed in previously in the book. This brief discourse largely reflects the author's viewpoint and current ideas about the subject, many of which originated from (or continued) the forerunner work by Krajcinovic (ref. Krajcinovic 1996). SDM as a discipline is far from being an established subject but by now it has a clear footprint. As such, the goal of this work is solely to portray the state of the art and expose novel ideas about SDM for consideration and discussion by the larger solid mechanics community. Some of these concepts – yet at their infancy – may possess the potential to develop into innovative, successful engineering tools. Designing the materials from the microstructure is already a current research and industry trend to achieve superior damage tolerance, partly driven by widespread usage of composites and more recently by nanotechnology. Another battle ground for SDM is indeed represented by the design of small (micro- and nano-sized) mechanical systems, such as MEMS and NEMS, where sample-size effects and discreteness are pervasive aspects of the design.

Appendix 1: Notes About Numerical Solution Schemes

All lattice simulations were performed with computer programs purposely developed in house (using C++, Fortran, Matlab). Two different solution schemes were pursued to solve for the nodal displacements of the lattice under several loading configurations, namely: molecular dynamics (MD) and direct stiffness method of finite elements (FE). The MD and FE methods each offer advantages and disadvantages. The MD and FE methods each offer advantages and disadvantages. The data obtained from the simulations are primarily the quasi-static force response $\overline{F} - \overline{u}$ and the cumulative curve $n - \overline{u}$. The FE requires more memory-storage of the MD model but offers greater accuracy, particularly in the softening. During damage nucleation, both approaches perform well but at the onset of damage localization the MD becomes less efficient and accurate. The increased compliance and highly heterogeneous distribution of kinetic and potential energies within the lattice are amongst the causes for slower convergence. MD is better suited to study the dynamical response of the lattice (given an intrinsic time scale) rather than the quasi-static response, where the simulation time to reach the equilibrated configuration might be unacceptable. The problem is very sensitive to lattice size and can have drastic effects on the accuracy of softening data for large lattice. Conversely, accuracy is not a concern for the FE method but there are other critical issues, such as the local instabilities described below. However, if a large number of inelastic transformations (Rinaldi et al. (2006a)) as well as strain-rate effects need to be included in the model, explicit solvers like MD offer greater flexibility than FE and are sometimes the only viable choice. A brief description of each technique follows.

Finite Elements
A standard "direct stiffness" solver belonging to the finite elements methods (FE) was implemented purposely to run damage-controlled simulations and generate information for the study of the transition and avalanches (ref. Rinaldi et al. (2006a)). The structure of the static linear elastic FE problem is described by the matrix equation $\mathbf{K}_L \mathbf{u}_L = \mathbf{F}_L$, where $\mathbf{K}_L$ is the system stiffness matrix of the lattice and $\{\mathbf{F}_L, \mathbf{u}_L\}$ are the nodal vectors of the external forces and displacements associated to the grains (the nodes). For the lattice the matrix $\mathbf{K}_L$ is obtained by the standard "assembly" procedure when both the orientation $\vartheta_{(ij)}$ and the scalar stiffness $k_{(ij)}$ of the *ij*-th spring (ref. footnote # 45) connecting the *i*-th and *j*-th nodes are known for all the Q "active" elements, where active refers to "pristine" or "broken

but compressed" springs. It is known that the global representation of the stiffness matrix of a spring element has four translational degrees of freedom (*dof*). The $\mathbf{K}_L$ is assembled by summing the augmented matrices $\mathbf{K}^{\mathrm{EL}}_{(ij)}$ of the Q active links, *i.e.* $\mathbf{K}_{\mathrm{L}} = \Sigma^{Q}\mathbf{K}^{\mathrm{EL}}_{(ij)}$. The ensemble of active links changes with the damage process and $\mathbf{K}_L$ must be updated at every damage step, when a new spring breaks or a broken one closes, becoming active again. Sparse solver techniques are adopted to optimize memory storage.

The non-standard part of this solver lays in the *stabilization check* that removes local instabilities caused by spring rupturing. Close to localization the lattice suffers indeed the inner formation of dandling edges and local mechanisms, as depicted in Figure 62, that renders $\mathbf{K}_L$ singular even after application of BCs. Therefore, unlike MD, unstable links must be searched and excluded from the assembly procedure. The *ij*-link in Figure 62(A) becomes dandling when $z_j = 1$ after the rupture of the *kj*-link. A kinematic mechanism (with 1 *dof*) forms locally if $z_i = 2$ and the *ij*-th and *ik*-th survivor springs of the *i*-th node are non consecutive (non-adjacent in the Voronoi graph), as happening in Figure 62(B) when the *ip*-link breaks. Both circumstances trouble direct solvers and the stability check increases significantly the run-time.

Damage-controlled simulations are achieved by searching the critical macro-configurations that determines the failure of the unbroken link with highest damage affinity $\mathcal{A}_{(ij)} = (1/2)k_{(ij)}\ell_{(ij)}(\varepsilon^2_{(ij)} - \varepsilon^{*\,2}_{(ij)})$, where $\varepsilon_{(ij)}$ and $\varepsilon^*_{(ij)}$ are the current strain and the critical tensile strain of the *ij*-th link respectively and $\{k_{(ij)}, \ell_{(ij)}\}$ are the stiffness and initial length. By letting $\overline{u}$ be an arbitrary macro-displacement imparted to the lattice, a trial displaced configuration is computed in a displacement-controlled fashion, with $u_{(ij)}$ and $\varepsilon_{(ij)}$ being the trial elongation and strain of the generic *ij*-th link on the microscale. Since $\varepsilon_{(ij)} \propto \overline{u}$ in the linear elastic framework, the next spring to fail is the one with the highest ratio $\lambda_{(ij)} = \varepsilon_{(ij)} / \varepsilon^*_{(ij)}$ between actual strain and critical strain from the quenched distribution, because it reaches first the failure condition $\lambda_{(ij)} = 1$ (*i.e.* $\mathcal{A}_{(ij)} = 0$) as $\overline{u}$ increases. Then, the corresponding lattice displacement $\overline{u}^*$, computed from the trial $\overline{u}$, is $\overline{u}^* = \overline{u} / \lambda^{MAX}_{(ij)}$. As a second step, $\mathbf{K}_L$ is updated by removing from the active set the *p*-th link with $\lambda_p = \lambda^{MAX}_{(ij)}$ and the new lattice configuration at $\overline{u}^*$ is computed, which allows assessing the load drop associated to new damage. Such a two-steps procedure is iterated until the lattice breaks down. A similar method was used by Delaplace et al. (1996) for fuse lattices. Data $\{\overline{F}, \overline{u}\}$ are collected only in correspondence to the critical macro-displacements $\overline{u}^*$ where new damage forms. The displacement-controlled curve, typically observed in real monotonic tensile tests, is

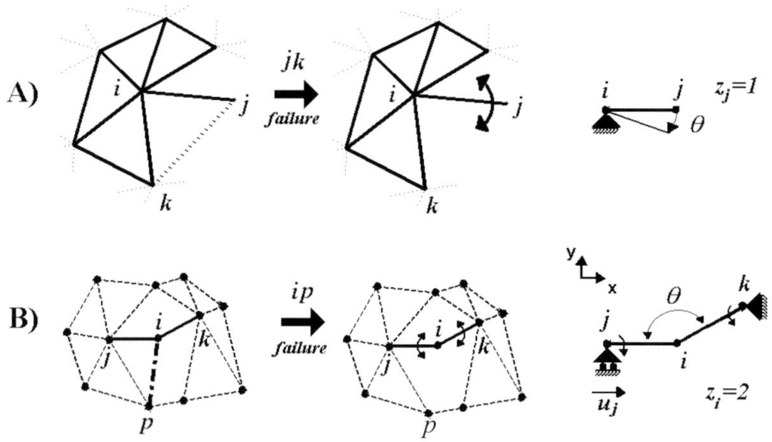

Figure 62. Formation of a dandling link from rupture of the *ij*-th link (A) and of a kinematical mechanism from rupture of *ip*-th link (B) in the primary lattice.

deduced from the envelope of the damage-controlled response by discarding the snap-back such that $\Delta\bar{u} \geq 0$ anywhere (ref. Figure 42).

Molecular Dynamics

Molecular dynamics (MD) is a numerical "technique used to solve the classical equations of motion for a system of N "molecules" interacting via a potential $\mathcal{V}$, *i.e.* a set of Newton's ordinary differential equations (ODE). The method is well known and has been used extensively in several branches of physical sciences (ref. Hoover (1986) , Allen and Tildesley (1994)). Nonetheless, MD is not frequently encountered in solid mechanics because the time-dependent problems are usually formulated by a continuum standpoint in terms of partial differential equations (PDE) and solvers based on variational principles (*e.g.* FE) are preferred. The dynamics of a conservative system of N point-particles of mass m_i with no rotational *dof* is governed by the following Cartesian equations

$$\ddot{\mathbf{r}}_i = -\nabla_{\mathbf{r}_i} \mathcal{V} / m_i = \mathbf{f}_i / m_i \qquad (i=1..N) \tag{127}$$

where $\mathbf{r}_i$ is the position vector and $\mathbf{f}_i$ is the force vector. The nature of the interaction between particles is not usually known and the formulation of the many-bodies potential function is, indeed, an extraordinary task for large ensembles. In most situations the particles are considered in pairs and $\mathcal{V}$ is approximated by the sum of the pair-potential function $\Phi(r_{ij})$ over all pairs (pair-wise additivity assumption)

$$\mathcal{V} = \frac{1}{2}\sum_{i}^{N}\sum_{j\neq i}^{N}\Phi\left(r_{ij}\right), \tag{128}$$

with $r_{ij} = \left|\mathbf{r}_{ij}\right| = \left|\mathbf{r}_i - \mathbf{r}_j\right|$. The Lennard-Jones 6-12 potential or the Morse potential are examples of classic diatomic potential used at the atomic scale. For lattice models of polycrystalline microstructures, where the nodes represent N^2 grains, the microscale interactions between two grains are assumed linear elastic and the pair potential is the quadratic form $\Phi=1/2k_{ij}\,(r_{ij}-r_{ij}^{0})^2$, with k_{ij} being the stiffness of the linear spring connecting the ij-th pair. Every grain interacts only with the nearest neighbors and the j-sum in Eq.(128) is limited to the coordination number z_i of the i-th grain

$$\mathcal{V} = \frac{1}{2}\sum_{i}^{N_0}\sum_{j\neq i}^{z_i}\left[\frac{1}{2}k_{ij}\left(r_{ij}-r_{ij}^{0}\right)^2\right]. \tag{129}$$

The actual coordination number depends on both damage and applied load, because the broken springs give no contribution to $\mathcal{V}$ if they are "open" (open microcracks) and are discarded. The asymmetry between compressive and tensile behaviors of the broken spring (Figure 3) is an inherent source on non-linearity on the macroscale as the secant stiffness is discontinuous when load reversal occurs. For grain size larger than, say, 1µm, quantum effects are negligible but not on the nano-scale or atomic-scale. Hence, even though the MD approach is usually valid, the simple quadratic potential cannot be used when inter-atomic interactions are predominant, such as in carbon-nanotubes (Belytschko et al. (2002)). The N^2 equations of motion for the lattice descend from Eq.(129) as

$$m_i\ddot{\mathbf{r}}_i = \mathbf{f}_i = -\nabla_{\mathbf{r}_i}\mathcal{V} = -\sum_{j\neq i}^{z_i}k_{ij}\left(r_{ij}-r_{ij}^{0}\right)\frac{\mathbf{r}_{ij}}{r_{ij}} \tag{130}$$

and are accompanied by an appropriate set of BCs. Algorithms for discrete-particle dynamics simulations are known from the mid '60s but only newer computers can handle systems of acceptable size. Amongst several ways to integrate Eq.(130), the original Verlet's algorithm was implemented in these simulations. The discretized form of the equations of motion is

$$\mathbf{r}_i(t+\Delta t) = \mathbf{r}_i(t) + \alpha\left[\mathbf{r}_i(t) - \mathbf{r}_i(t-\Delta t)\right] - \Delta t^2\sum_{j\neq i}^{z_i}\frac{k_{ij}}{m_i}\left(r_{ij}-r_{ij}^{0}\right)\frac{\mathbf{r}_{ij}}{r_{ij}}, \tag{131}$$

where $\alpha \leq 1$ is the damping parameter usually. The computation of the "next" configuration requires the storage of the two previous lattice configurations. Once the positions $\mathbf{r}_i(t)$ at time t are available, the accelerations $\mathbf{a}_i(t) = \ddot{\mathbf{r}}_i(t) = -\Sigma^{z_i}\,(k_{ij}/m_i\,(r_{ij}-r_{ij}^{\,0}))\mathbf{r}_{ij}/r_{ij}$ are computed first and the new positions

$\mathbf{r}_i\,(t+\Delta t)$ then. The steps are repeated identically for the next time increment once $\mathbf{r}_i\,(t+\Delta t)$ is shuffled back to be the new $\mathbf{r}_i\,(t)$. The computation of the accelerations is the most intensive part because requires an internal loop through the z_i neighbors to assemble the force vector acting on the *i*-th mass. Details about the iteration process are discussed by Allen and Tildesley (1994). For quasi-static loading the time, the mass, and the damping parameter are fictitious quantities that are not important by a physical standpoint in this case. A value α in the typical range $\alpha = 0.95 - 0.99$ (here $\alpha = 0.95$) is used to damp free oscillations and let the system reach the equilibrium configuration corresponding to the imposed BCs. The value of Δt affects the stability of the algorithm and was numerically set to 0.0001 to let the method converge for all lattice sizes. Convergence is reached when the kinetic energy of the lattice decreases below a given threshold. The pseudo-algorithm below summarize how the solver operates:

1. application of displacement increment ,
2. computation of accelerations $\ddot{\mathbf{r}}_i = \mathbf{F}_i / m_i = \Sigma^Q K_{ij} \cdot \Delta \mathbf{r}_{ji}$,
3. update of nodes positions,
4. computation of velocity and kinetic energy,
5. convergence check:
   a. exit if kinetic energy < specified tolerance,
   b. else iterate 2-5,
6. link rupture check:
   a. if rupture: removal from the network and iterate 2- 5,
   b. else application of new displacement increment – loop from 1,
7. exit if max displacement is reached or breakdown occurs.

Appendix 2: The One-way Correlation Hypothesis

With reference to the discussion in Section 3.3.4, although correlation does not generally imply causality, the perturbation of the microscale strain field from the pristine state due to damage obviously introduces a causal relationship between the rupturing processes in the diagonal and horizontal springs. This fact is referred to as "one-way correlation" in the third assumption of MODEL 2. A rigorous evaluation follows to support it. Preliminarily the concepts of correlation, autocorrelation, and *P*-value are recalled for convenience.

1) CORRELATION. In probability theory and statistics, correlation, also called correlation coefficient, is a numeric measure of the strength of linear relationship between two random variables. In statistical practice, correlation or co-relation refers to the departure of two variables from independence. In this broad sense there

are several coefficients, measuring the degree of correlation, adapted to the nature of data. The pair-correlation ρ_{XY} between two random variables X and Y with expected values $< X >$ and $< Y >$ and standard deviations σ_X and σ_Y is one common measure defined as:

$$\rho_{XY} = \frac{\operatorname{cov}(X,Y)}{\sigma_X \sigma_Y} = \frac{\langle XY \rangle - \langle X \rangle \langle Y \rangle}{\sqrt{\langle X^2 \rangle - \langle X \rangle^2}\sqrt{\langle Y^2 \rangle - \langle Y \rangle^2}} . \tag{132}$$

The points (X,Y) gather closer to the 45° line on a X-Y plot, called scatter plot, as the correlation increases.

2) AUTOCORRELATION. The autocorrelation of a discrete time series or a process X_t is simply the correlation of the process against a time-shifted version of itself. If X_t is second-order stationary with mean μ then this definition is

$$R(k) \simeq \frac{\langle (X_i - \langle X \rangle)(X_{i+k} - \langle X \rangle) \rangle}{\sigma_X^2} \tag{133}$$

where k is the time shift being considered (usually referred to as the lag). This function has the attractive property of being in the range $[-1, 1]$ with 1 indicating perfect correlation (the signals exactly overlap when time shifted by k) and -1 indicating perfect anti-correlation. Eq.(133) is a special case of the general Eq.(132).

3) P-VALUE. The assessment of the significance of the correlation coefficient rests on the P-value computed from testing the hypothesis of no correlation based on an asymptotic normal distribution of the correlation ρ_{XY}. The P-value is the probability of getting a correlation as large as the observed value by random chance, when the true correlation is zero. If this probability is small, say less than 0.01, then the correlation ρ_{XY} is significant, *i.e.* non zero (Montgomery et al. (2001)).

Under the general idea that two interacting links (broken or pristine) are likely near one another and that the perturbation fields are stronger in the vicinity of the ruptures (*e.g.* Figure 49), the one-way correlation between the two rupturing processes should show as a significant spatial correlation amongst pairs of diagonal and horizontal ruptures. For illustrative purposes the analysis is limited to the first 200 ruptures of the lattice $N = 48$ from Figure 32. These all occur in the hardening and account for 160 horizontals and 40 diagonals.

First, a relation is found between the 40 diagonals and the last preceding rupture (of whatever orientation), labeled "LAST". By taking the vector of the x coordinates of the 40 center points of the diagonal broken links as X in Eq.(132) and by letting $Y=X_{LAST}$ be the vector of the x coordinates of the preceding ruptures associated to X, the scatter plot in Figure 63(A) renders a marked correlation. The ρ_{XX} =0.58 is indeed significant based on a P-value smaller than 0.01. Similarly, when the y coordinates are examined, ρ_{YY} =0.44 is obtained. This observation suggests that the

40 diagonals are influenced by previous damage. The correlation coefficients are computed by pooling the 40 values associated to ruptures at 60° and −60°, which are marked differently with circles and crosses respectively to show their symmetric statistical behavior.

A second test examines the correlation between the 40 diagonals and the nearest horizontal (preexisting) ruptures, labeled "NHZ". The scatter plots render $(\rho_{XX}, \rho_{YY}) = (0.52, 0.53)$. Figure 63(B) shows the x coordinates only. This test highlights that the diagonals ruptures tend to form close to horizontal ruptures.

A third test highlights the correlation between the datasets LAST and NHZ. The scatter plot in Figure 63(C) demonstrates a very high correlation, of the order of 0.9, which suggests that LAST and NHZ might be essentially the same set. By accepting this likely interpretation and following a *transitive* logic, these three tests would prove a spatial and temporal correlation between each diagonal ruptures and the preceding horizontal rupture. The correlation emerges also from the histograms of "the distance from the previous rupture" (DPR) for Group 1 and Group 2 (not shown). In the former case the histogram resembles a uniform profile typical of non-correlated damage process, while in the latter case the histogram is highly skewed towards lower values.

Finally, other pair-wise comparisons (*e.g.* a diagonal rupture *vs.* other diagonal ones, or a horizontal rupture *vs.* other horizontal ones) suggest absence of autocorrelation at a significance $\alpha = 0.01$ (scatter plots not shown).

This statistical assessment, although based on pair correlations, supports MODEL 2. Higher order correlations could be informative and are left for future. In particular the possibility that a horizontal rupture (or a cluster of them) may be responsible for more than one diagonal should be acknowledged and evaluated.

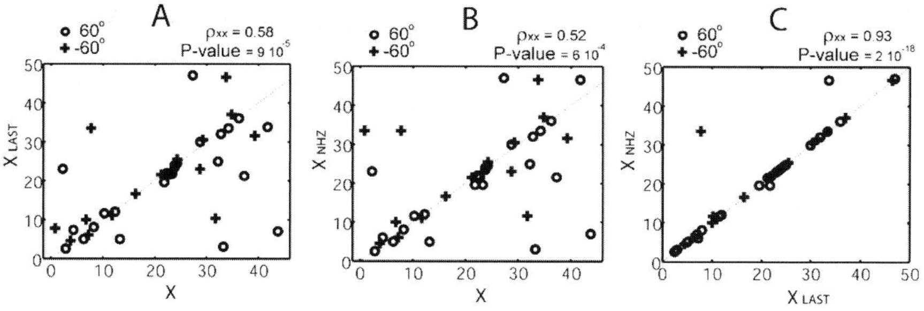

Figure 63. Scatter plots: (A) between the x-coordinates X of a diagonal rupture and the last previous rupture "LAST" (either horizontal or diagonal), (B) between the X and the nearest horizontal previous rupture "NHZ", (C) between "LAST" and "NHZ". Data refer to the first 40 ruptures registered for N=*48*. Circles and crosses corresponds to links with orientations 60° and −60° respectively. All correlations are significant at a significance level $\alpha = 0.01$.

References

Alava, M.J., Nukala, P.K.V.V., and Zapperi, S., (2006). Statistical models of fracture. *Advances in Physics*, Vol. 55, Nos. 3–4, 349–476.

Allen, M.P., and Tildesley, D.J., (1994). *Computer Simulation of Liquids*. Clarendon Press, Oxford, UK

Audoin, B., and Baste, S.,(1994). Ultrasonic evaluation of stiffness tensor changes and associated anisotropic damage in a ceramic matrix composite. *J. Appl. Mech.*, Vol. 61, 309-316.

Barabasi, A.-L., and Stanley, H.E., (1995). *Fractal Concepts in Surface Growth*. Cambridge Univ. Press, New York, NY.

Bazant, Z.P., (2000). Size effect. *Int. J. Solids Structures*, 37, Issues 1-2, 69-80.

Belytschko, T., Xiao, S.P., Schatz, G.C. and Ruoff, R.S., (2002). Atomistic simulations of nanotube fracture. *Phys. Rev. B*, Vol 65, 235430.

Belytschko, T., Loehnert, S., and Song, J.H., (2008). Multiscale aggregating discontinuities: a method for circumventing loss of material stability. *Int. J. Numer. Meth. Eng.*, 73, 869–894.

Betten, J., (1983). Damage tensors in continuum mechanics. *Jour. Mec. Theor. Appl.*, 2, 13–32.

Billington, E.W., and Tate, A., (1981). *The Physics of Deformation and Flaw*. Mc.Graw-Hill, Int. Book Comp., New York, NY.

Born, M., and Huang, K.,(1954). *Dynamical Theory of Crystal Lattices*. Oxford Univ. Press, New York NY.

Brunig, M., (2004). Eshelby stress tensor in large strain anisotropic damage mechanics. *Int. Jour. Mech. Sci.*, 46, 1763–1782.

Carpinteri, A., Lacidogna, G., Niccolini, G., and Puzzi, S., (2009). Morphological fractal dimension versus power-law exponent in the scaling of damaged media. *Int. J. Damage Mech.*, 18, 3, 259-282.

Chaboche, J.L. (1988). Continuum damage mechanics: Part I—General concepts. *Jour. Appl. Mech.*, 55,59–64.

Chaboche, J.L. (1988). Continuum damage mechanics: Part II—Damage growth, crack initiation, and crack growth. *J. Appl. Mech.*, 55, 65–72.

Chow, C.L., and Wang, J., (1987). An anisotropic theory of continuum damage mechanics for ductile fracture. *Eng. Frac. Mech.*, 27, 547–558.

Christopher, A., Mukul, K., and Wayne, K.E., (2003). Analysis of grain boundary networks and their evolution during grain boundary engineering. *Acta Mater.*, 51, 687–700.

Cordebois, J.P., and Sidorff, F., (1979). Damage Induced Anisotropy. *Colloque Euromech*, 115, Villard de Lans.

Davidge, R.W., (1979). *Mechanical Behavior of Ceramics*. Cambridge Univ. Press, Cambridge, UK.

Delaplace, A., Pijaudier-Cabot, G., and Roux, S., (1996). Localization of damage in discrete models and consequences on continuum modeling. *J. Mech. Phys. Solids*, Vol.44, Issue 1, 99-136.

Espinosa, H.D., and Zavatteri, P.D., (2003). A grain level model for the study of failure initiation and evolution in polycrystalline brittle materials. Part I: Theory and numerical implementation. *Mech. Mater.*, 35, 333–364.

Family, F., and Vicsek, T., (1985). Scaling of the active zone in the Eden process on percolation networks and the ballistic deposition model. *J. Phys. A*, 18, L75-L81.

Gerstle, K.H., Aschl, H., Bellotti, R., Bertacchi, P., Kotosovos, M.D., Ko, H.-Y., Linse, D., Newman, J.B., Rossi, P., Schickert, G., Taylor, M.A., Traina, L.A., Winkler, H., and Zimmerman, R. M.,(1980). Behavior of concrete under multiaxial stress states. *J. Engng. Mech. Div. ASCE 06*, 1383.

Goranson, U.G., (1993). Damage tolerance – facts and fiction. Boeing Commercial Airplane Group, U.S.A, 14th Plantema Memorial Lecture Presented at the *17th Symposium of the International on Aeronautical Fatigue*, Stockholm, Sweden.

Gouyet, J.-F., (1996). *Physics and Fractal Structures*. Masson, Paris.

Gurtin, M.E., (1975). *Handbuck der Physics*. Ed.: Truesdell, C., Vol. IV.

Hansen, A., Roux, S., and Herrmann, H.J., (1989). Rupture of central-force lattices. *J. Phys. France*, Vol. 50,733-744.

Hansen, A., Roux, S., and Hinrichen, E.L., (1990). Annealed Model for Breakdown Process. *Europhys. Lett.*, Vol. 13, 517-522.

Hansen, A., and Roux, S., (2001). Statistics tool box for damage and fracture. In: D. Krajcinovic and J. van Mier, Eds., *Damage and Fracture in Disordered Materials*, 17-102. Springer, Wien.

Hayhurst, D.R., (1972). Creep rupture under multiaxial state of stress. *J. Mech. Phys. Solids*, Vol 20, 381-392.

He, H., and Thorpe., M.F., (1985). Elastic properties of glasses. *Phys. Rev. Lett.*, Vol.54, 2107-2110.

Horii, H., and Nemat-Nasser, S., (1983). Overall moduli of solids with microcracks: load-induced anisotropy. *J. Mech. Phys. Solids*, 31, 155-171.

Hoover, W.G., (1986). *Molecular Dynamics*. Springer-Verlag, Berlin.

Hudson, J.A., Brown, E.T., and Fairhurst, C., (1971). Shape of the complete stress-strain curve for rock. In: *Stability of Rock Slopes. Proc. 13th Symp. on Rock Mech.*, E.J. Cording, Ed., pp. 773-795, Univ. of Illinois Urbana Press.

Jaeger, D., Schischka, J., Bagdahn, J., and Jaeger, R., (2009). Tensile testing of individual ultrathin electrospun Poly(L-lactic acid) fibers. *J. Appl. Polymer Sci.*, vol.114, 3774–3779.

Jagota, A., and Scherer, G.W., (1993). Viscosities and sintering rates of a two-dimensional granular composite. *J. Am. Ceram. Soc.*, Vol. 76, 3123-3135.

Ju, J.W., (1990). Isotropic and anisotropic damage variables in continuum damage mechanics. *J. Eng. Mech.*, 116, 2764–2770.

Kachanov, L.M., (1958). On the time to failure under creep conditions. *Izv. AN SSSR*, Otd. Tekhn. Nauk, No. 8, 26-31.

Kloster, M., Hansen, A., and Hemmer, P.C., (1997). Burst avalanches in solvable models of fibrous materials. *Phys. Rev. E*, vol. 56, 2615.

Krajcinovic, D., and Fonseka, G.U., (1981). The continuous damage theory of brittle materials. Part 1: general theory. *J. Appl. Mech.*, 48, 809–815.

Krajcinovic, D., and Fonseka, G.U., (1981). The continuous damage theory of brittle materials. Part 2: uniaxial and plane response modes. *J. Appl. Mech.*, 48, 816–824.
Krajcinovic, D., and Silva, M.A.G., (1982). Statistical aspects of the continuous damage theory. *Int. J. Solids Structures*, Vol. 18, 551-562.
Krajcinovic, D., (1983). Constitutive equations for damaging materials. *J. Appl. Mech.*, 50, 355–360.
Krajcinovic, D., and Basista, M., (1991). Rupture of central-force lattices revisited. *J. Phys. I*, Vol. 1, 241-245.
Krajcinovic, D., Lubarda, V., and Sumarac, D., (1993). Fundamental aspects of brittle cooperative phenomena – effective continua models. *Mech. Mater.*, 15, 99-115.
Krajcinovic, D., (1996). *Damage Mechanics*. North-Holland, Amsterdam, The Nederlands.
Krajcinovic, D., and Vujosevic, M., (1998). Strain localization - short to long correlation length transition. *Int. J. Solids Structures*, Vol. 35, Nos. 31-32, 4147-4166.
Krajcinovic, D., and Rinaldi, A., (2005). Thermodynamics and statistical physics of damage processes in quasi-ductile solids. *Mech. Mater.*, Vol 37. No 2-3: 299-315.
Krajcinovic, D., and Rinaldi, A., (2005). Statistical damage mechanics - 1. Theory. *J.Appl.Mech.*,72, 76-85.
Kreher, W. Pompe, W., (1989). Internal stress in heterogeneous solids. Akademie Verlag, Berlin.
Lemaitre, J., (1985). A continuous damage mechanics model for ductile fracture. *J. Eng. Mater. Tech.*, 107, 83–89.
Lemaitre, J., (1985). Coupled elasto-plasticity and damage constitutive equations. *Comp. Meth. Appl. Mech. Eng.*, 51, 31–49.
Mandelbrot, B.B., (1977). *Fractals: Form, Chance and Dimension*. Freeman, San Francisco.
Mariano, P.M., and Stazi, F.L., (2001). Strain localization in elastic microcracked bodies. *Comp. Meth. Appl. Mech. Eng.*; 190, 5657–77.
Mastilovic, S., and Krajcinovic, D., (1999). Statistical models of brittle deformation: Part II: computer simulations. *Int. J. Plasticity,* 15, 427-456.
Mastilovic, S., Krajcinovic, D., and Rinaldi, A., (2008). Ordering effect of kinetic energy on dynamic deformation of brittle solids. *Mech. Mater.*, 40, 407–417.
Mastilovic, S., (2008). *Investigation Of Dynamic Behavior Of Brittle Solids By Discrete Systems*. ISBN 978-86-85175-02-2 .
Moes, N., Dolbow, J., and Belytschko, T., (1999). A finite element method for crack growth without remeshing. *Int. J. Numer. Meth. Eng.*, 46, 131–150.
Monette, L., and Anderson, M.P., (1994). Elastic and fracture properties of the two-dimensional triangular and square lattices. *Modeling Simul. Mater. Sci. Eng.*, Vol. 2, 53-66.
Montgomery, D.C., Peck, E.A, and Vining, G.G., (2001). *Introduction to Linear Regression Analysis*. Wiley.
MRS Bullettin, (2008). Three-dimensional material science. *MRS Bullettin*, Vol.33, 6.
Murakami S., (1988). Mechanical modeling of material damage. *J. Appl. Mech.*, 55, 280–6.
Neville, A.M., (1981). *Properties of Concrete*. Prentice Hall. 3rd edition.
Okabe, A., Boots, B., Sugihara, K., and Chiu, S.N., (1999). *Spatial Tessellations, Concepts and Applications of Voronoi Diagrams*. John Wiley & Sons, New York, NY.

Ott, E., (2002). Chaos in dynamical systems, Cambridge, U.K., New York, Cambridge University Press, 2nd Ed.

Rice, J.R., (1978). Thermodynamics of the quasi-static growth of Griffith cracks. *J. Mech. Phys. Solids*, 26, 61-78.

Rinaldi, A., (2004). *Bridging the Scales with Statistical Damage Mechanics*. PhD Dissertation, Arizona State University.

Rinaldi, A., Peralta, P., Krajcinovic, D., and Lai, Y.C., (2006). Prediction of fatigue properties with discrete damage mechanics. *Int. J. Fatigue*, 28, 1069–1080.

Rinaldi, A., Mastilovic, S., and Krajcinovic, D., (2006). Statistical damage mechanics - 2. Constitutive relations. *Journal Theoretical and Applied Mechanics*, special issue on Damage Mechanics: Experiments and Modeling, 44, 3, 585-602.

Rinaldi, A., and Lai, Y-C., (2007). Damage theory of 2D disordered lattices: energetics and physical foundations of damage parameter. *Int. J. Plasticity*, 23, 1796-1825.

Rinaldi, A., Krajcinovic, D., and Mastilovic, S., (2007). Statistical damage mechanics and extreme value theory. *Int.J. Damage Mechanics*, 16,1.

Rinaldi, A., Krajcinovic, D., Peralta, P., and Lai, Y.C., (2008). Modeling polycrystalline microstructures with lattice models: a quantitative approach. *Mech. Mater.*, 40, 17-36.

Rinaldi, A., (2009). A rational model for 2D disordered lattices under uniaxial loading. *Int. J. Damage Mech.*, Vol. 18, No. 3, 233-257.

Scher, H., and Zallen, R., (1970). Critical density in percolation processes. *J. Chem. Phys.*, Vol. 53, 3759-3761.

Simsch, G., (1993). Influences on the stress-strain behavior of NSC (normal-strength concrete) and HSC (high strength concrete). *Darmstadt Concrete*, Vol. 8, 43-55.

Sornette, D., (1989). Elasticity and failure of a set of elements loaded in parallel. *J. Phys. A*, vol. 22, L243.

Toussaint, R., (2005). Statistical mechanics of interacting fiber bundles. *Proceedings of the 11th International Congress on fracture* CD Rom (Ed.) .

Tvergaard V., (1990). Material failure by void growth to coalescence. *Advances in Appl. Mech.*, 27, 83–151.

Van Mier, J.G.M., (2000). Measurement of damage parameters of brittle disordered media like concrete and rock. In: *Damage And Structure Of Disordered Materials*. Edited by D. Krajcinovic and J. Van Mier, pp. 135-217, Springer Wien.

Vicsek, T., (1989). *Fractal Growth Phenomena*. World Scientific.

Voyiadjis, G.Z., and Kattan, P.I., (1992). A plasticity-damage theory for large deformation of solids—I. theoretical formulation. *Int. J. Eng. Sci.*, 30, 1089–1108.

Weibull, W., (1939). A statistical theory of the strength of materials. *Proc. Royal Swedish Academy of Eng. Sci.*, Vol. 151, 1-45.

Weiner, J.H., (1983). *Statistical Mechanics of Elasticity*. Wiley - Interscience Publ., J. Wiley & Sons, New York, NY.

Zallen, R., (1983). *The Physics of Amorphous Solids*. J.Wiley & Sons, New York, NY.

PART III

Application of the Thermodynamical Theory of Elasto–Viscoplasticity in Modern Manufacturing Processes

by

Piotr Perzyna

Application of the Thermodynamical Theory of Elasto–Viscoplasticity in Modern Manufacturing Processes

Piotr Perzyna
Institute of Fundamental Technological Research,
Polish Academy of Sciences,
Pawińskiego 5B, 02–106 Warsaw, Poland
pperzyna@ippt.gov.pl

Abstract

The main objective of the lecture is to show broad application of the thermodynamic theory of elasto–viscoplasicity for the description of important problems in modern manufacturing processes, and particularly for meso-, micro-, and nano-mechanical issues. This description is particularly needed for the investigation by using the numerical methods how to avoid unexpected plastic strain localization and localized fracture phenomena in new manufacturing technology.

In the first part the development of thermo–elasto–viscoplastic constitutive model of a material which takes into consideration the induced anisotropy effects as well as observed contribution to strain rate effects generated by microshear banding is presented.

The model is developed within the thermodynamic framework of a unique, covariance constitutive structure with a finite set of the internal state variables. A set of internal state variables consists of one scalar and two tensors, namely the equivalent inelastic deformation $\in^p$, the second order microdamage tensor $\boldsymbol{\xi}$ with the physical interpretation that $(\boldsymbol{\xi} : \boldsymbol{\xi})^{1/2} = \xi$ defines the volume fraction porosity and the residual stress tensor (the back stress) $\boldsymbol{\alpha}$. The equivalent inelastic deformation $\in^p$ describes the dissipation effects generated by viscoplastic flow phenomena, the microdamage tensor $\boldsymbol{\xi}$ takes into account the anisotropic intrinsic microdamage mechanisms on internal dissipation and the back stress tensor $\boldsymbol{\alpha}$ aims at the description of dissipation effects caused by the second kind of induce anisotropy approximated by the kinematic hardening mechanism.

To describe the contribution to strain rate effects generated by microshear banding we propose to introduce certain scalar function which affects the relaxation time T_m in the viscoplastic flow rule. The relaxation time is used as a regularization parameter. Fracture criterion based on the evolution of the anisotropic intrinsic microdamage is formulated. The fundamental features of the proposed constitutive theory have been carefully discussed.

The objective of the second part is to discuss very efficient procedure of the numerical investigation of localized fracture in inelastic solids generated by impact–loaded adiabatic processes. Particular attention is focused on the proper description of a ductile mode of fracture propagating along the shear band for high impact velocities. This procedure of investigation is based on utilization the finite difference and finite element methods for regularized thermo-elasto-viscoplastic model of damaged material.

Particular problems have been considered as follows:

(i) Localization and localized fracture phenomena in inelastic solids under cyclic dynamic loadings;

(ii) Numerical analysis of macrocrack propagation along a bimaterial interface under dynamic loadings processes;

(iii) Numerical investigation of dynamic shear bands in inelastic solids as a problem of mesomechanics;

(iv) Numerical investigation of localized fracture phenomena in inelastic solids generated by impact–loaded adiabatic processes.

The physical and experimental motivation for each problem have been presented. The identification procedure for the material functions and constants involved in the constitutive equations is developed basing on the experimental observations.

Qualitative comparison of numerical results with experimental observation data has been presented. The numerical results obtained have proven the usefulness of the thermo–elasto–viscoplastic theory in the investigation of dynamic shear band propagations and localized fracture.

1. Prologue

Experimental observations concerning investigation of dynamic loading processes have shown that formation of microshear bands influences the evolution of microstructure of material. We can conclude that microshear banding contributes to viscoplastic strain rate effects. On the other hand analysis of recent experimental observations concerning investigations of fracture phenomena under dynamic loading processes suggests that there are two kinds of induced anisotropy: (i) the first caused by the residual type stress produced by the heterogeneous nature of the finite plastic deformation in polycrystalline solids; (ii) the second the fracture induced anisotropy generated by the evolution of the microdamage mechanism. It is noteworthy to stress that both these induced anisotropy effects are coupled.

In recent years there has been great activity in the field of continuum mechanics to describe anisotropic effects on fracture phenomena. This trend in research has been particularly valuable in the continuum damage mechanics, cf. Boehler (1990) and Voyiadjis, Ju nad Chaboche (1998). Analysis of recent experimental results concerning investigations of fracture phenomena under dynamic loading processes suggests that there are three main reasons for anisotropic effects: (i) The strain induced anisotropy is caused by the residual type stresses which result from the heterogeneous nature of the plastic deformation in polycrystalline materials. Experimental evidence indicates that yield surfaces exhibit anisotropic hardening. Subsequent yield surfaces are both translated and deformed in stress space. In phenomenological description this kind of anisotropy is modelled by the shift of the yield surface in stress space. This shift of the yield surface might be described by the residual stress tensor (the back stress) $\boldsymbol{\alpha}$ introduce as the internal state variable. (ii) The anisotropy caused by the formation of localized shear bands and the evolution of microdamage process along the localized regions. (iii) The anisotropy induced by the evolution of the intrinsic microdamage process. Experimental observations have shown that in the microdamage process the generated anisotropy is a consequence of rather random phenomena connected with some directional property of the formation of microcracks. This anisotropic effect is very much affected by the crystallographic structure of a material as well as by small fluctuations (due to wave interactions) of the applied stress tensor at particular point of a body during dynamic process.

The main objective of Chapter 2 is the development of thermo–elasto–viscoplastic constitutive model of a material which, takes into consideration observed contribution to strain rate effects generated by microshear banding, as well as both mentioned earlier induced anisotropy effects.

The model is developed within the thermodynamic framework of the rate type covariance constitutive structure with a finite set of the internal state variables. A set of internal state variables consists of one scalar and two tensors, namely equivalent inelastic deformation $\in^p$, the second order microdamage tensor $\boldsymbol{\xi}$ with the physical interpretation that $(\boldsymbol{\xi} : \boldsymbol{\xi})^{1/2} = \xi$ defines the volume fraction porosity and the residual stress (the back stress) tensor $\boldsymbol{\alpha}$. The equivalent inelastic deformation $\in^p$ describes the dissipation effects generated by viscoplastic flow phenomena, the microdamage tensor $\boldsymbol{\xi}$ takes into account the anisotropic intrinsic microdamage mechanisms on internal dissipation and the back stress tensor $\boldsymbol{\alpha}$ aims at the description of dissipation effects caused by the kinematic hardening, cf. Perzyna and Voyiadjis (2005).

To describe suitably the influence of both induced anisotropy effects and the stress triaxiality the new kinetic equations for the microdamage tensor $\boldsymbol{\xi}$ and for the back stress tensor $\boldsymbol{\alpha}$ are proposed. The relaxation time is used as a regularization parameter. To describe the contribution to strain rate effect generated by microshear banding we propose to introduce certain scalar function which affects the relaxation time T_m in the viscoplastic flow equation (cf. Nowak, Perzyna and Pęcherski(2007)). By assuming that the relaxation time tends to zero the thermo–elasto–plastic (rate independent) response of the damage material can be obtained. Fracture criterion based on the evolution of the anisotropic intrinsic microdamage is formulated. Very important part of this constitutive modelling is the identification procedure for the material functions and constants involved in the constitutive equations proposed. This procedure has been developed in the papers Glema, Łodygowski, Nowak, Perzyna and Voyiadjis (2005) and Glema, Łodygowski, Perzyna and Sumelka (2006).

Constitutive modelling for dynamic cyclic loadings is developed in Chapter 3.

Recent experimental observations for cycle fatigue damage mechanics at high temperature and dynamic loadings of metals suggest that the intrinsic microdamage process does very much dependent on the strain rate and the wave shape effects and is mostly developed in the regions where the plastic deformation is localized. The microdamage kinetics interacts with thermal and load changes to make failure of solids a highly rate, temperature and history dependent, nonlinear process. Such effects have been observed by Sidey and Coffin (1979). They investigated the oxygen–free high conductivity (OFHC) copper at 673 K using unequal strain rates to produce the wave shape. For OFHC copper, the fatigue lifetimes decreased by an order of magnitude as the tensile–going strain rate was reduced from 1.7×10^{-3} s^{-1} to 1.7×10^{-5} s^{-1} with the constant cyclic period. It was observed that accompanying this reduction in lifetime, the fracture mode changed from a transgranular fracture for the fast–slow wave shape to an intergranular single–crack fracture for equal ramp rates to interior cavitation for the slow–fast test.

A general constitutive model of elasto–viscoplastic damaged polycrystalline solids is developed within the thermodynamic framework of the rate type covariance structure with finite set of the internal state variables. A set of the internal state variables is assumed and interpreted such that the theory developed takes account of the effects as follows: (i) plastic non–normality; (ii) plastic strain induced anisotropy (kinematic hardening); (iii) softening generated by microdamage mechanisms (nucleation, growth and coalescence of microcracks); (iv) thermomechanical coupling (thermal plastic softening and thermal expansion); (v) rate sensitivity; (vi) plastic spin. To describe suitably the time and temperature dependent effects observed experimentally and the accumulation of the plastic deformation and damage during dynamic cyclic loading process the kinetics of microdamage and the kinematic hardening law have been modified. The relaxation time is used as a regularization parameter. By assuming that the relaxation time tends to zero, the rate independent elastic–plastic response is obtained.

Fracture criterion based on the evolution of porosity is used.

The main objective of Chapter 4 is the investigation of localization and localized fatigue fracture phenomena in thermo–viscoplastic flow processes under cyclic dynamic loadings.

The viscoplastic regularization procedure assures the stable integration algorithm by using the finite difference method. Particular attention is focused on the well–posedness of the evolution problem (the initial–boundary value problem) as well as on its numerical solutions.

The identification procedure is developed and the investigation of stability and convergence of the numerical method based on the finite difference approximation is presented.

Utilizing the finite difference method for regularized elasto–viscoplastic model, the numerical investigation of the three–dimensional dynamic adiabatic deformation in a particular body under cyclic loading condition is presented. Particular example has been considered.

A dynamic adiabatic cyclic loading process for a thin steel plate with sharp notch is investigated. To the upper edge of the plate is applied cyclic constraint realized by rigid rotation of the edge of the plate while the lower edge is supported rigidly. Small localized region, distributed asymmetrically near the tip of the notch, which undergoes significant deformation and temperature rise has been determined. Its evolution until occurrence of fatigue fracture has been simulated. The propagation of the macroscopic fatigue damage crack within the material of the plate is investigated. It has been found that the length of the macroscopic fatigue damage crack distinctly depends on the wave shape of the assumed loading cycle.

The results obtained are in accord with the experimental observations performed by Sidey and Coffin (1979).

In Chapter 5 numerical analysis of macrocrack propagation along a bimaterial interface under dynamic loading processes is presented. A general constitutive model of elasto–viscoplastic damaged polycrystalline solids is used.

The identification procedure is developed basing on the experimental observations.

The finite difference method for regularized elasto–viscoplastic model is used. The edge–cracked bimaterial specimen is considered. In the initial configuration, the height of the specimen is equal to 30 cm, width is 12.5 cm and the length of the initial crack is equal to 2.5 cm. The length of the boundary over which impact is applied is equal to 5 cm, the rise time is fixed at 0.1 μs and the impact velocity is varied. The impact area is localized symmetrically or asymmetrically to the shorter axis of the specimen (symmetrical line of the interface band). Considering the available data of recent experimental observation (cf. Rosakis, Samudrala and Coker (1999)) that have been carried out for relatively thin specimen the plane stress or plane strain conditions are assumed. The material of the specimen is AISI 4340 steel, while PMMA is the cohesive band, both modelled by thermo–elasto–viscoplastic constitutive equation with effects of isotropic hardening and softening generated by microdamage mechanisms and thermomechanical coupling. Fracture criterion based on the evolution of microdamage is assumed. Both, isothermal and adiabatic processes are considered.

Particular attention is focused on the investigation of the interactions and reflections of stress waves and the influence of these waves on the propagation of macrocrack within the interface band. The propagation of the macroscopic crack within the material band for both symmetrical and asymmetrical impact cases has been investigated. It has been found that macrocrack–tip speeds vary from the shear wave speed to the dilatational wave speed of the material and is higher than the Rayleigh surface wave speed. This result is in accord with the experimental observations performed by Rosakis, Samudrala and Coker (1999).

In technological processes fracture can occur as a result of an adiabatic shear band localization generally attributed to a plastic instability generated by thermal softening and intrinsic microdamage mechanisms within a material and thermal softening during plastic deformation.

Recent experimental observations have shown that the shear band procreates in a region of a body deformed where the resistance to plastic deformation is lower and the predisposition for localized shear band formation is higher. It has been found experimentally that in dynamic processes the shear band regions behave differently than adjacent zones. Within the shear band region the deformation process is characterized by very large strains (shear band strains over 100%) and very high strain rates ($10^4 - 10^6$ s^{-1}). The strain rate sensitivity of a material becomes very important feature of the shear band region.

The main objective of Chapter 6 is to discuss very efficient procedure of the numerical investigation of the propagation of shear bands in inelastic solids generated by impact–loaded adiabatic processes.

This procedure of investigation is based on utilization the finite element method and ABAQUS system for regularized thermo–elasto–viscoplastic constitutive model of damaged material.

A general constitutive model of thermo–elasto–viscoplastic polycrystalline solids with a finite set of internal state variables is used. The set of internal state variables consists of one scalar, namely equivalent inelastic deformation. The equivalent inelastic deformation can describe the dissipation effects generated by viscoplastic flow phenomena. The relaxation time is used as a regularization parameter.

The identification procedure for the material functions and constants involved in the constitutive equations is developed basing on the experimental observations of adiabatic shear bands in a HY–100 steel presented by Cho, Chi and Duffy (1988).

As a numerical example we consider dynamic shear band propagation in an asymmetrically impact–loaded prenotched thin plate. The plate is made of a HY–100 steel. A notch (260 μm wide) is further extended by 2 mm and is situated symmetrically on the edge. The specimen is supported at four points and is initially stationary. The constant velocity $V_0 = 38$ m/s is imposed for projectile. The projectile comes into contact with the specimen over the width of 50 mm.

The impact loading is simulated by a velocity boundary condition which are the results of dynamic contact problem. The velocity imposed in specimen in front of projectile increases during the process. The separation of the projectile from the specimen, resulting from wave reflections within the projectile and the specimen, occurs in the phenomenon. All surface areas have traction free boundary conditions except where the velocity boundary condition is applied. We idealize the initial boundary value problem observed experimentally in Guduru, Rosakis and Ravichandran (2001), by assuming the velocity boundary condition and different material of the specimen. The discretization parameters are assumed in such a way, to solve the problem of mesomechanics properly. The dimension of the accepted mesh is of the order 20 μm. A thin shear band region of finite width which undergoes significant deformation and temperature rise has been determined. Shear band advance, shear band velocity and the development of the temperature field as a function of time have been determined. Qualitative comparison of numerical results with experimental observation data has been presented. The numerical results obtained have proven the usefulness of the thermo–elasto–viscoplastic theory in the numerical investigation of dynamic shear band propagation.

The main objective of Chapter 7 is to discuss very efficient procedure of the numerical investigation of localized fracture in inelastic solids generated by impact–loaded adiabatic processes. Particular attention is focused on the proper description of a ductile mode of fracture propagating along the shear band for high impact velocities.

This procedure of investigation is based on utilization the finite difference for regularized thermo–elasto–viscoplastic constitutive model of damaged material.

A general constitutive model of thermo–elasto–viscoplastic damaged polycrystalline solids with a finite set of internal state variables is used. The set of internal state variables consists of two scalar, namely equivalent inelastic deformation and volume fraction porosity. The equivalent inelastic deformation can describe the dissipation effects generated by viscoplastic flow phenomena and the volume fraction porosity takes into account the microdamage evolution effects. The relaxation time is used as a regularization parameter. Fracture criterion based on the evolution of porosity is assumed.

The identification procedure for the material functions and constants involved in the constitutive equations is developed basing on the experimental observations of adiabatic shear bands in an AISI 4340 steel presented by Chakrabarti and Spretnak (1975), cf. also Cho, Chi and Duffy (1988) and Chi et al. (1988).

As a numerical example we consider dynamic shear band propagation and localized fracture in an asymmetrically impact–loaded prenotched thin plate. The plate is made of an AISI 4340 steel. A notch (50 mm wide) is situated symmetrically on the edge. The constant velocity $V_0 = 20$ m/s is imposed for projectile. The projectile comes into contact with the specimen over the width of 50 mm.

The impact loading is simulated by a velocity boundary condition which are the results of dynamic contact problem. The velocity imposed in specimen in front of projectile increases during the process. The separation of the projectile from the specimen, resulting from wave reflections within the projectile and the specimen, occurs in the phenomenon. All surface areas have traction free boundary conditions except where the velocity boundary condition is applied. We idealize the initial boundary value problem observed experimentally in Guduru, Rosakis and Ravichandran (2001), by assuming the velocity boundary condition and different material of the specimen. The discretization parameters are assumed in such a way, to solve the problem of mesomechanics properly. The dimension of the accepted mesh is of order 40 μm. A thin shear band region of finite width which undergoes significant deformation and temperature rise has been determined. Its evolution until occurrence of final fracture has been simulated. Shear band advance as a function of time, and the evolution of the Mises stress, equivalent plastic deformation, temperature, the microdamage and the crack path in the fracture region have been determined. Qualitative comparison of numerical results with experimental observation data has been presented. The numerical results obtained have proven the usefulness of the thermo–elasto–viscoplastic theory in the numerical investigation of dynamic shear band propagation and localized fracture.

2. The Thermodynamical Theory of Elasto–Viscoplasticity

2.1 Physical Foundations

2.1.1 Physical origin of elastic–viscoplastic response of solids

The high – rate deformation of face–centered cubic (f.c.c.) metals, such as copper, aluminum, lead and nickel has been recently extensively studied (cf. review paper by Follansbee (1986)). It has been shown that the apparent strain rate sensitivity of f.c.c. metals has two origins: that associated with the finite velocity of dislocations, and that connected with the evolution of the dislocation substructure. The first of these two components – the instantaneous rate sensitivity – is related to the wait – times associated with thermally activated dislocation motion. The second component has more to do with the relative importance of dislocation generation and annihilation at different strain rates, and shall be referred to as the strain rate history effect.

The rate and temperature dependence of the flow stress of metal crystals can be explained by different physical mechanisms of dislocation motion. The microscopic processes combine in various ways to give several groups of deformation mechanisms, each of which can be limited to the particular range of temperature and strain rate changes, cf. Perzyna (2005).

It will be profitable for further considerations to discuss the interaction of the thermally activated and phonon dumping mechanisms.

2.1.2 Interaction of the thermally activated and phonon damping mechanisms

If a dislocation is moving through the rows of barriers, then its velocity can be determined by the expression (cf. Fig.2.1)

$$\mathrm{v} = AL^{-1}/(t_S + t_B), \tag{2.1}$$

where AL^{-1} is the average distance of dislocation movement after each thermal activation, t_S is the time a dislocation spent at the obstacle, and t_B is the time of travelling between the barriers.

The shearing rate in single slip is given by the relationship (cf. Kumar and Kumble (1969), Teodosiu and Sidoroff (1976) and Perzyna (1977, 1988, 2002, 2005))

$$\dot{\epsilon}^p = \frac{1}{T_{mT}} \langle \exp\{U[(\tau - \tau_\mu)Lb]/k\vartheta\} + BAL^{-1}\nu/(\tau - \tau_B)b\rangle^{-1} \tag{2.2}$$

where U is the activation energy (Gibbs free energy), b denotes the Burgers vector, k is the Boltzman constant, ϑ actual absolute temperature, B denotes the dislocation drag coefficient, ν is the frequency of vibration of the dislocation, τ_μ denotes the athermal stress and τ_B is attributed to the stress needed to overcome the forest dislocation barriers to the dislocation motion and is called the back stress.

The two relaxation times

$$\frac{1}{T_{mT}} \frac{b\tau_B}{BAL^{-1}\nu} = \frac{\alpha b^2 \tau_B}{B} = \frac{1}{T_{mD}}, \tag{2.3}$$

and two effective resolved shear stresses

$$\tau_T^* = \tau - \tau_\mu \quad \text{and} \quad \tau_D^* = \tau - \tau_B \tag{2.4}$$

are separately defined for the thermally activated and phonon damping mechanisms, respectively.

If the time t_B taken by the dislocation to travel between the barriers in a viscous phonon medium is negligible when compared with the time t_S spent at the obstacle, then

$$\mathrm{v} = \frac{AL^{-1}}{t_S} \tag{2.5}$$

and we can focus our attention on the analysis of the thermally activated process.

When the ratio t_B/t_S increases then the dislocation velocity (2.1) can be approximated by the expression

$$\mathrm{v} = \frac{AL^{-1}}{t_B} \tag{2.6}$$

for the phonon damping mechanism.

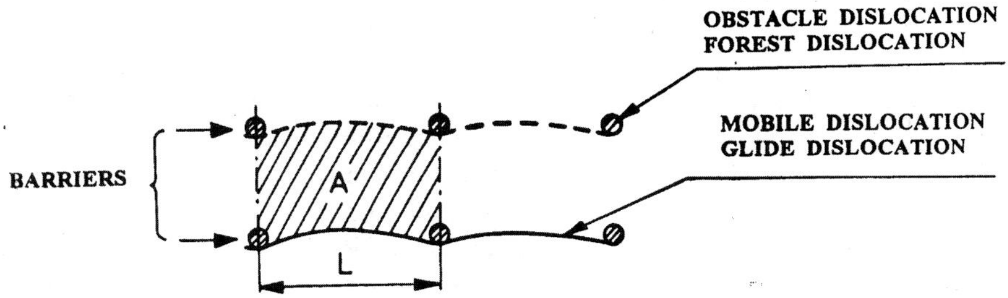

Figure 2.1: Intersection of forest dislocations

It is noteworthy to stress that the relaxation time varies when the mechanism of the dislocation motion changes from the thermally activated to the phonon damping.

2.2 Experimental Motivation for Microshear Banding Effects

2.2.1 Experimental observations of single crystal behaviour

Rashid et al. (1992) performed an experimental study of highly heterogeneous deformations in copper single crystals to investigate the importance of the dislocation substructure. Their experimental results are mainly qualitative in nature, and include optical photographs and micrographs of the deformed specimens, and scanning and transmission electromicrographs of the substructure. The results obtained seem to support the earlier findings related to the strain rate history dependence of the substructure evolution. The higher total dislocation density observed in the notch region of the dynamically deformed specimen, as compared to the same region in the quasi–statically deformed specimen, reflects the higher shear stress attained in the dynamic case.

When ductile single crystals of metals are finitely deformed, they display highly heterogeneous deformation, e.g. when crystals are stretched in tension, they can neck and then develop macroscopic bands of localized shearing, cf. Fig. 2.2.

Experimental observations of Chang and Asaro (1980, 1981), Lisiecki et al. (1982) and Spitzig (1981) for copper, aluminum – copper and nitrogenated Fe–Ti–Mn single crystals investigated in uniaxial tension tests have shown that in the first stage of the process, a crystal specimen undergoes uniform extension in single slip. At the point when the load–engineering strain trajectory reaches its maximum, a crystal specimen exhibits slight amounts of very diffuse necking, cf. Figs. 2.3, 2.4 and 2.5. The neck is usually symmetric in shape indicating that double slip is operative within it, cf. Lisiecki et al. (1982). At this stage of the tensile process, the gross plastic deformations are localized to the diffusely necked region and the thermomechanical coupling effects begin to play a crucial role. That is why in this region of the specimen the tensile process has to be considered as adiabatic. With continued extension the macroscopic, adiabatic shear bands have soon developed within the diffusely necked region.

It has been experimentally observed that at the inception of the macroscopic, adiabatic shear band, the direction of the band is slightly different from the detected coarse slip bands or slip traces. In other words, the

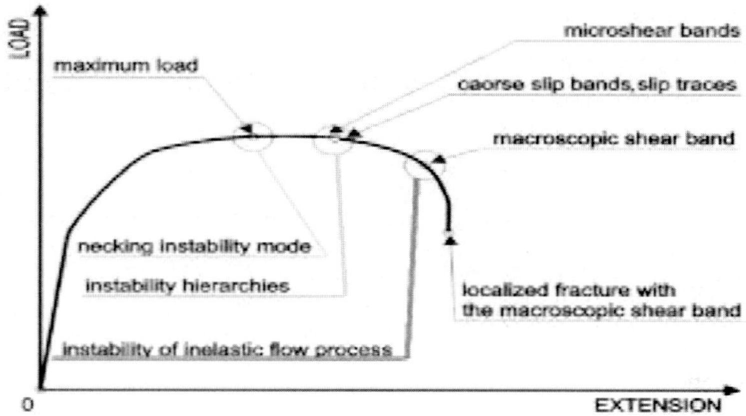

Figure 2.2: One dimensional loading tests

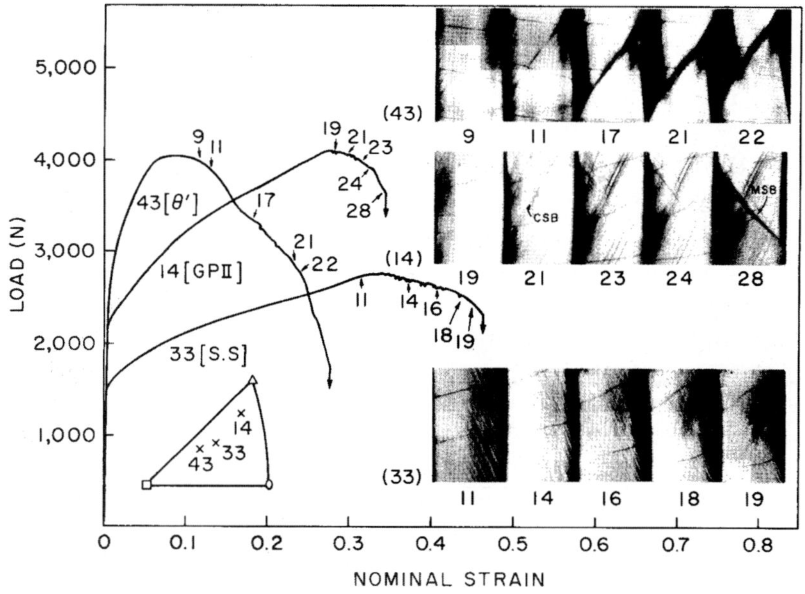

Figure 2.3: Load versus engineering strain curves for various ageing treatments. Numbered photos correspond to the indicated points on the load–strain curves (after Chang and Asaro (1981))

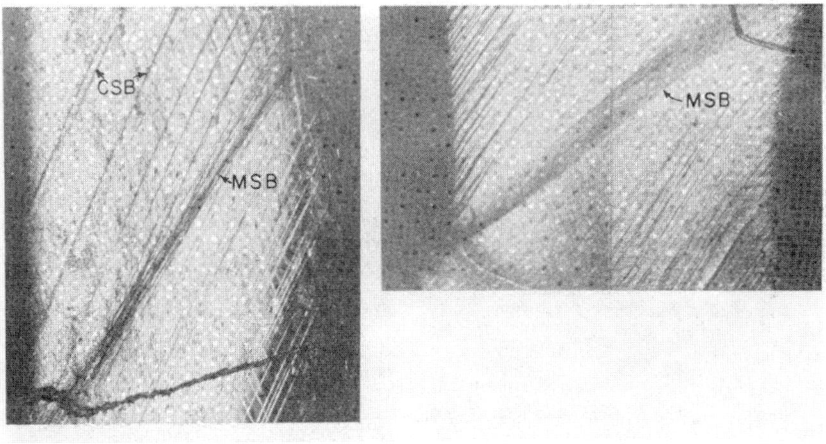

Figure 2.4: Coarse slip band (CSB) and macroscopic shear bands (MSB) in (a) GPII tested at 77 K and (b) a θ' strengthened crystal tested at 298 K (after Chang and Asaro (1981))

macroscopic shear bands are not aligned with the active slip systems in the crystal's matrix but are misaligned by angle δ, cf. Figs. 2.4 and 2.5.

2.2.2 Heuristic considerations

From the analysis of the experimental investigations of localized shearing in single crystals performed by Chang and Asaro (1980, 1981), Spitzig (1981), Lisiecki et al. (1982) and Rashid et al. (1982) we can follow the events in the order in which things naturally happen within a gauge length of the specimen during the uniaxial test, cf. Figs 2.2 and 2.6.

In the first stage of the adiabatic inelastic flow process a crystal specimen (a system) undergoes uniform extension and slip takes place. When control parameters are changed over a wide range, our system may run through a hierarchy of instabilities and accompanying cooperative phenomena. When we look at a microscopic level we observe that a crystal is well ordered, and is self organized in microscopic shear band pattern.

At the point when the load–engineering strain trajectory reaches its maximum, i.e. when the criterion of the onset of the localization by necking mode is satisfied, a crystal specimen exhibits slight amount of very diffuse symmetric necking.

With continued extension the instability of inelastic flow process takes

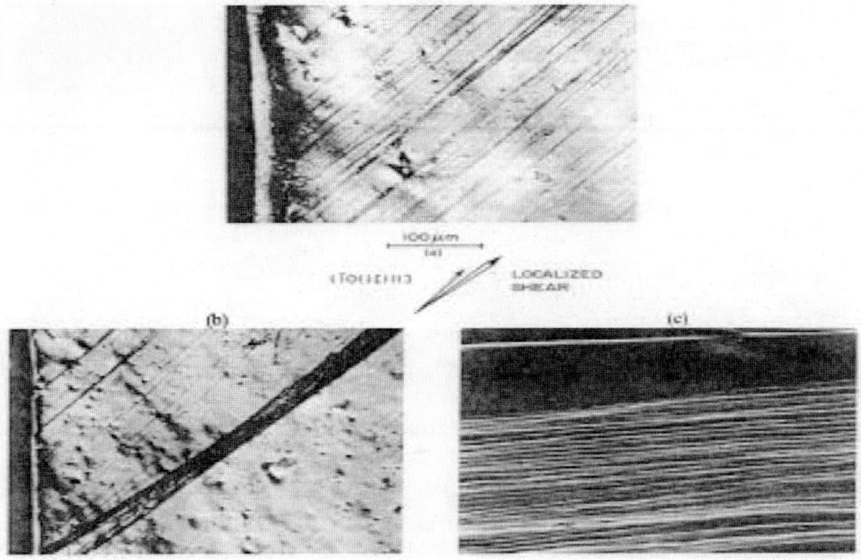

Figure 2.5: Slip traces and localized shear on the surface of nitrogenated Fe–Ti–Mn crystal of orientation D deformed at 295 K: (a) initial deformation until necking began, (b) subsequent deformation after removal of neck and localized shear bands from initial deformation, (c) slip traces within localized shear band in (b) (after Spitzig (1981))

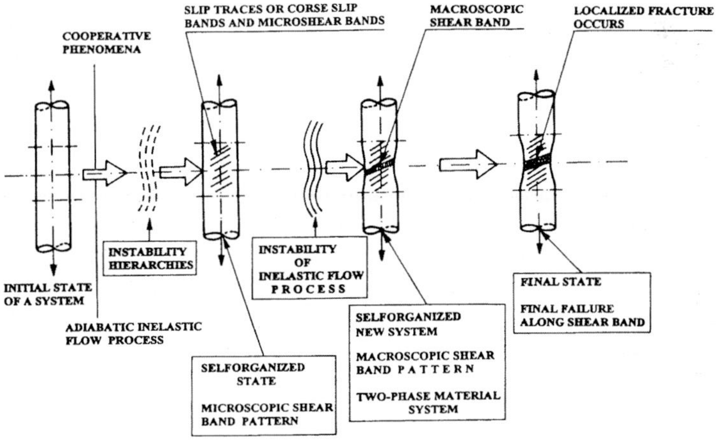

Figure 2.6: Subsequent states of adiabatic inelastic flow process of single crystal

place and we observe on macroscopic level the formation of adiabatic shear band pattern within the diffusely necked region. A system is self organized to a new – two–phase material system (cf. Fig. 2.6). This is mainly due to different substructure and its evolution within the regions of adiabatic shear bands when compared with the substructure in the attached zones, cf. Fig. 2.6. Final separation occurs by a ductile failure mechanism along the shear band, cf. Figs. 2.2 and 2.6.

2.2.3 Viscoplastic model of single crystals

The main idea of the viscoplastic flow mechanism is to accomplish in one model the description of behaviour of single crystals valid for the entire range of strain rate changes. In other words, the main conception is to encompass the interaction of the thermally activated and phonon damping mechanisms.

To achieve this aim the empirical overstress function Φ has been introduced and the strain rate is postulated in the form as follows (cf. Perzyna (1988))

$$\dot{\in}^p = \frac{1}{T}\langle\Phi\left[\frac{\tau}{\tau_Y(\in^p, \vartheta, \beta, \zeta)} - 1\right]\rangle \mathrm{sgn}\tau, \tag{2.7}$$

where T is the relaxation time, $\langle\cdot\rangle$ denotes the Macauley bracket and τ_Y is the static yield stress function. In this model the static yield stress function depends on the inelastic strain $\in^p$, temperature ϑ, the density of obstacle dislocations β and the concentration of point defects ζ.

It is noteworthy that the empirical overstress function Φ can be determined basing on available experimental results performed under dynamic loading.

2.2.4 Strain rate sensitivity

In previous sections fundamental features of finite deformation, rate dependent plastic flow of crystalline solids were discussed from microscopic and macroscopic phenomenological points of view. Particular viscoplastic flow model was proposed to predict deformation textures and large strain, temperature and rate dependent and strain hardening behaviour of polycrystals from the known behaviour of single crystals.

The possibility of making such a prediction rests on the tacit assumption that the mechanisms of plastic deformation in aggregates are substantially identical with those observed in single crystals.

Experimental justifications of the thermally activated and phonon damping mechanisms as well as the discussion of their range rate and temperature

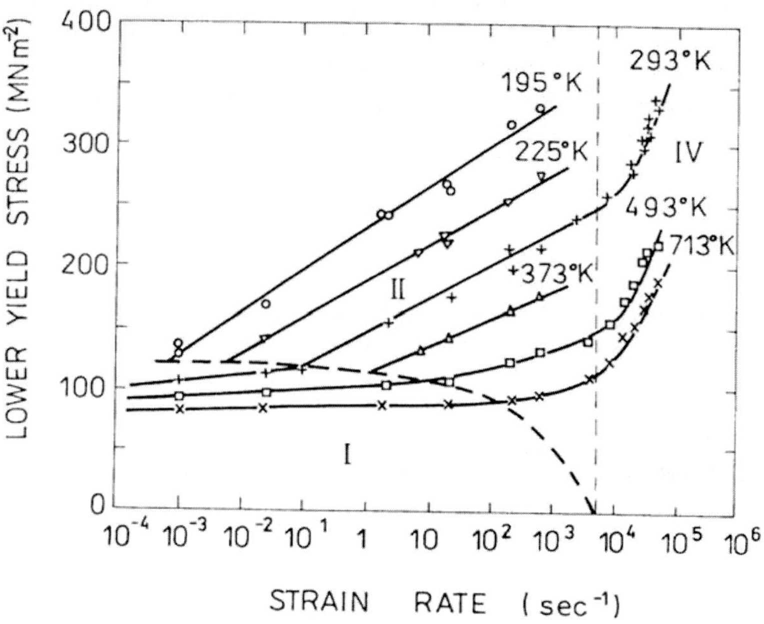

Figure 2.7: Variation of lower yield stress with strain rate, at constant temperature. After Campbell and Ferguson (1970)

changes for particular materials have been given in many papers. Particular importance for our purposes have results obtained by Campbell and Ferguson (1970). In their paper an account is given of experiments in which the shear flow stress of mild steel was measured at temperature from 195 to 713 K and strain rate from 10^{-3} to $4 \cdot 10^4$ s^{-1}. The flow stress at lower yield is plotted in Fig. 2.7 as shear stress against the logarithm of shear strain rate, for the various temperatures used throughout the investigation.

For the purpose of the discussion which follows, it is convenient to divide the curves into three regions, each corresponding to a certain range of strain rate which is a function of the temperature. Following Rosenfield and Hahn (1966) these will be referred to as region I, II and IV. These regions are indicated in Fig. 2.7.

In region I the flow stress shows a small temperature and strain rate sensitivity, the latter decreasing with increasing temperature. Prestraining increases the flow stress but has little effect on the rate sensitivity of the flow stress, $(\partial\tau/\partial\ln\dot{\in}^p)_\vartheta$, at room temperature. The dominant factor in region I seems to be the long–range internal stress fields due to dislocations, precipitate particles, grain boundaries etc.

In region II the flow stress shows greater rate and temperature sensitivities. From a survey of their own and previous work, Rosenfield and Hahn (1966) concluded that in this region the rate sensitivity $(\partial\tau/\partial\ln\dot{\in}^p)_\vartheta$ is independent of temperature and strain rate. However, the data of Campbell and Ferguson (1970) show a consistent increase in $(\partial\tau/\partial\ln\dot{\in}^p)_\vartheta$ as temperature is reduced.

It has been suggested by Campbell and Ferguson (1970) that the flow behaviour throughout region II can be explained by the thermal activation of dislocation motion.

Since the relaxation time T_{mT} is related to the dislocation structure it may be governed by the deformation history, rather than a function of the state variables $\in^p$, τ and ϑ.

The experimental data obtained by Campbell and Ferguson (1970) for mild steel in region II are properly interpreted by the linear approximation of the thermally activated theory.

Region IV is characterized by a rapid increase in semi– logarithmic rate sensitivity $(\partial\tau/\partial\ln\dot{\in}^p)_\vartheta$ with increasing strain rate, this parameter being approximately independent of temperature in the range 293 to 713 K.

In Fig. 2.8 the experimental data of Campbell and Ferguson (1970) for region IV are replotted using a linear strain–rate scale, and it is seen that, within the accuracy of measurement, they can be represented by straight lines at all three temperatures and all three values of pre–strain. While the slopes of these lines show only a small dependence on temperature, their intercepts on the stress axis vary greatly with temperature.

According to the interpretation presented by Campbell and Ferguson (1970), the intercepts in Fig. 2.8 are determined by the temperature–dependent barrier stress $\tau_\mu + U_\circ/v^*$ ($U_\circ$ denotes the activation energy for intersection at zero effective stress and v^* the activation volume), at which the strain rate reaches $1/T_{mT}$. When the applied stress exceeds this barrier stress, the time required to activate a dislocation past the short–range barriers of obstacles is negligible, hence its velocity is controlled by dissipation of energy as it moves through the lattice. Assuming that this dissipation is of a linear viscous nature, the excess stress τ_D will be proportional to the strain rate $\dot{\in}^p$, i.e.

$$\tau_D = \eta\dot{\in}^p, \tag{2.8}$$

where η denotes the macroscopic viscosity. Equating τ_D to the difference between the applied stress and the barrier stress $\tau_B = \tau_\mu + U_\circ/v^*$, we obtain

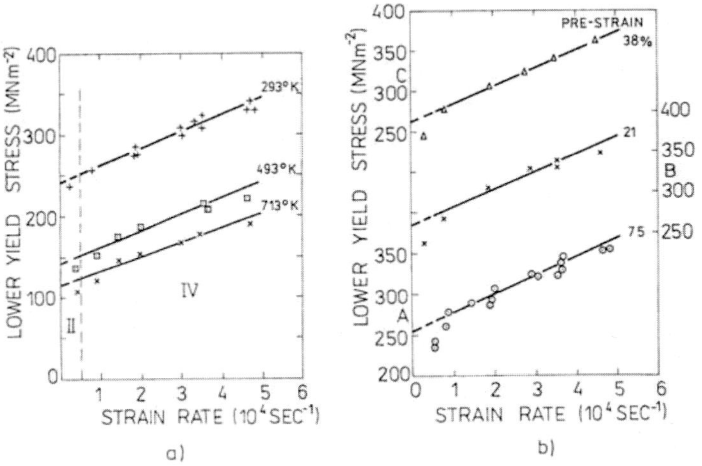

Figure 2.8: Variation of lower yield stress with strain rate (region IV). (a) Zero pre–strain; temperature 293, 493, 713 K. (b) Pre-strain 7.5, 21, 38%; temperature 293 K. After Campbell and Ferguson (1970)

for region IV

$$\tau = \tau_B + \eta \dot{\epsilon}^p. \tag{2.9}$$

Comparison (2.9) with $(2.3)_2$ under the condition that $\tau = \tau_B(1 + T_{mD}\dot{\epsilon}^p)$ gives

$$\eta = T_{mD}\tau_B = \frac{B}{\alpha b^2}. \tag{2.10}$$

The values of η can be obtained from the slops of the lines of Fig. 2.8.

For room temperature and zero pre–strain we have the relaxation time as follows

$$T_{mD} = \frac{\eta}{\tau_B} = \frac{2.1 \cdot 10^3}{250 \cdot 10^6}\text{s} = 0.84 \cdot 10^{-5}\text{s}.$$

For 493 K and zero pre–strain

$$T_{mD} = \frac{2.0 \cdot 10^3}{150 \cdot 10^6}\text{s} = 1.3 \cdot 10^{-5}\text{s}.$$

For 713 K and zero pre–strain

$$T_{mD} = \frac{1.8 \cdot 10^3}{120 \cdot 10^6}\text{s} = 1.5 \cdot 10^{-5}\text{s}.$$

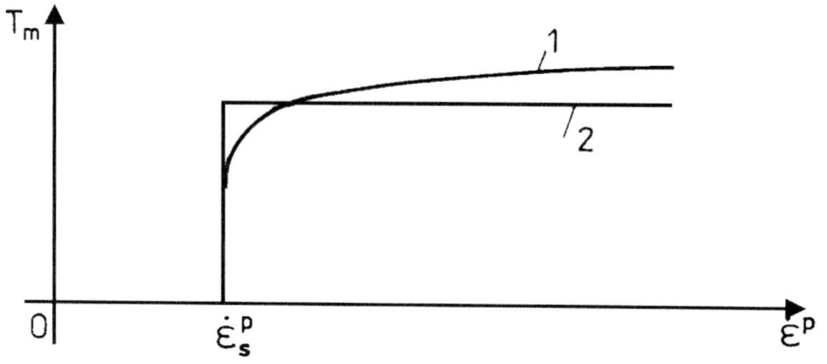

Figure 2.9: Variation of the relaxation time with strain rate for mild steel (1) real case; (2) assumption for classical elasto–viscoplasticity

Thus, in the phonon viscosity damping region IV the relaxation time is a function of temperature and is not sensitive to pre–stressing (for room temperature).

For region II the relaxation time T_{mT} is obtained as constant value

$$T_{mT} = 2 \cdot 10^{-4}\mathrm{s},$$

while in region IV the relaxation time T_{mD} is temperature dependent and can change from $T_{mD} = 0.8 \cdot 10^{-5}$s to $T_{mD} = 1.5 \cdot 10^{-5}$s.

For the viscoplastic model of polycrystalline solids the relaxation time T_m governs the viscoplastic flow in the entire range of strain rate changes and for classical models it is generally assumed as constant value. However, in real case it has to be a function of the rate of inelastic deformation $\dot{\epsilon}^p$, and can be determined based on experimental data, cf. Fig. 2.9.

Another possible idea has been presented by Perzyna (1980), namely that

$$T_m = \frac{\phi}{\gamma}, \tag{2.11}$$

where ϕ is the control function and γ is the temperature dependent viscosity coefficient. The dimensionless control function ϕ is assumed to depend on strain rate. Thus we have

$$T_m = \frac{1}{\gamma(\vartheta)} \phi \left(\frac{\dot{\epsilon}^p}{\dot{\epsilon}^p_s} - 1 \right). \tag{2.12}$$

In Fig. 2.10 the theoretical results obtained by Perzyna (1980) are compared with experimental data of Campbell and Ferguson (1970) for room

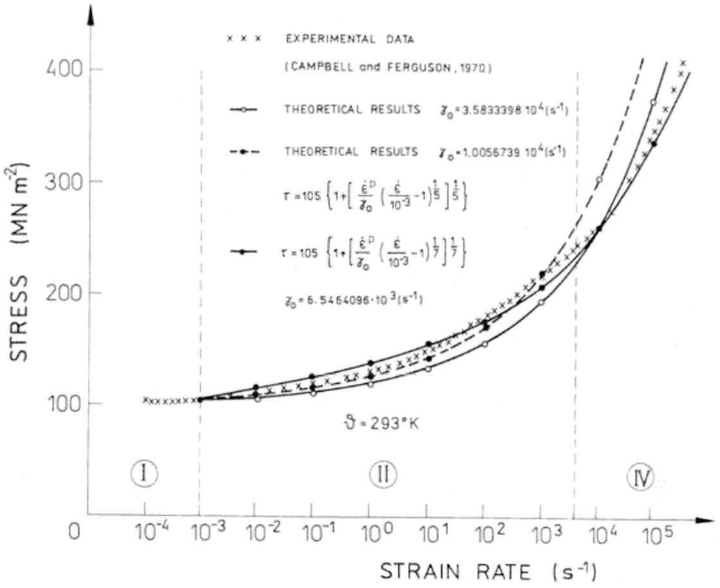

Figure 2.10: Comparison of theoretical description with experimental results

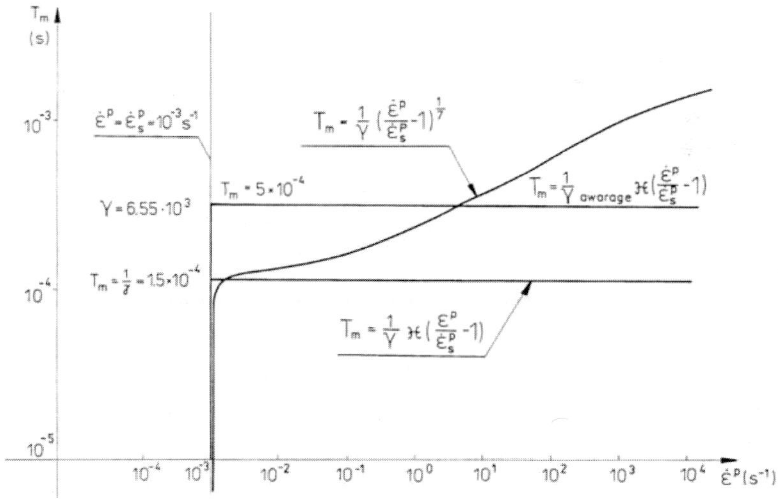

Figure 2.11: Variation of the relaxation time with strain rate for various assumption

temperature (293 K). Taking the best fitting curve we have

$$T_m = \frac{1}{6.55 \cdot 10^3} \left(\frac{\dot{\in}^p}{10^{-3}} - 1 \right)^{\frac{1}{7}} \text{(s)}. \tag{2.13}$$

Figure 2.11 shows the plots of the relaxation time (2.13) and (2.12) for $\phi \equiv \mathcal{H}(\cdot)$, i.e. for ϕ assumed as the Heaviside function.

Dowling, Harding and Campbell (1970) investigated the strain–rate sensitivity of the yield and flow stress of aluminum, copper, brass and mild steel over a range of strain rates from 10^{-3} to $4 \cdot 10^4$ s^{-1}. These experimental data can also be used for determination of the overstress function Φ and the relaxation time T_m for these metals.

Basing on the analysis of experimental observations for single crystals and polycrystalline metals we can conclude that microshear banding influences very much a substructure of a material under consideration. As a result of this it contributes to viscoplastic strain rate effects. We have also shown that the relaxation time T_m depends on the active microshear band fraction f_{ms} and on the rate of equivalent viscoplastic deformation $\dot{\in}^p$.

2.3 Experimental Motivation for Induced Anisotropy Effects

For our considerations of anisotropic effects, the metallurgical aspects of fracture and spalling have great importance. Effects of parameters such as grain size and substructure, crack/void initiation sites, micromechanical aspects of growth, and coalescence can influence fracture and spalling mechanisms.

To show the complexity of the anisotropy phenomena in microdamage process we shell discuss several available experimental results.

The most popular dynamical experiment[1] in the investigation of the fracture phenomenon in metals is a plate–impact configuration system. This experimental system consists of two plates, a projectile plane plate impacts against a target plane. This is a good example of a dynamic deformation process. If impact velocity is sufficiently high the propagation of a plastic wave through the target is generated. The reflection and interaction of waves result a net tensile pulse in the target plate. If this stress pulse has sufficient amplitude and sufficient time duration, it will cause separation of the material and spalling process.

[1]For a thorough discussion of the experimental and theoretical works in the field of dynamic fracture and spalling of metals please consult the review papers by Meyers and Aimone (1983) and Curran, Seaman and Shockey (1987), cf. also Meyers (1994).

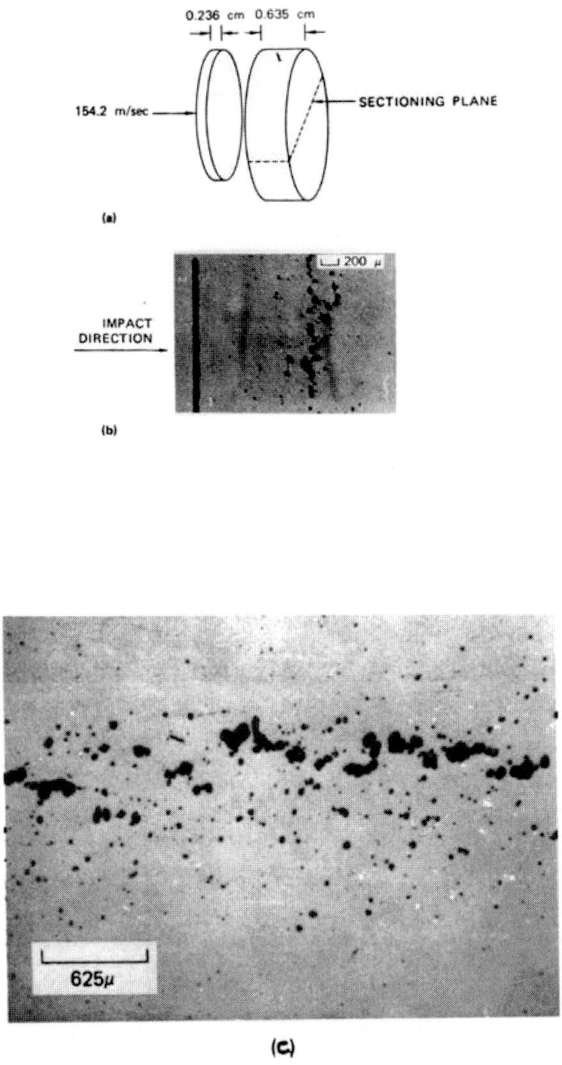

Figure 2.12: A cross section of an aluminum target plate that has undergone a planar impact by another aluminum plate; (a) impact configuration system, (b) observed ductile fracture on a cross section through target, (c) magnification of (b) (After Seaman, Curran and Shockey (1976))

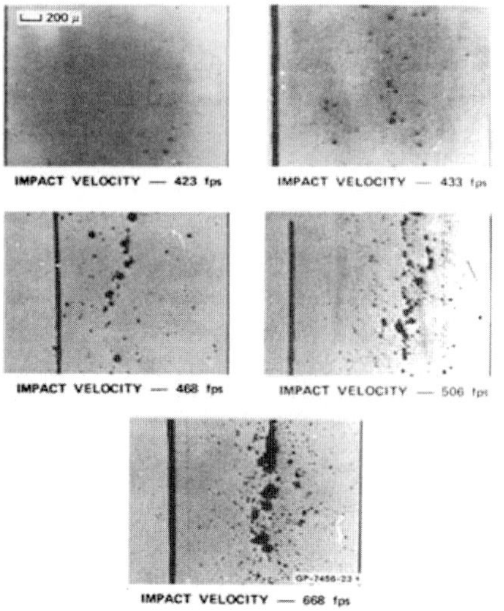

Figure 2.13: The final damage of the aluminum 1145 target plate for a constant shot geometry but for different impact velocities (After Barbee, Seaman, Crewdson and Curran (1972))

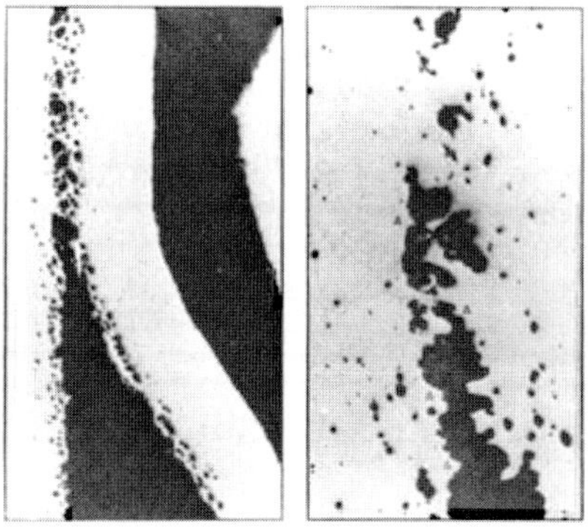

Figure 2.14: A sample of full separation of an aluminum target; (a) ductile crack propagation by void coalescence, (b) tip of ductile crack shown in (a), at higher magnification. Impact test performed by Seaman, Curran and Shockey (1976)

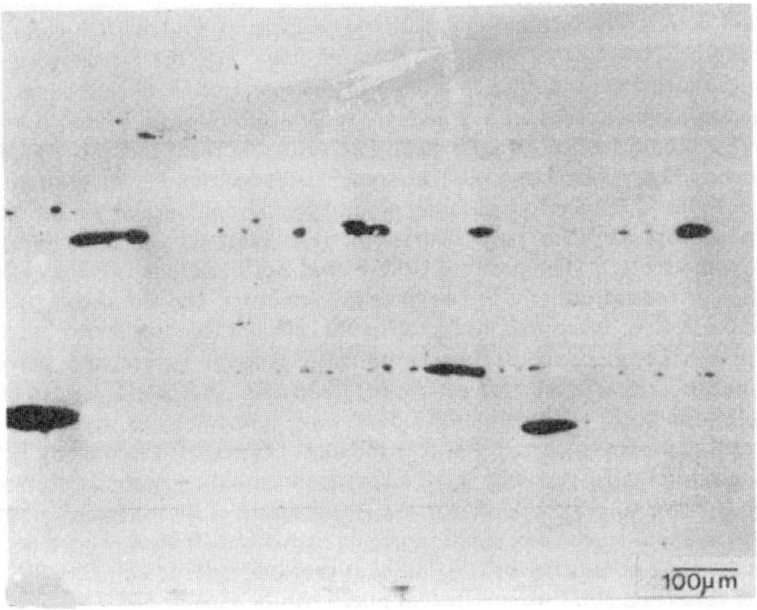

Figure 2.15: Spalling by ductile void formation in nickel (After Shockey, Seaman and Curran (1973))

The reason for choosing this particular kind of dynamical experiment is that postshot photomicrographic observations of the residual porosity are available, and the stress amplitude and pulse duration can be performed sufficiently great to produce substantial porosity and the spall of the target plate.

The experimental data presented by Seaman, Curran and Shockey (1976) illustrate damage phenomena and provide a common basis for considering damage criteria. They have used a plate–impact configuration system. Following the compression waves resulting from the impact, rarefaction waves have intersected near the middle of the target plate to cause damage in the form of nearly spherical voids. The heaviest damage is localized in a narrow zone, which is called the spall plane. Both the number and the size of voids decrease with distance from this zone. This type of damage is termed ductile fracture because of high ductility (ability to flow) required of the plate material, Fig. 2.12.

The final damage of the target plate (aluminium 1145) for a constant shot geometry but for different impact velocities has been performed by Barbee, Seaman, Crewdson and Curran (1972). The results suggest dependence

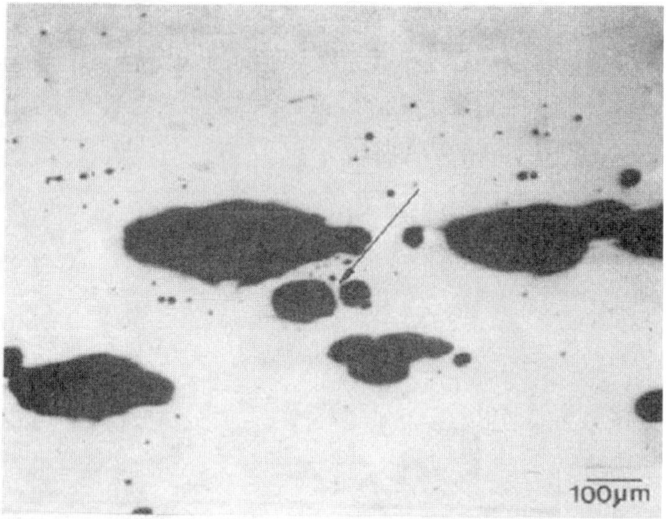

Figure 2.16: Void nucleation, growth, and coalescence observed in nickel (After Shockey, Seaman and Curran (1973))

of spalling process on the pulse amplitude, Fig. 2.13.

A sample of full separation is shown in Fig. 2.14, an aluminium target impacted by a plate has been damaged to the extent that full separation occurred near the center of the target, cf. Seaman, Curran and Shockey (1976). The authors suggested that this full separation appears as a macrocrack propagating through heavily damaged material. The macrocrack occurs as a result of coalescence of microvoids which is also visible in Fig. 2.14.

Spalling by ductile void formation in nickel is characterized by ellipsoidal cracks with large axes perpendicular to the direction of the applied tensile stress, cf. Fig. 2.15. A more advanced microdamage process has been shown in Fig. 2.16, where crack nucleation growth, and coalescence mechanisms in nickel have been observed.

An example of brittle fracture for Armco iron is presented in Fig. 2.17 (cf. Curran, Seaman and Shockey (1981)). It shows the polished cross section through plate impact specimen with very well visible cleavage (penny shape) microcracks. The damage, which appears as randomly oriented planar microcracks, depends on the impact velocity as well as on the duration of the tensile wave. The second property is directly observed from the results presented in Fig. 2.18 (cf. Curran, Seaman and Shockey (1977)). Use of a tapered flyer results in longer tensile impulses at the thicker end. As it is

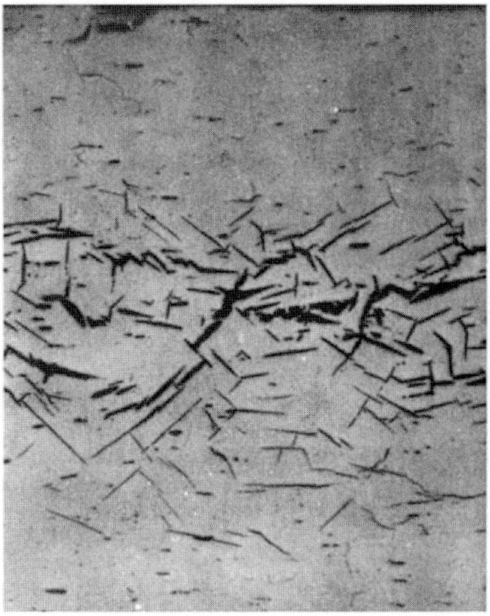

Figure 2.17: Internal cleavage (penny shape) microcracks caused by shock loading in the polished cross section of an Armco iron specimen (After Curran, Seaman and Shockey (1981))

shown in Fig. 2.18 these longer pulses lead to greater damage in the Armco iron target (the inset gives to approximate durations of the tensile pulses). The damage observed in this experiment is termed brittle, although the microcrack growth is much slower than elastic crack velocities, indicating considerable plastic flow at micro crack tips.

Adiabatic shear band localization during dynamic process can be another good example of anisotropic effects. Shear band form in one direction, which is determined by the state of stress and the properties of the material of a body, as well as by the boundary conditions.

Grebe, Pak and Meyers (1985) were conducted ballistic impact experiments on 12.5 mm thick commercial purity titanium and T–6pct Al–4pct V alloy plates using steel projectiles with 10.5 mm diameter. The impact velocities in their experiments varied between 578 m/s and 846 m/s. The microstructural damage mechanisms associated with shear band formation, shock wave and dynamic fracture were investigated by optical and scanning and transmission electron microscopy. The shear band were found along the two sides of the cross–section passing through the axis of the projectile.

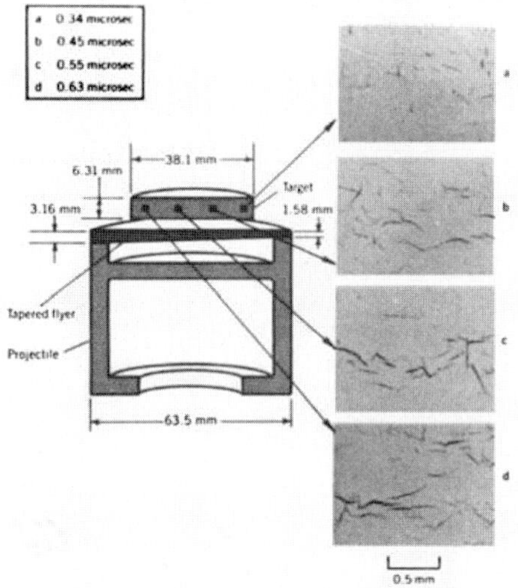

Figure 2.18: Tapered flyer impact experimental results for the Armco iron target (After Curran, Seaman and Shockey (1977))

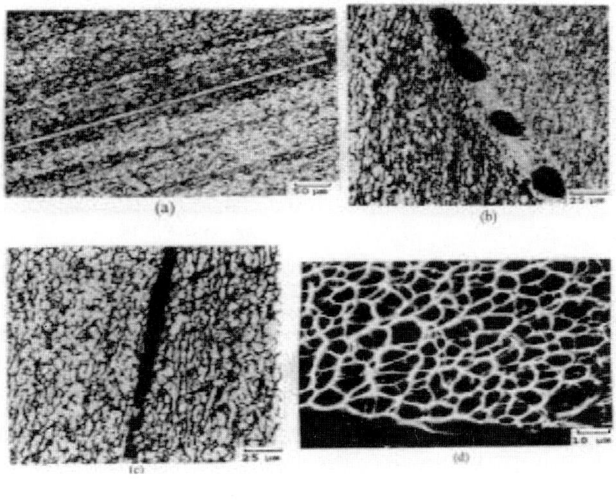

Figure 2.19: Shear band in Ti6Al4V target impacted at 846 m/s (After Grebe, Pak and Meyers (1985)). a) Single shear band; b) Microcracks in the shear band region; c) Elongated macrocracks along the shear band; d) Characteristic dimples observed in spall region

The measured shear band width in T6A14V varied between 1 and 10 μm. Observations of the onset of fracture along the shear band were also conducted. Spherical and ellipsoidal microcracks in T6A14V were found along the bands, Fig. 2.19. The mechanism of final failure in T6A14V is a simple propagation of a macrocrack along the damaged material within the shear band region. In the explanation of the phenomenon of fracture along shear band very important role has the microdamage process which consists of the nucleation, growth and coalescence of microvoids.

The investigations reported by Grebe, Pak and Meyers (1985) indicated that in dynamic processes the shear band regions behave differently than adjacent zones, cf. Fig. 2.19. Within the shear band region the deformation process is characterized by very large strains (shear strains over 100%) and very high strain rates ($10^3 - 10^5$ s^{-1}). The strain rate sensitivity of a material becomes very important feature of the shear band region and the micro–damage process is intensified.

2.4 Kinematics of finite deformation and fundamental definitions

2.4.1 Fundamental measures of total deformation

Our notation throughout is as follows: $\mathcal{B}$ and $\mathcal{S}$ are manifolds, points in $\mathcal{B}$ are denoted $\mathbf{X}$ and those in $\mathcal{S}$ by $\mathbf{x}$. The tangent spaces are written $T_{\mathbf{X}}\mathcal{B}$ and $T_{\mathbf{x}}\mathcal{S}$. Coordinate systems are denoted $\{X^A\}$ and $\{x^a\}$ for $\mathcal{B}$ and $\mathcal{S}$, respectively, with corresponding bases $\mathbf{E}_A$ and $\mathbf{e}_a$ and dual bases $\mathbf{E}^A$ and $\mathbf{e}^a$.

Let us take the Riemannian spaces on manifolds $\mathcal{B}$ and $\mathcal{S}$, i.e. $\{\mathcal{B}, \mathbf{G}\}$ and $\{\mathcal{S}, \mathbf{g}\}$, the metric tensors $\mathbf{G}$ and $\mathbf{g}$ are defined as follows $\mathbf{G} : T\mathcal{B} \to T^*\mathcal{B}$ and $\mathbf{g} : T\mathcal{S} \to T^*\mathcal{S}$, where $T\mathcal{B}$ and $T\mathcal{S}$ denote the tangent bundles of $\mathcal{B}$ and $\mathcal{S}$, respectively, and $T^*\mathcal{B}$ and $T^*\mathcal{S}$ their dual tangent bundles.

Let the metric tensor G_{AB} be defined by $G_{AB}(\mathbf{X}) = (\mathbf{E}_A, \mathbf{E}_B)_{\mathbf{X}}$, and similarly define g_{ab} by $g_{ab}(\mathbf{x}) = (\mathbf{e}_a, \mathbf{e}_b)_{\mathbf{x}}$, where $(\, , \,)_{\mathbf{X}}$ and $(\, , \,)_{\mathbf{x}}$ denote the standard inner products in $\mathcal{B}$ and $\mathcal{S}$, respectively.

Let

$$\mathbf{x} = \phi(\mathbf{X}, t) \tag{2.14}$$

be regular motion, then $\phi_t : \mathcal{B} \to \mathcal{S}$ is a C^1 actual configuration (at time t) of $\mathcal{B}$ in $\mathcal{S}$. The tangent of ϕ is denoted $\mathbf{F}$ and is called the deformation gradient of ϕ; thus $\mathbf{F} = T\phi$. For $\mathbf{X} \in \mathcal{B}$, we let $\mathbf{F}(\mathbf{X})$ denote the restriction of $\mathbf{F}$ to $T_{\mathbf{X}}\mathcal{B}$.

Thus

$$\mathbf{F}(\mathbf{X}, t) : T_{\mathbf{X}}\mathcal{B} \to T_{\mathbf{x}=\phi(\mathbf{X},t)}\mathcal{S} \tag{2.15}$$

is a linear transformation for each $\mathbf{X} \in \mathcal{B}$ and $t \in I \subset I\!R$. For each $\mathbf{X} \in \mathcal{B}$ there exists an orthogonal transformation $\mathbf{R}(\mathbf{X}) : T_{\mathbf{X}}\mathcal{B} \to T_{\mathbf{x}}\mathcal{S}$ such that $\mathbf{F} = \mathbf{R} \cdot \mathbf{U} = \mathbf{V} \cdot \mathbf{R}$. Notice that $\mathbf{U}$ and $\mathbf{V}$ operate within each fixed tangent space. We call $\mathbf{U}$ and $\mathbf{V}$ the right and left stretch tensor, respectively. For each $\mathbf{X} \in \mathcal{B}$, $\mathbf{U}(\mathbf{X}) : T_{\mathbf{X}}\mathcal{B} \to T_{\mathbf{X}}\mathcal{B}$ and for each $\mathbf{x} \in \mathcal{S}$, $\mathbf{V}(\mathbf{x}) : T_{\mathbf{x}}\mathcal{S} \to T_{\mathbf{x}}\mathcal{S}$.

The material (or Lagrangian) strain tensor $\mathbf{E} : T_{\mathbf{X}}\mathcal{B} \to T_{\mathbf{X}}\mathcal{B}$ is defined by

$$2\mathbf{E} = \mathbf{C} - \mathbf{I}, \quad (\mathbf{I} \text{ denotes the identity on } T_{\mathbf{X}}\mathcal{B}), \tag{2.16}$$

where

$$\mathbf{C} = \mathbf{F}^T \cdot \mathbf{F} = \mathbf{U}^2 = \mathbf{B}^{-1}. \tag{2.17}$$

The spatial (or Eulerian) strain tensor $\mathbf{e} : T_{\mathbf{x}}\mathcal{S} \to T_{\mathbf{x}}\mathcal{S}$ is defined by

$$2\mathbf{e} = \mathbf{i} - \mathbf{c}, \quad (\mathbf{i} \text{ denotes the identity on } T_{\mathbf{x}}\mathcal{S}), \tag{2.18}$$

where

$$\mathbf{c} = \mathbf{b}^{-1} \quad \text{and} \quad \mathbf{b} = \mathbf{F} \cdot \mathbf{F}^T = \mathbf{V}^2. \tag{2.19}$$

The various strain tensors can be redefined in terms of pull–back and push–forward operations. For the material strain tensor $\mathbf{E}$ and the spatial strain tensor $\mathbf{e}$ we have

$$\begin{aligned} \mathbf{E}^\flat &= \phi^*(\mathbf{e}^\flat), & E_{AB}(\mathbf{X}) &= e_{ab}(\mathbf{x}) F^a_A(\mathbf{X}) F^b_B(\mathbf{X}), \\ \mathbf{e}^\flat &= \phi_*(\mathbf{E}^\flat), & e_{ab}(\mathbf{x}) &= E_{AB}(\mathbf{X}) (\mathbf{F}(\mathbf{X})^{-1})^A_a (\mathbf{F}(\mathbf{X})^{-1})^B_b, \end{aligned} \tag{2.20}$$

where the symbol $\flat$ denotes the index lowering operator.

2.5 Finite elasto–viscoplastic deformation

Motivated by the micromechanics of single crystal plasticity we postulate a local multiplicative decomposition of the form

$$\mathbf{F}(\mathbf{X}, t) = \mathbf{F}^e(\mathbf{X}, t) \cdot \mathbf{F}^p(\mathbf{X}, t), \tag{2.21}$$

where $\mathbf{F}^{e^{-1}}$ is interpreted as the local deformation that releases the stresses from each neighborhood $\mathcal{N}(\mathbf{x}) \subset \phi(\mathcal{B})$ in the current configuration of the body, cf. Fig. 2.20.

Let us consider a particle X, which at time $t = 0$ occupied the place $\mathbf{X}$ in the reference (material) configuration $\mathcal{B}$, its current place at time t in the actual (spatial) configuration $\mathcal{S}$ is $\mathbf{x} = \phi(\mathbf{X}, t)$ and its position in the unloaded actual configuration $\mathcal{S}'$ is denoted by $\mathbf{y}$. Thus we have

$$\mathbf{F}^e : T_{\mathbf{y}}\mathcal{S}' \to T_{\mathbf{x}}\mathcal{S}, \quad \mathbf{F}^p : T_{\mathbf{X}}\mathcal{B} \to T_{\mathbf{y}}\mathcal{S}', \tag{2.22}$$

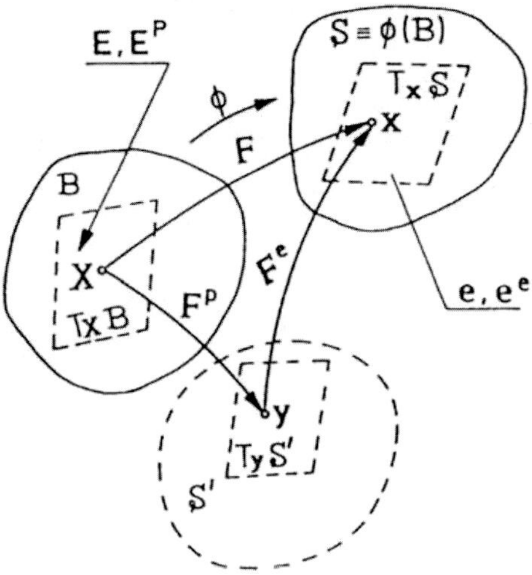

Figure 2.20: Schematic representation of the multiplicative decomposition of the deformation gradient

where $T_{\mathbf{y}}\mathcal{S}'$ denotes the tangent space in the unloaded actual configuration $\mathcal{S}'$. It is noteworthy that $\mathbf{F}^e$ and $\mathbf{F}^p$ defined by (2.22) are linear transformations.

We shall tread the tangent space $T_{\mathbf{y}}\mathcal{S}'$ as an auxiliary tool which helps to define the plastic strain tensors[2].

The plastic strain tensor $\mathbf{E}^p : T_{\mathbf{X}}\mathcal{B} \to T_{\mathbf{X}}\mathcal{B}$ is defined by

$$\mathbf{E}^p = \frac{1}{2}(\mathbf{C}^p - \mathbf{I}), \tag{2.23}$$

where

$$\mathbf{C}^p = \mathbf{F}^{p^T} \cdot \mathbf{F}^p = \mathbf{U}^{p^2} = \mathbf{B}^{p^{-1}} \quad \text{and} \quad \mathbf{E}^e \stackrel{\text{def}}{=} \mathbf{E} - \mathbf{E}^p. \tag{2.24}$$

Similarly the elastic strain tensor $\mathbf{e}^e : T_x\mathcal{S} \to T_x\mathcal{S}$ is defined by

$$\mathbf{e}^e = \frac{1}{2}(\mathbf{i} - \mathbf{c}^e), \tag{2.25}$$

where

$$\mathbf{c}^e = \mathbf{b}^{e^{-1}}, \quad \mathbf{b}^e = \mathbf{F}^e \cdot \mathbf{F}^{e^T} = \mathbf{V}^{e^2} \quad \text{and} \quad \mathbf{e}^p \stackrel{\text{def}}{=} \mathbf{e} - \mathbf{e}^e. \tag{2.26}$$

[2]For precise definition of the finite elasto–plastic deformation see Perzyna (1995) and Duszek–Perzyna and Perzyna (1998). Different approach to define the finite elasto–plastic deformation has been presented by Nemat–Nasser (1992).

It is noteworthy to compare the relation

$$\mathbf{F} = \mathbf{R} \cdot \mathbf{U} = \mathbf{V} \cdot \mathbf{R} \tag{2.27}$$

with

$$\mathbf{F} = \mathbf{F}^e \cdot \mathbf{F}^p = \mathbf{R}^e \cdot \mathbf{U}^e \cdot \mathbf{R}^p \cdot \mathbf{U}^p = \mathbf{V}^e \cdot \mathbf{R}^e \cdot \mathbf{V}^p \cdot \mathbf{R}^p. \tag{2.28}$$

The following commutative diagrams summarize the situation.

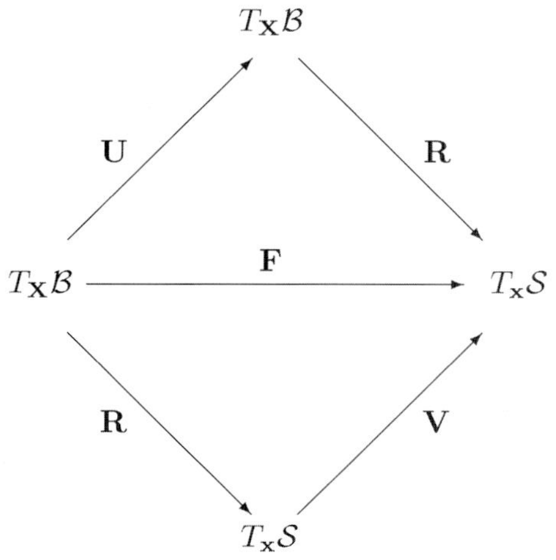

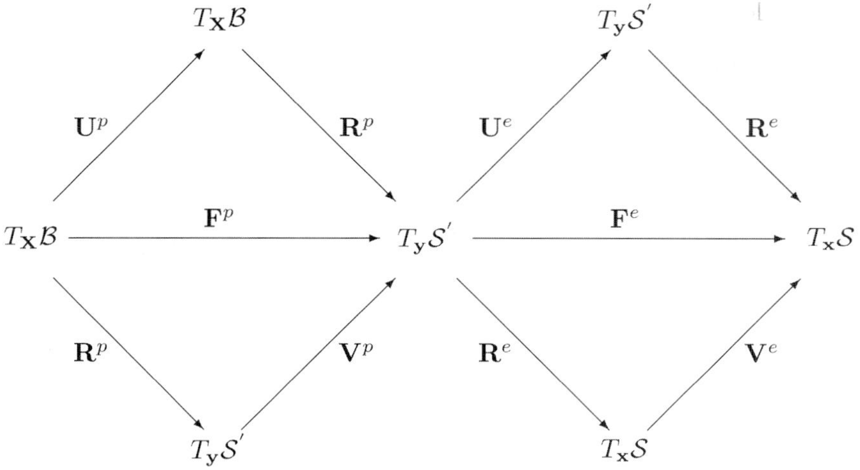

From the second diagram it is clear that the tangent space $T_{\mathbf{y}}\mathcal{S}^{'}$ is playing an auxiliary role indeed.

The plastic tensors $\mathbf{E}^p$and $\mathbf{e}^p$ operate within each fixed tangent space; that is $\mathbf{E}^p : T_{\mathbf{X}}\mathcal{B} \to T_{\mathbf{X}}\mathcal{B}$ and $\mathbf{e}^p : T_{\mathbf{x}}\mathcal{S} \to T_{\mathbf{x}}\mathcal{S}$.

We can show that the following relations are valid

$$\phi_*(\mathbf{E}^{p^\flat}) = \mathbf{e}^{p^\flat}, \qquad \phi^*(\mathbf{e}^{e^\flat}) = \mathbf{E}^{e^\flat}. \tag{2.29}$$

2.5.1 Rates of the deformation tensor

Let $\phi(\mathbf{X}, t)$ be a C^2 motion of $\mathcal{B}$. Then the spatial velocity is $\boldsymbol{\upsilon}_t = \mathbf{V}_t \circ \phi_t^{-1}$, where $\mathbf{V}_t = \frac{\partial \phi}{\partial t}$ is the material velocity, i.e. $\boldsymbol{\upsilon} : \mathcal{S} \times I \to T\mathcal{S}$, $I \subset I\!R$.

The collection of maps $\phi_{t,s}$ such that for each s and $\mathbf{x}$, $t \to \phi_{t,s}(\mathbf{x})$ is an integral curve of $\boldsymbol{\upsilon}$, and $\phi_{s,s}(\mathbf{x}) = \mathbf{x}$, is called the flow or evolution operator of $\boldsymbol{\upsilon}$, i.e.

$$\{\phi_{t,s} \mid \phi_{t,s} = \phi_t \circ \phi_s^{-1} : \phi_s(\mathcal{B}) \to \phi_t(\mathcal{B})\} \tag{2.30}$$

and

$$\phi_{t,s} \circ \phi_{s,r} = \phi_{t,r}, \qquad \phi_{t,t} = \text{identity} \tag{2.31}$$

for all $r, s, t \in I \subset I\!R$.

If $\mathbf{t}$ is a C^1 (possible time–dependent) tensor field on $\mathcal{S}$, then the Lie derivative of $\mathbf{t}$ with respect to $\boldsymbol{\upsilon}$ is defined by[3]

$$\mathrm{L}_{\boldsymbol{\upsilon}}\mathbf{t} = \left(\frac{d}{dt}\phi_{t,s}^*\mathbf{t}_t\right)\Big|_{t=s}. \tag{2.32}$$

If we hold t fixed in $\mathbf{t}_t$, we obtain the autonomous Lie derivative

$$\mathcal{L}_{\boldsymbol{\upsilon}}\mathbf{t} = \left(\frac{d}{dt}\phi_{t,s}^*\mathbf{t}_s\right)\Big|_{t=s}. \tag{2.33}$$

Thus

$$\mathrm{L}_{\boldsymbol{\upsilon}}\mathbf{t} = \frac{\partial \mathbf{t}}{\partial t} + \mathcal{L}_{\boldsymbol{\upsilon}}\mathbf{t}. \tag{2.34}$$

If $\mathbf{t} \in \mathbf{T}^r{}_s(\mathcal{S})$ (elements of $\mathbf{T}^r{}_s(\mathcal{S})$ are called tensors on $\mathcal{S}$, contravariant of order r and covariant of order s) then $\mathrm{L}_{\boldsymbol{\upsilon}}\mathbf{t} \in \mathbf{T}^r{}_s(\mathcal{S})$.

The spatial velocity gradient $\mathbf{l}$ is defined by

$$\mathbf{l} = D\boldsymbol{\upsilon} : T_{\mathbf{x}}\mathcal{S} \to T_{\mathbf{x}}\mathcal{S}, \quad \text{i.e.} \quad l^a_b = \upsilon^a\mid_b = \frac{\partial \upsilon^a}{\partial x^b} + \gamma^a_{bc}\upsilon^c, \tag{2.35}$$

where γ^a_{bc} denotes the Christoffel symbol for $\mathbf{g}$.

[3]The algebraic and dynamic interpretations of the Lie derivative have been presented by Abraham et al. (1988), cf. also Marsden and Hughes (1983).

The spatial velocity gradient $\mathbf{l}$ can be expressed as follows

$$\mathbf{l} = D\boldsymbol{\upsilon} = \dot{\mathbf{F}} \cdot \mathbf{F}^{-1} = \dot{\mathbf{F}}^{e} \cdot \mathbf{F}^{e^{-1}} + \mathbf{F}^{e} \cdot (\dot{\mathbf{F}}^{p} \cdot \mathbf{F}^{p^{-1}}) \cdot \mathbf{F}^{e^{-1}} = \mathbf{l}^{e} + \mathbf{l}^{p} = \mathbf{d} + \boldsymbol{\omega}\mathbf{d}^{e} + \boldsymbol{\omega}^{e} + \mathbf{d}^{p} + \boldsymbol{\omega}^{p}, \tag{2.36}$$

where $\mathbf{d}$ denotes the spatial rate of deformation tensor and $\boldsymbol{\omega}$ is called the spin.

Let us define the material (or Lagrangian) rate of deformation tensor $\mathbf{D}$ as follows

$$\mathbf{D}(\mathbf{X}, t) = \frac{\partial}{\partial t}\mathbf{E}(\mathbf{X}, t). \tag{2.37}$$

We have very important relation

$$\mathbf{d}^{\flat} = \mathrm{L}_{\boldsymbol{\upsilon}}\mathbf{e}^{\flat} = \phi_{*}\frac{\partial}{\partial t}(\phi^{*}\mathbf{e}^{\flat}) = \phi_{*}(\frac{\partial}{\partial t}\mathbf{E}^{\flat}) = \phi_{*}(\mathbf{D}^{\flat}). \tag{2.38}$$

On the other hand

$$\mathbf{d}^{\flat} = \mathrm{L}_{\boldsymbol{\upsilon}}\mathbf{e}^{\flat} = \mathrm{L}_{\boldsymbol{\upsilon}}\left[\frac{1}{2}\left(\mathbf{g} - \mathbf{b}^{-1}\right)\right]^{\flat} = \frac{1}{2}\mathrm{L}_{\boldsymbol{\upsilon}}\mathbf{g} = \frac{1}{2}\left(g_{cb}\upsilon^{c}\mid_{a} + g_{ac}\upsilon^{c}\mid_{b}\right)\mathbf{e}^{a}\otimes\mathbf{e}^{b}, \tag{2.39}$$

i.e. the symmetric part of the velocity gradient $\mathbf{l}$ (the symbol $\otimes$ denotes the tensor product).

The components of the spin $\boldsymbol{\omega}$ are given by

$$\omega_{ab} = \frac{1}{2}\left(g_{ac}\upsilon^{c}\mid_{b} - g_{cb}\upsilon^{c}\mid_{a}\right) = \frac{1}{2}\left(\frac{\partial \upsilon_{a}}{\partial x^{b}} - \frac{\partial \upsilon_{b}}{\partial x^{a}}\right), \tag{2.40}$$

and

$$\mathbf{d}^{e^{\flat}} = \mathrm{L}_{\boldsymbol{\upsilon}}\mathbf{e}^{e^{\flat}}, \qquad \mathbf{d}^{p^{\flat}} = \mathrm{L}_{\boldsymbol{\upsilon}}\mathbf{e}^{p^{\flat}}. \tag{2.41}$$

2.5.2 Rates of the stress tensors

The first Piola–Kirchhoff stress tensor P^{aA} is the two–point tensor obtained by performing a Piola transformation on the second index of the Cauchy stress tensor $\boldsymbol{\sigma}$, i.e.

$$P^{aA} = J(\mathbf{F}^{-1})^{A}_{b}\sigma^{ab}, \tag{2.42}$$

where J denotes the Jacobian of the deformation.

The second Piola–Kirchhoff stress tensor $\mathbf{S}$ is defined as follows

$$S^{AB} = (\mathbf{F}^{-1})^{A}_{a}P^{aB} = J(\mathbf{F}^{-1})^{A}_{a}(\mathbf{F}^{-1})^{B}_{b}\sigma^{ab} = (\mathbf{F}^{-1})^{A}_{a}(\mathbf{F}^{-1})^{B}_{b}\tau^{ab}, \tag{2.43}$$

i.e.

$$\mathbf{S} = \phi^*(\boldsymbol{\tau}), \tag{2.44}$$

where $\boldsymbol{\tau} = J\boldsymbol{\sigma}$ is called the Kirchhoff stress tensor.

The rate of the Kirchhoff stress tensor $\boldsymbol{\tau}$ is given by

$$\mathrm{L}_{\boldsymbol{\upsilon}}\boldsymbol{\tau} = \phi_* \frac{\partial}{\partial t}(\phi^*\boldsymbol{\tau}) = \phi_*(\frac{\partial}{\partial t}\mathbf{S}) = \mathbf{F}\cdot(\frac{\partial}{\partial t}\mathbf{S})\cdot\mathbf{F}^T \circ \phi_t^{-1}. \tag{2.45}$$

Let us define

$$\boldsymbol{\tau}_1 = \tau^{ab}\mathbf{e}_a \otimes \mathbf{e}_b \in \mathbf{T}^2{}_0(\mathcal{S}), \qquad \boldsymbol{\tau}_2 = \tau_a{}^b\mathbf{e}^a \otimes \mathbf{e}_b \in \mathbf{T}_1{}^1(\mathcal{S}),$$
$$\boldsymbol{\tau}_3 = \tau^a{}_b\mathbf{e}_a \otimes \mathbf{e}^b \in \mathbf{T}^1{}_1(\mathcal{S}). \tag{2.46}$$

Then

$$(\mathrm{L}_{\boldsymbol{\upsilon}}\boldsymbol{\tau}_1)^{ab} = \frac{\partial\tau^{ab}}{\partial t} + \frac{\partial\tau^{ab}}{\partial x^c}\upsilon^c - \tau^{cb}\frac{\partial\upsilon^a}{\partial x^c} - \tau^{ac}\frac{\partial\upsilon^b}{\partial x^c}. \tag{2.47}$$

is the rate associated with the name Oldroyd (cf. Oldroyd (1950)). The Zaremba–Jaumann rate (cf. Zaremba (1903a,b) and Jaumann (1911)) is defined as follows

$$\frac{1}{2}\left[(\mathrm{L}_{\boldsymbol{\upsilon}}\boldsymbol{\tau}_3)^a{}_c g^{cb} + g^{ac}(\mathrm{L}_{\boldsymbol{\upsilon}}\boldsymbol{\tau}_2)_c{}^b\right] = \frac{\partial\tau^{ab}}{\partial t} + \frac{\partial\tau^{ab}}{\partial x^c}\upsilon^c + \tau^{ad}\omega_d{}^b - \tau^{db}\omega^a{}_d. \tag{2.48}$$

2.5.3 Fundamental properties of the Lie derivatives

Let us take again $\mathbf{t} \in \mathbf{T}^r{}_s(\mathcal{S})$ a given time dependent spatial tensor field on $\mathcal{S}$ and let $\boldsymbol{\zeta}$ be a diffeomorphism of $\mathcal{S}$ to another manifold $\boldsymbol{\zeta}(\phi(\mathcal{B}))$, cf. Fig. 2.21. Any spatial tensor field $\mathbf{t} \in \mathbf{T}^r{}_s(\mathcal{S})$ is said to transform objectively under superposed diffeomorphism $\boldsymbol{\zeta}$ if it transforms according to the rule

$$\mathbf{t}' = \boldsymbol{\zeta}_*\mathbf{t}, \tag{2.49}$$

where $\boldsymbol{\zeta}_*$ is the push–forward operation.

Let $\boldsymbol{\upsilon}'$ be the velocity field of $\boldsymbol{\zeta}_t \circ \phi_t$. Then we have (cf. Marsden and Hughes, 1983)

$$\mathrm{L}_{\boldsymbol{\upsilon}'}\mathbf{t}' = \boldsymbol{\zeta}_*(\mathrm{L}_{\boldsymbol{\upsilon}}\mathbf{t}). \tag{2.50}$$

This means that objective tensors have objective Lie derivatives. It is noteworthy to recall here that rates which are objective with respect to diffeomorphism are called covariant.

The Oldroyd rate of the Kirchhoff stress tensor (2.47) is objective with respect to diffeomorphisms while the Zaremba–Jaumann rate (2.48) is objective with respect to isometries. The reason of it is caused by the fact that the operations of raising and lowering indices do not commute with Lie differentiation. This corollary has very important consequences for the formulation of the objective constitutive structures.

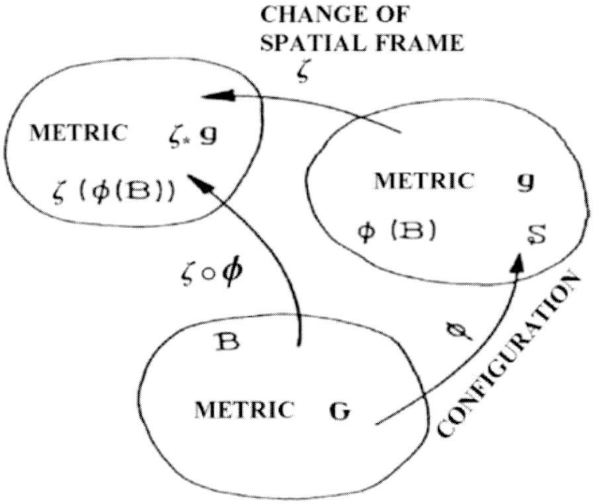

Figure 2.21: Schematic representation of the change of spatial frame generated by superposed spatial diffeomorphism

2.6 Thermo–Elasto–Viscoplasticity Constitutive Model

2.6.1 Constitutive postulate

Let us assume that: (i) conservation of mass, (ii) balance of momentum, (iii) balance of moment of momentum, (iv) balance of energy, (v) entropy production inequality hold.

We introduce the four fundamental postulates:

(i) Existence of the free energy function. It is assumed that the free energy function is given by

$$\psi = \hat{\psi}(\mathbf{e}, \mathbf{F}, \vartheta; \boldsymbol{\mu}), \tag{2.51}$$

where ϑ is temperature and $\boldsymbol{\mu}$ denotes a set of the internal state variables.
To extend the domain of the description of the material properties and particularly to take into consideration different dissipation effects we have to introduce the internal state variables represented by the vector $\boldsymbol{\mu}$.

(ii) Axiom of objectivity (spatial covariance). The constitutive structure should be invariant with respect to any diffeomorphism (any motion) $\zeta : \mathcal{S} \to \mathcal{S}$.

(iii) The axiom of the entropy production. For any regular motion of a body $\mathcal{B}$ the constitutive functions are assumed to satisfy the reduced dissipation inequality

$$\frac{1}{\rho_{Ref}}\boldsymbol{\tau} : \mathbf{d} - (\eta\dot{\vartheta} + \dot{\psi}) - \frac{1}{\rho\vartheta}\mathbf{q} \cdot \mathrm{grad}\vartheta \geq 0, \tag{2.52}$$

where η is the specific (per unit mass) entropy, and $\mathbf{q}$ denotes the heat flow vector field.

(iv) The evolution equation for the internal state variable vector $\boldsymbol{\mu}$ is assumed in the form as follows

$$\mathrm{L}_{\boldsymbol{\upsilon}}\boldsymbol{\mu} = \hat{\mathbf{m}}(\mathbf{e}, \mathbf{F}, \vartheta, \boldsymbol{\mu}), \tag{2.53}$$

where the evolution function $\hat{\mathbf{m}}$ has to be determined based on careful physical interpretation of a set of the internal state variables and analysis of available experimental observations.
The determination of the evolution function $\hat{\mathbf{m}}$ (in practice a finite set of the evolution functions) appears to be the main problem of the modern constitutive modelling.

2.6.2 Fundamental assumptions

For our practical purposes it is sufficient to assume that the internal state vector $\boldsymbol{\mu}$ has the form

$$\boldsymbol{\mu} = (\in^p, \boldsymbol{\xi}, \boldsymbol{\alpha}), \tag{2.54}$$

where $\in^p$ is the equivalent viscoplastic deformation, i.e.

$$\in^p = \int_0^t \left(\frac{2}{3}\mathbf{d}^p : \mathbf{d}^p\right)^{\frac{1}{2}} dt, \tag{2.55}$$

$\boldsymbol{\xi}$ denote the microdamage second order tensor, with the physical interpretation that $(\boldsymbol{\xi} : \boldsymbol{\xi})^{\frac{1}{2}} = \xi$ defines the volume fraction porosity and takes account for microdamaged effects and $\boldsymbol{\alpha}$ denotes the residual stress (the back stress) and aims at the description of the kinematic hardening effects.

Let us introduce the plastic potential function $f = f(\tilde{J}_1, \tilde{J}_2, \vartheta, \boldsymbol{\mu})$, where $\tilde{J}_1$, $\tilde{J}_2$ denote the first two invariants of the stress tensor $\tilde{\boldsymbol{\tau}} = \boldsymbol{\tau} - \boldsymbol{\alpha}$.

Let us postulate the evolution equations as follows

$$\mathbf{d}^p = \Lambda \mathbf{P}, \quad \boldsymbol{\omega}^p = \Lambda \boldsymbol{\Omega}, \quad \mathrm{L}_{\boldsymbol{\upsilon}} \boldsymbol{\xi} = \boldsymbol{\Xi}, \quad \mathrm{L}_{\boldsymbol{\upsilon}} \boldsymbol{\alpha} = \mathbf{A}, \tag{2.56}$$

where for elasto–viscoplastic model of a material we assume (cf. Perzyna (1963, 1966, 1971, 1995, 2005))

$$\Lambda = \frac{1}{T_m} \langle \Phi(\frac{f}{\kappa} - 1) \rangle, \tag{2.57}$$

T_m denotes the relaxation time for mechanical disturbances, the isotropic work–hardening–softening function κ is

$$\kappa = \hat{\kappa}(\in^p, \vartheta, \boldsymbol{\xi}), \tag{2.58}$$

Φ is the empirical overstress function, the bracket $\langle \cdot \rangle$ defines the ramp function,

$$\mathbf{P} = \left. \frac{\partial f}{\partial \boldsymbol{\tau}} \right|_{\boldsymbol{\xi}=const} \left(\left\| \frac{\partial f}{\partial \boldsymbol{\tau}} \right\| \right)^{-1}, \tag{2.59}$$

$\boldsymbol{\Omega}$, $\boldsymbol{\Xi}$ and $\mathbf{A}$ denote the evolution functions which have to be determined.

2.6.3 Microshear banding effects

To describe the microshear banding effects let us assume that the relaxation time T_m depends on the active microshear bands fraction f_{ms} and on the rate of equivalent viscoplastic deformation $\dot{\in}^P$ (cf. Pęcherski (1998) and Nowak, Perzyna and Pęcherski (2007)), i.e.

$$T_m = T_m \left(f_{ms}, \dot{\in}^P \right). \tag{2.60}$$

Additionally we introduce the simplification as follows

$$T_m = T_m^0 \phi_1 \left(f_{ms} \right) \phi_2 \left(\dot{\in}^P \right). \tag{2.61}$$

For example, for mild steel (cf. considerations in Section 3.3) we can postulate

$$\phi_1 \left(f_{ms} \right) = \left[1 - f_{ms} \left(\in^P \right) \right] = \left[1 - f_{ms}^0 \frac{1}{1 + \exp \left(a - b \in^P \right)} \right], \tag{2.62}$$

and

$$\phi_2 \left(\dot{\in}^P \right) = \left(\frac{\dot{\in}^P}{\dot{\in}_s^P} - 1 \right)^{\frac{1}{7}}, \tag{2.63}$$

where f^0_{ms}, a and b are material parameters.

Finally we have

$$T_m = T^0_m \left[1 - f^0_{ms} \frac{1}{1 + \exp\left(a - b \in^P\right)}\right] \left(\frac{\dot{\in}^P}{\dot{\in}^P_s} - 1\right)^{\frac{1}{7}}. \tag{2.64}$$

2.6.4 Constitutive assumption for the plastic spin

Let us postulate that $\boldsymbol{\Omega}$ has the form[4] (cf. Dafalias, 1983 and Loret, 1983)

$$\boldsymbol{\Omega} = \eta^* (\boldsymbol{\alpha} \cdot \mathbf{P} - \mathbf{P} \cdot \boldsymbol{\alpha}), \tag{2.65}$$

where η^* denotes the scalar valued function of the invariants of the tensors $\boldsymbol{\alpha}$, $\boldsymbol{\xi}$ and $\mathbf{P}$, and may depend on temperature ϑ.

To take into consideration the observed time dependent effects it is advantageous to use the proposition of the description of the intrinsic microdamage process presented by Perzyna (1986), Duszek–Perzyna and Perzyna (1994), and Dornowski and Perzyna (2000).

2.6.5 Anisotropic intrinsic microdamage mechanisms

Let us assume that the intrinsic microdamage process consists of the nucleation and growth mechanisms.[5]

Based on the heuristic suggestions and taking into account the influence of the stress triaxiality and anisotropic effects on the nucleation and growth mechanisms we assume the evolution equation for the microdamage tensor $\boldsymbol{\xi}$ as follows

$$\begin{aligned} \mathrm{L}_{\boldsymbol{\upsilon}} \boldsymbol{\xi} = & \frac{\partial h^*}{\partial \boldsymbol{\tau}} \frac{1}{T_m} \langle \Phi [\frac{\tilde{I}_n}{\tau_n(\boldsymbol{\xi}, \boldsymbol{\alpha}, \vartheta, \in^p)} - 1] \rangle + \\ & \frac{\partial g^*}{\partial \boldsymbol{\tau}} \frac{1}{T_m} \langle \Phi [\frac{\tilde{I}_g}{\tau_{eq}(\boldsymbol{\xi}, \boldsymbol{\alpha}, \vartheta, \in^p)} - 1] \rangle . \end{aligned} \tag{2.66}$$

The tensorial function $\frac{\partial h^*}{\partial \boldsymbol{\tau}}$ describes the mutual microcrack interaction for nucleation mechanism, while the tensional function $\frac{\partial g^*}{\partial \boldsymbol{\tau}}$ represents the mu-

[4] For a thorough discussion of a concept of the plastic spin and its constitutive description in phenomenological theories for macroscopic large plastic deformations please consult the critical review paper by Van der Giessen (1991).

[5] Recent experimental observation results (cf. Shockey, Seaman and Curran (1985)) have shown that coalescence mechanism can be treated as nucleation and growth process on a smaller scale.

tual microcrack interaction for growth process, τ_n and τ_{eq} denote the threshold stresses for microcrack nucleation and growth, respectively.

$$\tilde{I}_n = a_1 \tilde{J}_1 + a_2 \sqrt{\tilde{J}_2'} + a_3 \left(\tilde{J}_3'\right)^{\frac{1}{3}} \tag{2.67}$$

defines the stress intensity invariant for nucleation, a_i $(i = 1, 2, 3)$ are the material constants, $\tilde{J}_2'$ and $\tilde{J}_3'$ are the second and third invariants of the stress deviator $\tilde{\boldsymbol{\tau}}' = (\boldsymbol{\tau} - \boldsymbol{\alpha})'$,

$$\tilde{I}_g = b_1 \tilde{J}_1 + b_2 \sqrt{\tilde{J}_2'} + b_3 \left(\tilde{J}_3'\right)^{\frac{1}{3}}, \tag{2.68}$$

defines the stress intensity invariant for growth and b_i $(i = 1, 2, 3)$ are the material constants. This determines the evolution function $\boldsymbol{\Xi}$.

In the evolution equation(2.66) the functions

$$h = \hat{h}\left(\boldsymbol{\tau}, \vartheta, \in^P, \boldsymbol{\xi}, \boldsymbol{\alpha}\right) \tag{2.69}$$

$$g = \hat{g}\left(\boldsymbol{\tau}, \vartheta, \in^P, \boldsymbol{\xi}, \boldsymbol{\alpha}\right) \tag{2.70}$$

play the fundamental role, and have to be determined based on available experimental observation data. We also introduced the denotations as follows

$$\frac{\partial h^*}{\partial \boldsymbol{\tau}} = \frac{\partial \hat{h}}{\partial \boldsymbol{\tau}} \left(\left\|\frac{\partial \hat{h}}{\partial \boldsymbol{\tau}}\right\|\right)^{-1}, \qquad \frac{\partial g^*}{\partial \boldsymbol{\tau}} = \frac{\partial \hat{g}}{\partial \boldsymbol{\tau}} \left(\left\|\frac{\partial \hat{g}}{\partial \boldsymbol{\tau}}\right\|\right)^{-1}. \tag{2.71}$$

The threshold stress τ_n and τ_{eq} for microcrack nucleation and growth, respectively, can be assumed as the material functions in the form

$$\tau_n = \tau_n\left(\boldsymbol{\xi}, \boldsymbol{\alpha}, \vartheta, \in^P\right), \qquad \tau_{eq} = \tau_{eq}\left(\boldsymbol{\xi}, \boldsymbol{\alpha}, \vartheta, \in^P\right). \tag{2.72}$$

2.6.6 Kinematic hardening

To determine the evolution function $\mathbf{A}$ we shall follow some results obtained by Duszek and Perzyna (1991). The kinematic hardening evolution law takes the form

$$\mathrm{L}_{\boldsymbol{\upsilon}} \boldsymbol{\alpha} = \frac{1}{T_m} \langle \Phi(\frac{f}{\kappa} - 1) \rangle \left[r_1 \mathbf{P} + r_2 \frac{\mathbf{P} : \mathbf{Q}}{\tilde{\boldsymbol{\tau}} : \mathbf{Q} + r_3 \boldsymbol{\xi} : \mathbf{Q}} (\tilde{\boldsymbol{\tau}} + r_3 \boldsymbol{\xi}) \right], \tag{2.73}$$

where r_1, r_2 and r_3 are the material coefficients and

$$\mathbf{Q} = \left[\frac{\partial f}{\partial \boldsymbol{\tau}} + \left(\frac{\partial f}{\partial \boldsymbol{\xi}} - \frac{\partial \kappa}{\partial \boldsymbol{\xi}} \right) : \frac{\partial \boldsymbol{\xi}}{\partial \boldsymbol{\tau}} \right] \left\| \frac{\partial f}{\partial \boldsymbol{\tau}} + \left(\frac{\partial f}{\partial \boldsymbol{\xi}} - \frac{\partial \kappa}{\partial \boldsymbol{\xi}} \right) : \frac{\partial \boldsymbol{\xi}}{\partial \boldsymbol{\tau}} \right\|^{-1}. \tag{2.74}$$

The kinetic law (2.73) represents the linear combination of the Prager and Ziegler kinematic hardening rules and additionally depends linearly on the microdamage tensor $\boldsymbol{\xi}$.

2.6.7 Thermodynamic restrictions and rate type constitutive relations

Suppose the axiom of the entropy production holds. Then the constitutive assumption (2.51) and the evolution equations (2.56) lead to the results as follows

$$\boldsymbol{\tau} = \rho_{Ref}\frac{\partial\hat{\psi}}{\partial\mathbf{e}}, \qquad \eta = -\frac{\partial\hat{\psi}}{\partial\vartheta}, \qquad -\frac{\partial\hat{\psi}}{\partial\boldsymbol{\mu}}\cdot \mathrm{L}_{\boldsymbol{\upsilon}}\boldsymbol{\mu} - \frac{1}{\rho\vartheta}\mathbf{q}\cdot \mathrm{grad}\vartheta \geq 0. \tag{2.75}$$

The rate of internal dissipation is determined by

$$\begin{aligned}\vartheta\hat{i} &= -\frac{\partial\hat{\psi}}{\partial\boldsymbol{\mu}}\cdot \mathrm{L}_{\boldsymbol{\upsilon}}\boldsymbol{\mu} \qquad (2.76)\\ &= -\left[\frac{\partial\hat{\psi}}{\partial \in^P}\sqrt{\frac{2}{3}} + \frac{\partial\hat{\psi}}{\partial\boldsymbol{\alpha}} : \left(r_1\mathbf{P} + r_2\frac{\mathbf{P}:\mathbf{Q}}{\tilde{\boldsymbol{\tau}}:\mathbf{Q} + r_3\boldsymbol{\xi}:\mathbf{Q}}(\tilde{\boldsymbol{\tau}} + r_3\boldsymbol{\xi})\right)\right]\Lambda\\ &\quad -\frac{\partial\hat{\psi}}{\partial\boldsymbol{\xi}} : \boldsymbol{\Xi}.\end{aligned}$$

Operating on the stress relation $(2.75)_1$ with the Lie derivative and keeping the internal state vector constant, we obtain

$$\mathrm{L}_{\boldsymbol{\upsilon}}\boldsymbol{\tau} = \mathcal{L}^e : \mathbf{d} - \mathcal{L}^{th}\dot{\vartheta} - [(\mathcal{L}^e + \mathbf{g}\boldsymbol{\tau} + \boldsymbol{\tau}\mathbf{g} + \mathcal{W}) : \mathbf{P}]\frac{1}{T_m}\langle\Phi(\frac{f}{\kappa} - 1)\rangle, \tag{2.77}$$

where

$$\mathcal{L}^e = \rho_{Ref}\frac{\partial^2\hat{\psi}}{\partial\mathbf{e}^2}, \qquad \mathcal{L}^{th} = -\rho_{Ref}\frac{\partial^2\hat{\psi}}{\partial\mathbf{e}\partial\vartheta},$$
$$\mathcal{W} = \eta^*[(\mathbf{g}\boldsymbol{\tau} - \boldsymbol{\tau}\mathbf{g}) : (\boldsymbol{\alpha}\mathbf{g} - \mathbf{g}\boldsymbol{\alpha})]. \tag{2.78}$$

Substituting $\dot{\psi}$ into the energy balance equation and taking into account the results $(2.75)_3$ and (2.76) gives

$$\rho\vartheta\dot{\eta} = -\mathrm{div}\mathbf{q} + \rho\vartheta\hat{i}. \tag{2.79}$$

Operating on the entropy relation $(2.75)_2$ with the Lie derivative and substituting the result into (2.79) we obtain

$$\rho c_p\dot{\vartheta} = -\mathrm{div}\mathbf{q} + \vartheta\frac{\rho}{\rho_{Ref}}\frac{\partial\boldsymbol{\tau}}{\partial\vartheta} : \mathbf{d} + \rho\chi^*\boldsymbol{\tau} : \mathbf{d}^p + \rho\chi^{**}\mathbf{K} : \mathrm{L}_{\boldsymbol{\upsilon}}\boldsymbol{\xi}, \tag{2.80}$$

where the specific heat

$$c_p = -\vartheta\frac{\partial^2\hat{\psi}}{\partial\vartheta^2}, \qquad \chi^{**}\mathbf{K} = -\left(\frac{\partial\hat{\psi}}{\partial\boldsymbol{\xi}} - \vartheta\frac{\partial^2\hat{\psi}}{\partial\vartheta\partial\boldsymbol{\xi}}\right) \tag{2.81}$$

and χ^* and χ^{**} are the irreversibility coefficients.

2.6.8 Fracture criterion based on the evolution of microdamage

We base the fracture criterion on the evolution of the microdamage tensor $\boldsymbol{\xi}$.

Let us assume that for $(\boldsymbol{\xi} : \boldsymbol{\xi})^{\frac{1}{2}} = \xi^F$ catastrophe takes place (cf. Perzyna (1984)), that is

$$\kappa = \hat{\kappa}(\in^p, \vartheta, \boldsymbol{\xi})|_{(\boldsymbol{\xi}:\boldsymbol{\xi})^{\frac{1}{2}}=\xi^F} = 0. \tag{2.82}$$

It means that for $(\boldsymbol{\xi} : \boldsymbol{\xi})^{\frac{1}{2}} = \xi^F$ the material loses its carrying capacity. The condition (2.82) describes the main feature observed experimentally that the load tends to zero at the fracture point.

In is noteworthy that the isotropic hardening–softening material function $\hat{\kappa}$ proposed in Eq. (2.58) should satisfy the fracture criterion (2.82).

2.7 Analysis of the Fundamental Features of the Model

2.7.1 Invariance with respect to diffeomorphism

Our description of the thermo–elasto–viscoplastic constitutive structure is invariant with respect to any diffeomorphism. It means that the constitutive structure is invariant to any superposed motion. Such constitutive structure is called covariant, cf. Marsden and Hughes (1983).

The covariance property of the constitutive structure has been achieved due to the assumption that the rates of the deformation tensors and the stress tensors (as well as all vectors and tensors) are defined based on the Lie derivative. Of course, the covariance description has very important consequence for proper mathematical investigations of some phenomena which can be discussed in the solution of the evolution problems. It will be also very crucial for proper description of meso–, micro–, and nano–mechanical problems and particularly in their numerical solutions.

2.7.2 Finite plastic deformation and plastic spin effects

The kinematics of finite elasto–viscoplastic deformation is based on notions of the Riemann space on manifolds and tangent spaces, cf. Marsden and Hughes (1983). A multiplicative decomposition of the deformation gradient into elastic and viscoplastic parts is assumed. The viscoplastic spin has been also taken into account in the constitutive structure.

Due to these assumptions we obtain the consistent description of finite elasto–viscoplastic deformations. The main effect generated by the viscoplastic spin can be observed from the rate type constitutive relation for the Kirchhoff stress tensor $\boldsymbol{\tau}$ (cf. Eq. (2.77)). In this equation the additional term $\frac{1}{T_m}\langle\Phi(\frac{f}{\kappa}-1)\rangle\mathcal{W} : \mathbf{P}$ describes the influence of the viscoplastic spin.

2.7.3 Plastic non–normality

Plastic non–normality (i.e. $\mathbf{P} \neq \mathbf{Q}$ is generated by the influence of microdamage mechanisms. This effect is clear from the comparison of $\mathbf{P}$ (cf. Eq. (2.59)) and $\mathbf{Q}$ (cf. Eq. (2.74)). It has very important influence on the description of localization and localized fracture phenomena.

2.7.4 Softening effects generated by microdamage mechanisms

By introducing the internal state variable $\boldsymbol{\xi}$, i.e. the microdamage tensor in the constitutive structure the description of the intrinsic microdamage process has been achieved. From the available experimental observations we learn that there are three main mechanisms responsible for the evolution of microdamage, namely nucleation, growth and coalescence of microcracks. Taking advantage of the conjecture that coalescence mechanism can be treated as nucleation and growth process on a small scale we simplified the description of the intrinsic microdamage by making allowance only for the nucleation and growth mechanisms.

From the fundamental rate equation for temperature ϑ (cf. Eq. (2.80)) we observe that the microdamage mechanism generates dissipation effects. On the other hand from the form of the isotropic work–hardening–softening function κ (cf. Eq. (2.58)) we see the direct description of softening effects caused by the intrinsic microdamage process.

It is noteworthy to add that the fundamental fracture criterion (cf. Eq. (2.82)) is also based on the evolution of microdamage.

2.7.5 Anisotropic effects in microdamage mechanisms

The induced anisotropy generated by the microdamage machanisms is described by the internal state variable $\boldsymbol{\xi}$, i.e. the microdamage tensor.

Kinetic equation for $\boldsymbol{\xi}$ takes into account the influence of the stress triaxiality and anisotropic effects on the nucleation and growth mechanisms.

2.7.6 Plastic deformation induced anisotropic effects

Experimental observations of plastic flow processes have shown that finite plastic deformations induced the effect of anisotropy of a material. To describe this effect the residual stress (the back stress) $\boldsymbol{\alpha}$ has been introduced to the constitutive structure as the internal state variable. The aim of this is to take into account the kinematic hardening phenomenon, which can be treated as first approximation of the description of plastic deformation induced anisotropy.

The evolution equation postulated for the back stress $\boldsymbol{\alpha}$ represents the linear combination of the Prager and Ziegler kinematic hardening rules and additionally depends on the microdamage tensor $\boldsymbol{\xi}$.

This kind of anisotropy generated by the kinematic hardening law will have very important influence on the localized fracture phenomena in thermo–viscoplastic flow processes under cyclic dynamic loadings.

2.7.7 Thermomechanical couplings (thermal plastic softening and thermal expansion)

An analysis of thermomechanical coupling effects can be based on the investigation of the evolution problem formulated for a thermo–elasto–viscoplastic model. Thermal expansion effect influences the fundamental, adiabatic, elastodynamic matrix $\mathbb{E}$ for an adiabatic process, so it can have very important influence on the propagation and interaction of the deformation waves.

On the other hand the thermal plastic softening as typical dissipative effect influences an adiabatic process through the nonlinear function $\mathbf{f}$. Of course, it will have a lot of influence on the localization phenomena as well as on the fracture phenomena.[6]

2.7.8 Influence of stress triaxiality on the evolution of microdamage

The introduced modification of the kinetics of microdamage leads to the description of the influence of stress triaxiality. This has been done by assumption that the stress intensity invariants for nucleation and growth depend on the three invariants of stress tensor, cf. Eqs. (2.67) and (2.68).

It has been shown by Dornowski and Perzyna (1999, 2000, 2002) that this modification helps to describe the accumulation of damage during dynamic loading process.

2.7.9 Rate sensitivity

To analyse the influence of the strain rate effect let as write the evolution constitutive equation $(2.56)_1$ in the form as follows

$$\|\mathbf{d}^P\| = \frac{1}{T_m}\langle\Phi(\frac{f}{\kappa} - 1)\rangle\|\mathbf{P}\|, \tag{2.83}$$

[6]For a thorough analytical discussion of the thermal plastic softening and thermal expansion effects on the localization of plastic deformation please consult the papers by Duszek–Perzyna and Perzyna (1994, 1998).

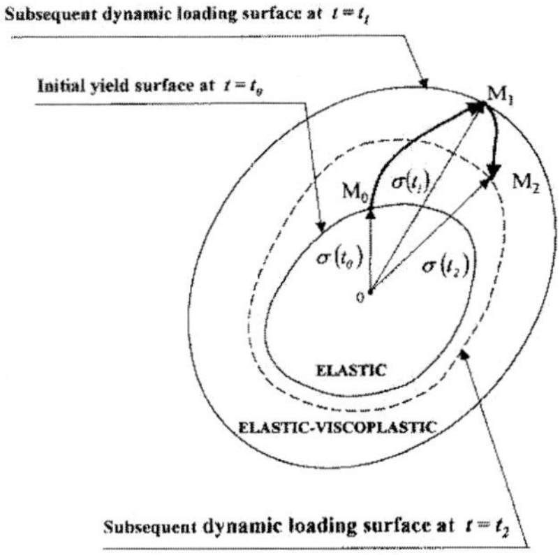

Figure 2.22: Dynamic loading surface for the elasto–viscoplastic response

where $\|\cdot\|$ denotes the norm of the vectors in the 6−dimensional space. The equation (2.83) yields the following dynamical yield criterion

$$f = \hat{\kappa}(\in^P, \vartheta, \boldsymbol{\xi}) \left\{ 1 + \Phi^{-1} \left[T_m^0 \phi_1(f_{ms}) \phi_2 \left(\dot{\in}^P \right) \frac{\|\mathbf{d}^P\|}{\|\mathbf{P}\|} \right] \right\}. \tag{2.84}$$

This relation may be interpreted as a description of actual change of the yield surface during the thermodynamical process. This change is caused by isotropic and anisotropic work–hardening–softening effects, by the evolution of substructure and dependence on temperature and by the influence of the strain rate effects. It is noteworthy to add that the relation (2.84) constitutes a basis for experimental investigations which seek to examine the theoretical assumptions, cf. Fig. 2.22.

2.7.10 Length–scale sensitivity of the constitutive model

The constitutive equations for a thermo–elastic–viscoplastic model introduce implicitly a length–scale parameter into the dynamic initial–boundary value problem, i.e.

$$l = \beta c T_m \quad \text{or} \quad l = \beta c T_m^0 \phi_1(f_{ms}) \phi_2(\dot{\in}^P), \tag{2.85}$$

and as example for mild steel we have

$$l = \beta c T_m^0 \left[1 - f_{ms}^0 \frac{1}{1 + \exp(a - b \in^P)} \right] \left(\frac{\dot{\in}^P}{\dot{\in}_s^P} - 1 \right)^{\frac{1}{7}}, \tag{2.86}$$

where T_m is the relaxation time for mechanical disturbances, and is directly related to the viscosity of the material, c denotes the velocity of the propagation of the elastic waves in the problem under consideration, and the proportionality factor β depends on the particular initial–boundary value problem and may also be conditioned on the microscopic properties of the material.

The relaxation time T_m can be viewed either as a microstructural parameter to be determined from experimental observations or as a mathematical regularization parameter.

It is noteworthy to stress that the length–scale sensitivity of the constitutive structure is of great importance for proper description of mesomechanical problems.

2.7.11 Dissipation effects

We observe that for an adiabatic process **dissipation effects** are generated by following phenomena: (i) viscoplastic flow; (ii) viscoplastic deformation induced anisotropy; and (iii) microdamage mechanisms.

For an adiabatic process ($\mathbf{q} = 0$) Eq. (2.80) takes the form

$$c_p \dot{\vartheta} = \vartheta \frac{1}{\rho_{Ref}} \frac{\partial \boldsymbol{\tau}}{\partial \vartheta} : \mathbf{d} + \chi * \boldsymbol{\tau} : \mathbf{d}^P + \chi^{**} \mathbf{K} : \mathrm{L}_{\boldsymbol{\upsilon}} \boldsymbol{\xi}. \tag{2.87}$$

When the nondissipative term is neglected, then Eq. (2.87) takes the form

$$c_p \dot{\vartheta} = \chi^* \boldsymbol{\tau} : \mathbf{d}^p + \chi^{**} \mathbf{K} : \mathrm{L}_{\boldsymbol{\upsilon}} \boldsymbol{\xi}. \tag{2.88}$$

From Eq. (2.88) we can compute the irreversibility coefficient χ^*. It gives

$$\chi^* = \frac{c_p \dot{\vartheta} - \chi^{**} \mathbf{K} : \mathrm{L}_{\boldsymbol{\upsilon}} \boldsymbol{\xi}}{\boldsymbol{\tau} : \mathbf{d}^P}. \tag{2.89}$$

For $\chi^{**} = 0$, i.e. when the influence of the intrinsic microdamage mechanism is not taken into consideration, Eq. (2.89) takes the form

$$\chi^* = \frac{c_p \dot{\vartheta}}{\boldsymbol{\tau} : \mathbf{d}^P}. \tag{2.90}$$

For this particular case the irreversibility coefficient χ^* has a simple interpretation as the heat rate conversion to plastic work rate fraction. However,

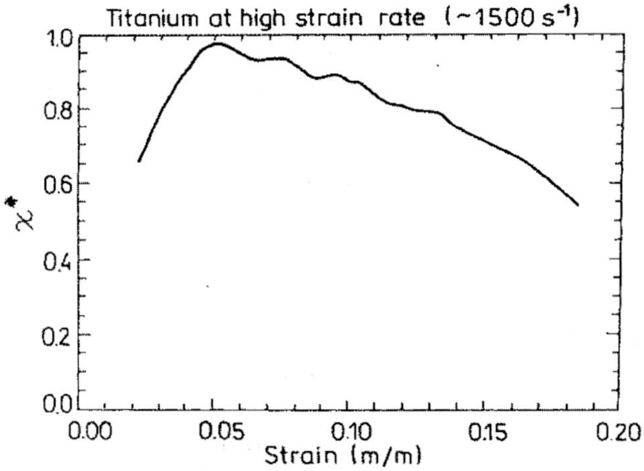

Figure 2.23: The irreversibility coefficient χ^* versus strain calculated for Ti–6Al–4V titanium using the average of the temperature of the two detectors (After Mason et al. (1994))

Eq. (2.89) shows that the remaining work rate is attributed to the energy rate lost for microdamage effects.

When modelling thermomechanical behaviour of materials, χ^* is usually assumed to be a constant in the range $0.85 - 0.95$ (a practice that dates back to the work of Taylor and Quinney (1934)).

Recent experimental investigations performed by Mason et al. (1994) by using a Kolsky (split Hopkinson) pressure bar and a high–speed infrared detector array have clearly shown that this assumption may not be correct for all metals, cf. Fig. 2.23.

The reason for this considerable discrepancy is clearly visible from Eq. (2.89). The rate of the stored energy implied by the evolution of microdamage is responsible for the decreasing of χ^* (e.g. as it has been observed for Ti–6Al–4V deformed at high strain rates).

2.7.12 Dispersion effects

It is very well known fact that the stress wave propagation in an elasto–viscoplastic medium has the dispersive nature.

The dispersion effect is very crucial for the development of the regularization procedure for the rate independent plastic flow evolution problems, cf. Section 2.7.14.

Of course, the dispersion effects will also influence very much the initiation and development of localization phenomena. A thorough analysis of these consequences has been presented in Glema, Łodygowski and Perzyna (2000, 2001, 2003).

2.7.13 Regularization of the evolution problem

Very recently, it has been widely recognized to consider an elastic–viscoplastic model of a material as a regularization method for solving mesh–dependent problems of plasticity. In these regularized initial–boundary value problems, wave propagation phenomena play a fundamental role. An elastic–viscoplastic model introduces dissipative as well as dispersive nature for the propagated waves. The dispersion property implies that in the elastic–viscoplastic medium any initial disturbance can break up into a system of group of oscillations or wavelets. On the other hand, the dissipation property causes the amplitude of a harmonic wavetrain to decay with time. In the evolution problem considered in such dissipative and dispersive medium, the stress deformation due to wave reflections and interactions are not uniformly distributed, and this kind of heterogeneity can lead to strain localization in the absence of geometrical or material imperfections.

2.7.14 Synergetic effects generated by cooperative phenomena

The question arises as follows: can cooperative phenomena described by a thermo–elasto–viscoplastic model lead to some synergetic effects. This fundamental question has not easy answer. However, some investigations of the influence of synergetic effects on localization phenomena have been already presented, cf. for single crystals: Duszek–Perzyna and Perzyna (1993, 1996), Duszek–Perzyna, Korbel and Perzyna (1997), Perzyna and Korbel (1996, 1998) and Perzyna (1998, 2002); and for polycrystalline solids: Duszek–Perzyna and Perzyna (1991, 1994, 1995, 1998) and Perzyna (1994, 1995, 1998). Numerical investigation of synergetic effects for localization and fracture phenomena has been given in Duszek–Perzyna and Perzyna (1994, 1998) and Łodygowski and Perzyna (1997).

3. Constitutive Modelling for Dynamic Cyclic Loadings

3.1 Experimental and physical motivations

Sidey and Coffin (1979) investigated the mechanisms of fatigue as they bear on fatigue damage at elevated temperature when time and strain rate effects are important. This regime is often referred to as that of time–dependent fatigue, cf. Dornowski and Perzyna (2000).

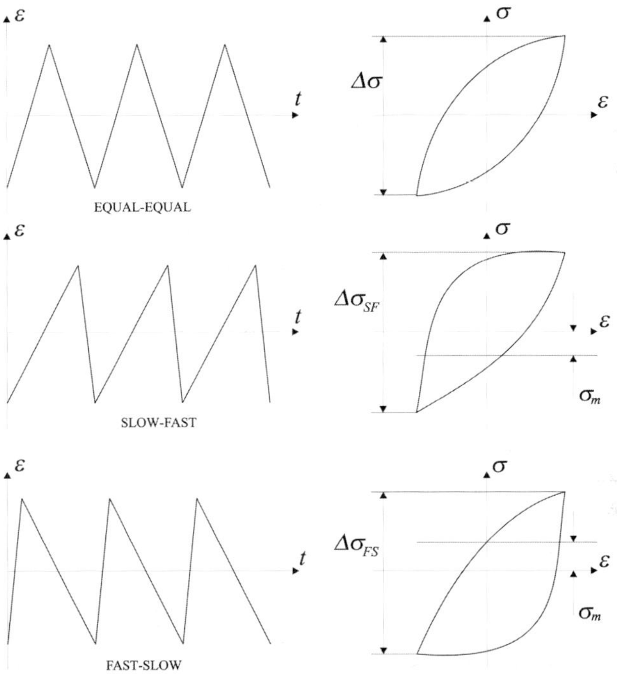

Figure 3.1: Wave shapes and resulting hysteresis loops for equal and unequal forward and reserve strain rates (After Sidey and Coffin (1979))

Test on oxygen-free high conductivity (OFHC) copper at 673 K are examined using unequal strain rates to produce the wave shape. Some typical wave shape are shown in Fig. 3.1, they include equal–equal, slow–fast and fast–slow. Specimens with a gage length of 12.70 mm and diameter of 6.35 mm were used. Strain–controlled fatigue tests were carried out at 673 K in air with a total strain range of 1.0 percent. In each test the cycle time was kept constant at 600 s but the tensile and compressive strain rates were varied so that a study of wave–shape effects could be made. In the case of the unbalanced loops, the ratio of fast to slow strain rates was fixed at 100 to 1 with the slow strain rate being 1.7×10^{-5} s^{-1}. A strain rate of 3.3×10^{-5} s^{-1} was used in equal ramp rate tests. After failure, the gage–length was sectioned longitudinally for metallographic observations.

Table 3.1: Effect of wave–shape on the number of cycles to failure of OFHC copper at 673 K

Cycle	Slow–Fast	Equal	Fast–Slow
N_f	104	380	1138
$t_f(h)$	17	63	190

Table 3.1 shows the number of cycles to failure and the time to failure under the various testing conditions. It can be seen that when the total cycle time is kept constant the number of cycles to failure decreases as the tensile–going strain rate decreases.

Figure 3.2 shows the failure crack in the fast–slow test. The crack path is transgranular and has started from the surface. Many transgranular surface cracks had initiated and grown for depths up to 0.5 mm but presumably these had ceased growing when one crack became dominant. No internal intergranular cracks were observed.

In contrast, failure in the slow–fast test was intergranular (Fig. 3.3). Near the fracture edge, extensive intergranular cracks can be seen which have been opened out by the final failure. Many of these cracks were wedge type, but at higher magnifications linked cavities could be seen. It was noted that cavitation was present throughout the gage length. Thus the failure was typical of that for creep fracture with most of the cracks being orientated at right angles to the applied stress direction. The fracture path in the equal strain rate test was intergranular (Fig. 3.4). Also internal and short surface intergranular cracks were observed. The surface cracks were about one grain in depth and less numerous than in the slow–fast specimen. There was evidence of cavities either near the fracture or in the bulk of the specimen.

The slow strain rate tension test failed by the propagation of an intergranular crack. Microscopically the specimen was very similar to the equal ramp test with intergranular wedge cracks being situated near the fracture surface and no cracking in the bulk region.

Thus, metallography of the specimen indicated that the decrease in fatigue life was associated with a change in the fracture mode from trans-

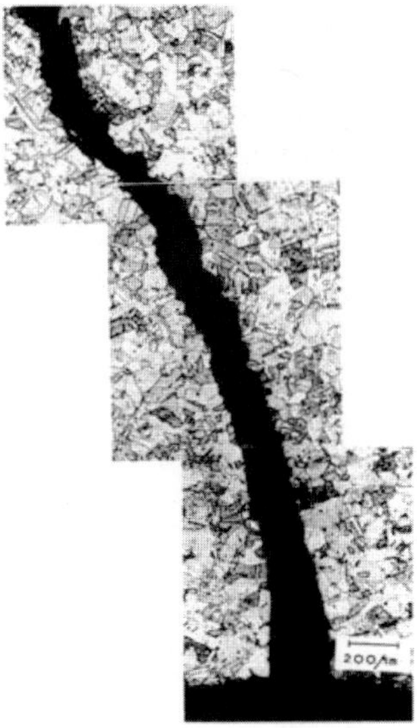

Figure 3.2: Main failure crack in OFHC copper after fast–slow cycling showing transgranular fracture mode (After Sidey and Coffin (1979))

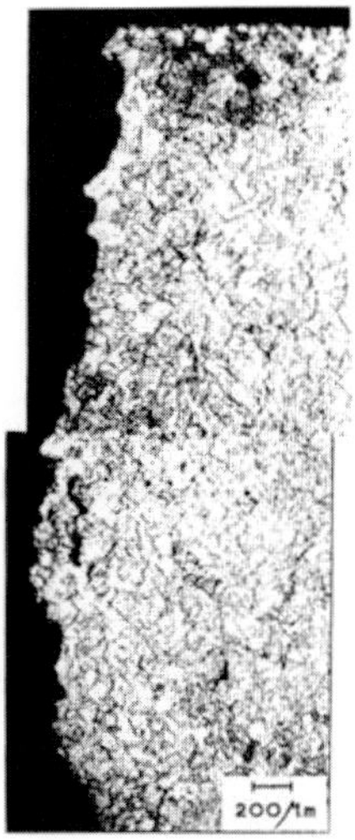

Figure 3.3: Fracture edge of slow–fast cycle specimen showing intergranular fracture path and surface and interior intergranular cracks (After Sidey and Coffin (1979))

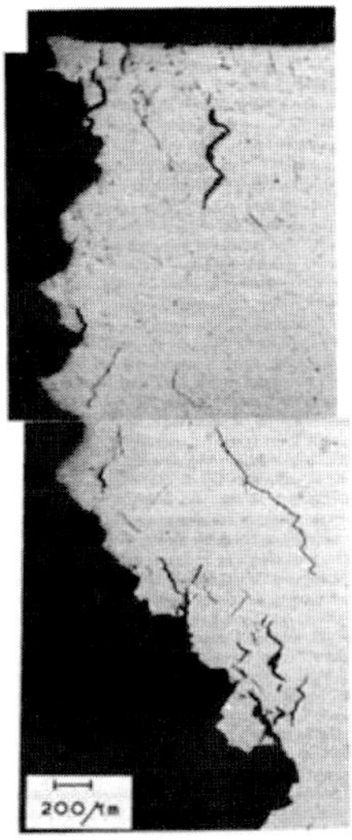

Figure 3.4: Equal ramp cycle specimen in unetched condition showing interlinkage of wedge cracks near the fracture edge (After Sidey and Coffin (1979))

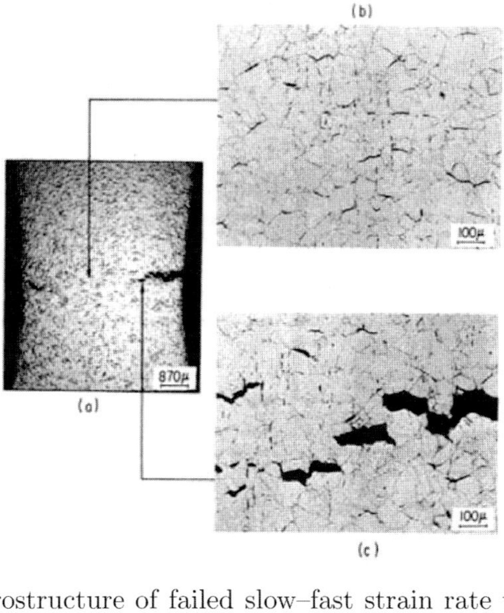

Figure 3.5: Microstructure of failed slow–fast strain rate test on AISI 304 stainless steel at 923 K in air (After Sidey and Coffin (1979))

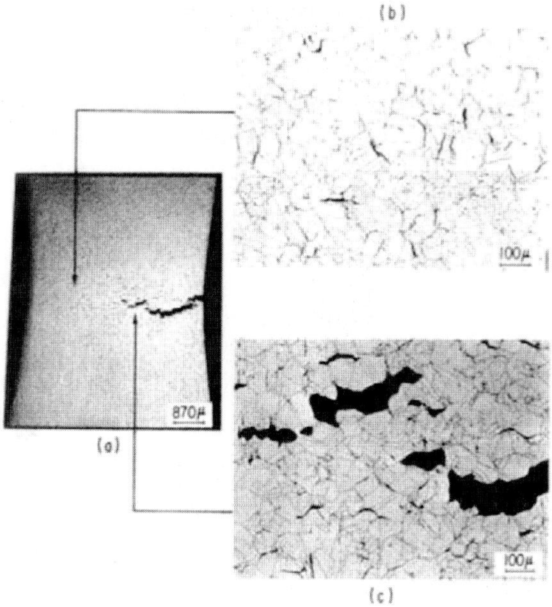

Figure 3.6: Microstructure of failed slow–fast strain rate test on AISI 304 stainless steel at 923 K in vacuum (After Sidey and Coffin (1979))

granular to intergranular cracking. Cavity damage occurred only when the tensile–going strain rate was less than the compressive–going strain rate.

Similar results have been obtained for the stainless steel (AISI 304) tested at 923 K in air and vacuum, cf. Sidey and Coffin (1979). Factography revealed that both the air and vacuum slow–fast tests failed by intergranular fracture, and, in addition, extensive interior grain boundary cracking was observed, as shown in Figs. 3.5 and 3.6. This indicates that interior damage processes are unaffected by the external environment, but appear to be strongly influenced by the wave shape. They found that for equal–equal and fast–slow tests the tension hold test has a degrading effect on life, but not as severe as that of the slow–fast test. Metallographic and factographic findings have shown that the slow–fast tests developed intergranular cracks and grain–boundary cavities, while the fracture surface for the equal strain rate and fast–slow tests were transgranular.

3.2 Fundamental assumptions

A number of plasticity models have been recently proposed for cyclic loadings, cf. Auricchio, Taylor and Lubliner (1992), Auricchio and Taylor (1995), Chaboche (1986), Dafalias and Popov (1976), Duszek and Perzyna (1991), Mróz (1967), Ristinmaa (1995), Van der Giessen (1989) and Wang and Ohno (1991).

However none of these theories is able to describe properly the mechanism of fatigue damage when time and strain rate effects are important. Such effects have been observed by Sidey and Coffin (1979). They investigated oxygen–free high conductivity (OFHC) copper at 673 K using unequal strain rates to produce the wave shape. It has been observed that the intrinsic micro–damage process does very much depend on the strain rate effects, cf. chapter 2.

The main objective is the development of the thermo–elasto–viscoplasticity theory of damaged polycrystalline solids which takes into account the time and temperature dependent effects observed experimentally and the accumulation of the plastic deformation and damage during dynamic cyclic loading processes.

To describe these effects we intend to modify a constitutive model of thermo–elasto–viscoplastic damaged polycrystalline solids developed by Duszek–Perzyna and Perzyna (1994). The main modification concerns the kinetics of microdamage and the kinematic hardening law.

For our practical purposes it is sufficient to assume that the internal state vector $\boldsymbol{\mu}$ has the form as follows

$$\boldsymbol{\mu} = (\in^p, \xi, \boldsymbol{\alpha}). \tag{3.1}$$

The evolution equations for $\mathbf{d}^p$, $\boldsymbol{\omega}^p$ and $\boldsymbol{\alpha}$ are postulated as in Chapter 2 (cf. Eqs. (2.56)) and for the porosity ξ in the form

$$\dot{\xi} = \Xi. \tag{3.2}$$

3.3 Intrinsic micro–damage process

An analysis of the experimental observations for cycle fatigue damage mechanisms at high temperature of metals performed by Sidey and Coffin (1979) suggests that the intrinsic micro–damage process does very much depend on the strain rate effects as well as on the wave shape effects. In the tests in which duration of extension stress was larger than duration of compression stress (in single cycle) decreasing of the fatigue lifetime was observed and fracture mode changed from a transgranular fracture for the fast–slow wave shape to an intergranular single–crack fracture for equal ramp rates to interior cavitation for the slow–fast test.

To take into consideration these observed time dependent effects it is advantageous to use the proposition of the description of the intrinsic micro–damage process presented by Perzyna (1986a,b) and Duszek–Perzyna and Perzyna (1994) and Dornowski and Perzyna (1999, 2000, 2002).

Based on this heuristic suggestion and taking into account the influence of the stress triaxiality on the nucleation mechanism we postulate for rate dependent plastic flow

$$\left(\dot{\xi}\right)_{nucl} = \frac{1}{T_m} h^*(\xi, \vartheta) \left[\exp \frac{m^*(\vartheta) \mid \tilde{I}_n - \tau_n(\xi, \vartheta, \in^p) \mid}{k\vartheta} - 1 \right], \tag{3.3}$$

where k denotes the Boltzmann constant, $h^*(\xi, \vartheta)$ represents a void nucleation material function which is introduced to take account of the effect of microvoid interaction effect, $m^*(\vartheta)$ is a temperature dependent coefficient, $\tau_n(\xi, \vartheta, \in^p)$ is the porosity, temperature and equivalent plastic strain dependent threshold stress for microvoid nucleation,

$$\tilde{I}_n = a_1 \tilde{J}_1 + a_2 \sqrt{\tilde{J}_2'} + a_3 \left(\tilde{J}_3'\right)^{\frac{1}{3}} \tag{3.4}$$

defines the stress intensity invariant for nucleation, a_i $(i = 1, 2, 3)$ are the material constants, $\tilde{J}_1$ denotes the first invariant of the stress tensor $\tilde{\boldsymbol{\tau}} = \boldsymbol{\tau} - \boldsymbol{\alpha}$, $\tilde{J}_2'$ and $\tilde{J}_3'$ are the second and third invariants of the stress deviator $\tilde{\boldsymbol{\tau}}' = (\boldsymbol{\tau} - \boldsymbol{\alpha})'$.

For the growth mechanism we postulate (cf. Johnson (1981); Perzyna (1986a,b); Perzyna and Drabik (1989, 2008))

$$\left(\dot{\xi}\right)_{grow} = \frac{1}{T_m} \frac{g^*(\xi, \vartheta)}{\kappa_0} \left[\tilde{I}_g - \tau_{eq}(\xi, \vartheta, \in^p) \right], \tag{3.5}$$

where $T_m \kappa_0$ denotes the dynamic viscosity of a material, $g^*(\xi, \vartheta)$ represents a void growth material function and takes account for void interaction, $\tau_{eq}(\xi, \vartheta, \in^p)$ is the porosity, temperature and equivalent plastic strain dependent void growth threshold stress,

$$\tilde{I}_g = b_1 \tilde{J}_1 + b_2 \sqrt{\tilde{J}_2'} + b_3 \left(\tilde{J}_3'\right)^{\frac{1}{3}}, \tag{3.6}$$

defines the stress intensity invariant for growth and b_i $(i = 1, 2, 3)$ are the material constants.

Finally the evolution equation for the porosity ξ has the form

$$\dot{\xi} = \frac{h^*(\xi,\vartheta)}{T_m}\left[\exp\frac{m^*(\vartheta) \mid \tilde{I}_n - \tau_n(\xi,\vartheta,\in^p) \mid}{k\vartheta} - 1\right] + \frac{g^*(\xi,\vartheta)}{T_m\kappa_0}\left[\tilde{I}_g - \tau_{eq}(\xi,\vartheta,\in^p)\right]. \tag{3.7}$$

This determines the evolution function Ξ.

To have consistent theory of elasto–viscoplasticity we can replace the exponential function in the nucleation term and the linear function in the growth term by the empirical overstress function Φ, then the evolution equation for the porosity ξ takes the form as follows

$$\dot{\xi} = \frac{1}{T_m}h^*(\xi,\vartheta)\langle\Phi[\frac{\tilde{I}_n}{\tau_n(\xi,\vartheta,\in^p)} - 1]\rangle + \frac{1}{T_m}g^*(\xi,\vartheta)\langle\Phi[\frac{\tilde{I}_g}{\tau_{eq}(\xi,\vartheta,\in^p)} - 1]\rangle. \tag{3.8}$$

3.4 Kinematic hardening

For a constitutive model describing the behaviour of a material under cyclic loading processes the crucial role plays the evolution equation for the back stress $\boldsymbol{\alpha}$, which is responsible for the description of the induced plastic strain anisotropy effects.

We shall follow some fundamental results obtained by Duszek and Perzyna (1991a). Let us postulate

$$\mathrm{L}_{\boldsymbol{\upsilon}}\boldsymbol{\alpha} = A(\mathbf{d}^p, \tilde{\boldsymbol{\tau}}, \vartheta, \xi). \tag{3.9}$$

Making use of the tensorial representation of the function A and taking into account that there is no change of $\boldsymbol{\alpha}$ when $\tilde{\boldsymbol{\tau}} = 0$ and $\mathbf{d}^p = 0$ the evolution law (3.9) can be written in the form (cf. Truesdell and Noll (1965))

$$\begin{aligned}\mathrm{L}_{\boldsymbol{\upsilon}}\boldsymbol{\alpha} \;=\;& \eta_1\mathbf{d}^p + \eta_2\tilde{\boldsymbol{\tau}} + \eta_3\mathbf{d}^{p^2} + \eta_4\tilde{\boldsymbol{\tau}}^2 + \eta_5\left(\mathbf{d}^p\cdot\tilde{\boldsymbol{\tau}} + \tilde{\boldsymbol{\tau}}\cdot\mathbf{d}^p\right)\\ &+\eta_6\left(\mathbf{d}^{p^2}\cdot\tilde{\boldsymbol{\tau}} + \tilde{\boldsymbol{\tau}}\cdot\mathbf{d}^{p^2}\right) + \eta_7\left(\mathbf{d}^p\cdot\tilde{\boldsymbol{\tau}}^2 + \tilde{\boldsymbol{\tau}}^2\cdot\mathbf{d}^p\right)\\ &+\eta_8\left(\mathbf{d}^{p^2}\cdot\tilde{\boldsymbol{\tau}}^2 + \tilde{\boldsymbol{\tau}}^2\cdot\mathbf{d}^{p^2}\right),\end{aligned} \tag{3.10}$$

where $\eta_1, \ldots, \eta_8$ are functions of the basic invariant of $\mathbf{d}^p$ and $\tilde{\boldsymbol{\tau}}$, the porosity parameter ξ and temperature ϑ.

A linear approximation of the general evolution law (3.10) leads to the result

$$\mathrm{L}_{\boldsymbol{\upsilon}}\boldsymbol{\alpha} = \eta_1\mathbf{d}^p + \eta_2\tilde{\boldsymbol{\tau}}. \tag{3.11}$$

This kinetic law represents the linear combination of the Prager and Ziegler kinematic hardening rules, cf. Prager (1955) and Ziegler (1959).

To determine the connection between the material functions η_1 and η_2 we take advantage of the geometrical relation (cf. Duszek and Perzyna (1991a))

$$(\mathrm{L}_{\boldsymbol{\upsilon}}\boldsymbol{\alpha} - r\mathbf{d}^p) : \mathbf{Q} = 0, \tag{3.12}$$

where

$$\mathbf{Q} = \left[\frac{\partial f}{\partial \boldsymbol{\tau}} + \left(\frac{\partial f}{\partial \xi} - \frac{\partial \kappa}{\partial \xi}\right)\frac{\partial \xi}{\partial \boldsymbol{\tau}}\right] \left\|\frac{\partial f}{\partial \boldsymbol{\tau}} + \left(\frac{\partial f}{\partial \xi} - \frac{\partial \kappa}{\partial \xi}\right)\frac{\partial \xi}{\partial \boldsymbol{\tau}}\right\|^{-1}, \tag{3.13}$$

and r denotes the new material function.

The relation (3.12) leads to the result

$$\eta_2 = \frac{1}{T_m}\langle\Phi(\frac{f}{\kappa} - 1)\rangle\left[r(\xi, \vartheta) - \eta_1\right]\frac{\mathbf{P} : \mathbf{Q}}{\tilde{\boldsymbol{\tau}} : \mathbf{Q}}. \tag{3.14}$$

Finally the kinematic hardening evolution law takes the form

$$\mathrm{L}_{\boldsymbol{\upsilon}}\boldsymbol{\alpha} = \frac{1}{T_m}\langle\Phi(\frac{f}{\kappa} - 1)\rangle\left[r_1(\xi, \vartheta)\mathbf{P} + r_2(\xi, \vartheta)\frac{\mathbf{P} : \mathbf{Q}}{\tilde{\boldsymbol{\tau}} : \mathbf{Q}}\tilde{\boldsymbol{\tau}}\right], \tag{3.15}$$

where

$$r_1(\xi, \vartheta) = \eta_1, \qquad r_2(\xi, \vartheta) = r - \eta_1. \tag{3.16}$$

It is noteworthy to add that the developed procedure can be used as general approach for obtaining various particular kinematic hardening laws. As an example let us assume that the evolution A in (3.9) instead of $\mathbf{d}^p$ and $\tilde{\boldsymbol{\tau}}$ depends on $\mathbf{d}^p$ and $\boldsymbol{\alpha}$ only (cf. Agah–Tehrani et al. (1987)). Then instead of (3.15) we obtain

$$\mathrm{L}_{\boldsymbol{\upsilon}}\boldsymbol{\alpha} = \frac{1}{T_m}\langle\Phi(\frac{f}{\kappa} - 1)\rangle\left[\zeta_1(\xi, \vartheta)\mathbf{P} - \zeta_2(\xi, \vartheta)\boldsymbol{\alpha}\right], \tag{3.17}$$

where

$$\zeta_1 = r_1, \qquad \zeta_2 = -r_2(\xi, \vartheta)\frac{\mathbf{P} : \mathbf{Q}}{\boldsymbol{\alpha} : \mathbf{Q}}. \tag{3.18}$$

When the infinitesimal deformations and rate independent response of a material are assumed and the intrinsic micro–damage effects are neglected then the kinematic hardening law (3.17) reduces to that proposed by Armstrong and Frederick (1966).

The kinematic hardening law (3.17) leads to the nonlinear stress–strain relation with the characteristic saturation effect. The material function $\zeta_1(\xi, \vartheta)$ for $\xi = \xi_0$ and $\vartheta = \vartheta_0$ can be interpreted as an initial value of the kinematic hardening modulus and the material function $\zeta_2(\xi, \vartheta)$ determines the character of the nonlinearity of kinematic hardening. The particular forms of the functions ζ_1 and ζ_2 have to take into account the degradation nature of the influence of the intrinsic micro–damage process on the evolution of anisotropic hardening.

3.5 Thermodynamic restrictions and rate type constitutive relations

Suppose the axiom of the entropy production holds. Then the constitutive assumption (2.51) and the evolution equations $(2.56)_{1,2}$, (3.8) and (3.15) lead to the results as follows

$$\boldsymbol{\tau} = \rho_{Ref}\frac{\partial\hat{\psi}}{\partial\mathbf{e}}, \qquad \eta = -\frac{\partial\hat{\psi}}{\partial\vartheta}, \qquad -\frac{\partial\hat{\psi}}{\partial\boldsymbol{\mu}}\cdot \mathrm{L}_{\boldsymbol{\upsilon}}\boldsymbol{\mu} - \frac{1}{\rho\vartheta}\mathbf{q}\cdot\mathrm{grad}\vartheta \geq 0. \tag{3.19}$$

The rate of internal dissipation is determined by

$$\begin{aligned}\vartheta\hat{i} = -\frac{\partial\hat{\psi}}{\partial\boldsymbol{\mu}}\cdot \mathrm{L}_{\boldsymbol{\upsilon}}\boldsymbol{\mu} = &-\left[\tfrac{\partial\hat{\psi}}{\partial\in^p}\sqrt{\tfrac{2}{3}} + \tfrac{\partial\hat{\psi}}{\partial\boldsymbol{\alpha}} : \left(r_1\mathbf{P} + r_2\tfrac{\mathbf{P}:\mathbf{Q}}{\tilde{\boldsymbol{\tau}}:\mathbf{Q}}\tilde{\boldsymbol{\tau}}\right)\right]\Lambda \\ &-\frac{\partial\hat{\psi}}{\partial\xi}\Xi.\end{aligned} \tag{3.20}$$

Operating on the stress relation $(3.19)_1$ with the Lie derivative and keeping the internal state vector constant, we obtain (cf. Duszek–Perzyna and Perzyna (1994))

$$\mathrm{L}_{\boldsymbol{\upsilon}}\boldsymbol{\tau} = \mathcal{L}^e : \mathbf{d} - \mathcal{L}^{th}\dot{\vartheta} - [(\mathcal{L}^e + \mathbf{g}\boldsymbol{\tau} + \boldsymbol{\tau}\mathbf{g} + \mathcal{W}) : \mathbf{P}]\frac{1}{T_m}\langle\Phi(\frac{f}{\kappa} - 1)\rangle, \tag{3.21}$$

where

$$\begin{gathered}\mathcal{L}^e = \rho_{Ref}\frac{\partial^2\hat{\psi}}{\partial\mathbf{e}^2}, \qquad \mathcal{L}^{th} = -\rho_{Ref}\frac{\partial^2\hat{\psi}}{\partial\mathbf{e}\partial\vartheta}, \\ \mathcal{W} = \eta^*[(\mathbf{g}\boldsymbol{\tau} - \boldsymbol{\tau}\mathbf{g}) : (\boldsymbol{\alpha}\mathbf{g} - \mathbf{g}\boldsymbol{\alpha})].\end{gathered} \tag{3.22}$$

Substituting $\dot{\psi}$ into the energy balance equation and taking into account the results $(3.19)_3$ and (3.20) gives

$$\rho\vartheta\dot{\eta} = -\mathrm{div}\mathbf{q} + \rho\vartheta\hat{i}. \tag{3.23}$$

Operating on the entropy relation $(3.19)_2$ with the Lie derivative and substituting the result into (3.23) we obtain

$$\rho c_p\dot{\vartheta} = -\mathrm{div}\mathbf{q} + \vartheta\frac{\rho}{\rho_{Ref}}\frac{\partial\boldsymbol{\tau}}{\partial\vartheta} : \mathbf{d} + \rho\chi^*\boldsymbol{\tau} : \mathbf{d}^p + \rho\chi^{**}\dot{\xi}, \tag{3.24}$$

where the specific heat

$$c_p = -\vartheta \frac{\partial^2 \hat{\psi}}{\partial \vartheta^2} \tag{3.25}$$

and the irreversibility coefficients χ^* and χ^{**} are determined by

$$\begin{aligned}
\chi^* &= -\left[\left(\frac{\partial \hat{\psi}}{\partial \in^p} - \vartheta \frac{\partial^2 \hat{\psi}}{\partial \vartheta \partial \in^p}\right)\sqrt{\frac{2}{3}}\right. \\
&\quad \left. + \left(\frac{\partial \hat{\psi}}{\partial \boldsymbol{\alpha}} - \vartheta \frac{\partial^2 \hat{\psi}}{\partial \vartheta \partial \boldsymbol{\alpha}}\right) : \left(r_1 \mathbf{P} + r_2 \frac{\mathbf{P} : \mathbf{Q}}{\tilde{\boldsymbol{\tau}} : \mathbf{Q}} \tilde{\boldsymbol{\tau}}\right)\right] \frac{1}{\boldsymbol{\tau} : \mathbf{P}}, \\
\chi^{**} &= -\left(\frac{\partial \hat{\psi}}{\partial \xi} - \vartheta \frac{\partial^2 \hat{\psi}}{\partial \vartheta \partial \xi}\right).
\end{aligned} \tag{3.26}$$

3.6 Fracture criterion based on the evolution of porosity

We base the fracture criterion on the evolution of the internal state variable ξ.

Let us assume that for $\zeta = \xi^F$ catastrophe takes place (cf. Perzyna (1984)), that is

$$\kappa = \hat{\kappa}(\in^p, \vartheta, \xi)|_{\zeta} = 0. \tag{3.27}$$

It means that for $\zeta = \xi^F$ the material loses its carrying capacity. The condition (3.27) describes the main feature observed experimentally that the load tends to zero at the fracture point.

In is noteworthy that the isotropic hardening–softening material function $\hat{\kappa}$ proposed in Eq. (2.58) should satisfy the fracture criterion (3.27).

3.7 Rate independent plastic response

We have to confess that the results of Section 3.5 are almost the same as those obtained in Section 2.6.7.

3.7.1 Rate independent plastic response as a limit case

Let us assume that the relaxation time $T_m = 0$, then from $(2.56)_1$ and (2.57) we have the yield criterion

$$f - \kappa = 0. \tag{3.28}$$

The coefficient Λ can now be determined from the consistency condition

$$\dot{f} - \dot{\kappa} = 0, \tag{3.29}$$

which yields

$$\Lambda = \langle \frac{1}{H}\{\mathbf{Q} : [\dot{\boldsymbol{\tau}} - (\mathbf{d}\cdot\boldsymbol{\alpha} + \boldsymbol{\alpha}\cdot\mathbf{d})] + \pi\dot{\vartheta}\}\rangle, \tag{3.30}$$

where

$$\begin{aligned} H &= H^* + H^{**}, \\ H^* &= \sqrt{\frac{2}{3}}\left[\frac{\partial\hat{\kappa}}{\partial \in^p} - \left(\frac{\partial f}{\partial\xi} - \frac{\partial\hat{\kappa}}{\partial\xi}\right)\frac{\partial\xi}{\partial \in^p}\right]\left\|\frac{\partial f}{\partial\boldsymbol{\tau}} + \right. \\ &\left.\left(\frac{\partial f}{\partial\xi} - \frac{\partial\hat{\kappa}}{\partial\xi}\right)\frac{\partial\xi}{\partial\boldsymbol{\tau}}\right\|^{-1}, \\ H^{**} &= r\mathbf{P} : \mathbf{Q}, \\ \pi &= \left[\left(\frac{\partial f}{\partial\xi} - \frac{\partial\hat{\kappa}}{\partial\xi}\right)\frac{\partial\xi}{\partial\vartheta} + \frac{\partial f}{\partial\vartheta} - \frac{\partial\hat{\kappa}}{\partial\vartheta}\right] \\ &\left\|\frac{\partial f}{\partial\boldsymbol{\tau}} + \left(\frac{\partial f}{\partial\xi} - \frac{\partial\hat{\kappa}}{\partial\xi}\right)\frac{\partial\xi}{\partial\boldsymbol{\tau}}\right\|^{-1}. \end{aligned} \tag{3.31}$$

The relations (3.30) and (3.31) suggest the interpretation of H^* and H^{**} as the isotropic and anisotropic hardening modulus rates, respectively.

Finally, we have for rate independent plastic response

$$\mathbf{d}^p = \langle \frac{1}{H}\{\mathbf{Q} : [\dot{\boldsymbol{\tau}} - (\mathbf{d}\cdot\boldsymbol{\alpha} + \boldsymbol{\alpha}\cdot\mathbf{d})] + \pi\dot{\vartheta}\}\rangle\mathbf{P}. \tag{3.32}$$

3.7.2 Rate independent intrinsic micro–damage process

For rate independent response when $T_m = 0$ the evolution equation for porosity (3.8) yields

$$\tilde{I}_n = \tilde{I}_g = \tau_n(\xi, \vartheta, \in^p) = \tau_{eq}(\xi, \vartheta, \in^p) = \kappa. \tag{3.33}$$

To obtain the evolution equation for porosity parameter ξ we shall follow the procedure developed by Duszek–Perzyna and Perzyna (1994). The result (3.33) leads to the evolution equation as follows

$$\dot{\xi} = A_1\dot{\tilde{J}}_1 + A_2\dot{\tilde{J}}_2' + A_3\dot{\tilde{J}}_3' + B\dot{\in}^p + C\dot{\vartheta}, \tag{3.34}$$

where A_1, A_2, A_3, B and C are determined material functions. The evolution equation (3.34) can be directly related to the Gurson proposition modified by Tvergaard (cf. Gurson (1975)), Mear and Hutchinson (1985), Tvergaard (1987), Duszek and Perzyna (1988)) by the assumptions

$$\begin{aligned} A_1\dot{\tilde{J}}_1 &= k_1\dot{\tilde{J}}_1, \qquad A_2 = A_3 = 0, \\ B\dot{\in}^p &= k_2\tilde{\boldsymbol{\tau}} : \mathbf{d}^p + k_3\mathbf{g} : \mathbf{d}^p. \end{aligned} \tag{3.35}$$

Introducing the assumptions (3.35) into (3.34) gives

$$\dot{\xi} = k_1 \dot{\tilde{J}}_1 + k_2 \tilde{\boldsymbol{\tau}} : \mathbf{d}^p + k_3 \mathbf{g} : \mathbf{d}^p + C\dot{\vartheta}. \tag{3.36}$$

The first two terms in (3.36) are responsible for the description of the nucleation, the first due to the cracking of the second–phase particles and the second is generated by the debonding of the second–phase particles from the matrix material as the plastic work progressively increases. The third term describes the growth process and is assumed to be controlled only by the plastic flow phenomena. The last term can be interpreted as the description of the nucleation induced by the transient change of temperature. For processes in which temperature change is smooth and small this term can be neglected (i.e. we can assume $C = 0$).

3.7.3 Fundamental rate type constitutive equations

The rate equation (3.21) can also be written in the form as follows

$$\mathrm{L}_{\boldsymbol{\upsilon}} \boldsymbol{\tau} = \mathcal{L}^e : \mathbf{d} - (\mathcal{L}^e : \mathbf{P} + \mathbf{b})\Lambda - \mathcal{L}^{th}\dot{\vartheta}, \tag{3.37}$$

where

$$\mathbf{b} = (\mathbf{P} + \boldsymbol{\Omega}) \cdot \boldsymbol{\tau} + \boldsymbol{\tau} \cdot (\mathbf{P} - \boldsymbol{\Omega}) \tag{3.38}$$

(cf. here the results for single crystals presented by Hill and Rice (1972) and Duszek–Perzyna and Perzyna (1993a)).

Substituting in (3.37) Λ and $\boldsymbol{\Omega}$ from Eqs (3.30) and (2.65) we finally obtain the rate type constitutive equation for the Kirchhoff stress tensor $\boldsymbol{\tau}$ in the form

$$\mathrm{L}_{\boldsymbol{\upsilon}} \boldsymbol{\tau} = \mathcal{L} : \mathbf{d} - \mathcal{M}\dot{\vartheta}, \tag{3.39}$$

where the fundamental elasto–plastic matrix $\mathcal{L}$ and the thermal tensor $\mathcal{M}$ are defined as follows

$$\begin{aligned} \mathcal{L} &= \mathcal{L}^e - \frac{(\mathcal{L}^e + \mathbf{g}\boldsymbol{\tau} + \boldsymbol{\tau}\mathbf{g} + \mathcal{W}) : \mathbf{PQ} : (\mathcal{L}^e + \tilde{\boldsymbol{\tau}}\mathbf{g} + \mathbf{g}\tilde{\boldsymbol{\tau}})}{H + \mathbf{P} : (\mathcal{L}^e + \mathbf{g}\boldsymbol{\tau} + \boldsymbol{\tau}\mathbf{g}) : \mathbf{Q} + \mathbf{P} : \mathcal{W} : \mathbf{Q}}, \\ \mathcal{M} &= \mathcal{L}^{th} - \frac{(\mathcal{L}^e + \mathbf{g}\boldsymbol{\tau} + \boldsymbol{\tau}\mathbf{g} + \mathcal{W}) : \mathbf{P}(\mathbf{Q} : \mathcal{L}^{th} - \pi)}{H + \mathbf{P} : (\mathcal{L}^e + \mathbf{g}\boldsymbol{\tau} + \boldsymbol{\tau}\mathbf{g}) : \mathbf{Q} + \mathbf{P} : \mathcal{W} : \mathbf{Q}}. \end{aligned} \tag{3.40}$$

It is noteworthy to point out here that the fundamental elasto–plastic matrix $\mathcal{L}$ determined by $(3.40)_1$ is nonsymmetric. There are three reasons for the nonsymmetry of the fundamental elasto–plastic matrix $\mathcal{L}$, namely the kinematic hardening effects (i.e. the stress tensor $\tilde{\boldsymbol{\tau}} = \boldsymbol{\tau} - \boldsymbol{\alpha}$ arises in the second bracket in the numerator of the second term instead of $\boldsymbol{\tau}$), the plastic non–normality (i.e. $\mathbf{P} \neq \mathbf{Q}$) and the plastic spin (this effect is represented by the additional term $\mathcal{W}$. For the particular case when these three effects are neglected we have very important result, namely the fundamental elasto–plastic matrix $\mathcal{L}$ becomes symmetric.

The rate equation (3.24) for rate independent response takes the form

$$\begin{aligned}\rho c_p \dot{\vartheta} &= -\mathrm{div}\mathbf{q} + \vartheta \frac{\rho}{\rho_{Ref}} \frac{\partial \boldsymbol{\tau}}{\partial \vartheta} : \mathbf{d} + \rho \chi_1 \Lambda \\ &+ \rho \chi_2 \left[\mathrm{L}_{\boldsymbol{\upsilon}} \boldsymbol{\tau} : \mathbf{g} + (\mathbf{g} \cdot \tilde{\boldsymbol{\tau}} + \tilde{\boldsymbol{\tau}} \cdot \mathbf{g}) : \mathbf{d} \right], \qquad (3.41)\end{aligned}$$

where the irreversibility coefficients χ_1 and χ_2 are defined as follows

$$\begin{aligned}\chi_1 &= -\left[\left(\frac{\partial \hat{\psi}}{\partial \zeta} - \vartheta \frac{\partial^2 \hat{\psi}}{\partial \vartheta \partial \zeta} \right) \sqrt{\frac{2}{3}} + \right. \\ &\left(\frac{\partial \hat{\psi}}{\partial \boldsymbol{\alpha}} - \vartheta \frac{\partial^2 \hat{\psi}}{\partial \vartheta \partial \boldsymbol{\alpha}} \right) : \left(r_1 \mathbf{P} + r_2 \frac{\mathbf{P} : \mathbf{Q}}{\tilde{\boldsymbol{\tau}} : \mathbf{Q}} \tilde{\boldsymbol{\tau}} \right) \\ &\left. + \left(\frac{\partial \hat{\psi}}{\partial \boldsymbol{\xi}} - \vartheta \frac{\partial^2 \hat{\psi}}{\partial \vartheta \partial \boldsymbol{\xi}} \right) (k_2 \tilde{\boldsymbol{\tau}} : \mathbf{P} + k_3 \mathbf{g} : \mathbf{P}) \right], \\ \chi_2 &= -\left(\frac{\partial \hat{\psi}}{\partial \boldsymbol{\xi}} - \vartheta \frac{\partial^2 \hat{\psi}}{\partial \vartheta \partial \boldsymbol{\xi}} \right) k_1. \qquad (3.42)\end{aligned}$$

It is noteworthy that the rate type equations (3.39) and (3.40) take into account such effects as the plastic spin, plastic non–normality, plastic strain induced anisotropy (kinematic hardening, i.e. nonsymmetry of the fundamental matrix $\mathcal{L}$), covariant terms, micro–damage process (i.e. softening generated by microcrack nucleation and growth mechanisms), thermomechanical couplings (i.e. thermal plastic softening and thermal expansion) and of course due to cooperative phenomena the synergetic effects.

4. Localization and Localized Fracture Phenomena in Cyclic Dynamic Loading Processes

4.1 Formulation of the evolution problem

Find $\boldsymbol{\varphi}$ as function of t and $\mathbf{x}$ satisfying[7]

$$\left.\begin{array}{ll}\text{(i)} & \dot{\boldsymbol{\varphi}} = \mathcal{A}(t,\boldsymbol{\varphi})\boldsymbol{\varphi} + \mathbf{f}(t,\boldsymbol{\varphi});\\ \text{(ii)} & \boldsymbol{\varphi}(0) = \boldsymbol{\varphi}^0(\mathbf{x});\\ \text{(iii)} & \text{The boundary conditions.}\end{array}\right\} \tag{4.1}$$

The evolution problem (4.1) describes an adiabatic inelastic flow process provided

$$\boldsymbol{\varphi} = \begin{bmatrix} \boldsymbol{\upsilon} \\ \rho \\ \boldsymbol{\tau} \\ \boldsymbol{\alpha} \\ \xi \\ \vartheta \end{bmatrix},$$

$$\mathbf{f} = \begin{bmatrix} 0 \\ 0 \\ -\frac{1}{T_m}\langle\Phi(\frac{f}{\kappa}-1)\rangle\left[\left(\boldsymbol{\mathcal{L}}^e + \frac{\chi^*}{\rho c_p}\boldsymbol{\mathcal{L}}^{th}\boldsymbol{\tau} + \mathbf{g}\boldsymbol{\tau} + \boldsymbol{\tau}\mathbf{g}\right) : \mathbf{P}\right] - \frac{\chi^{**}\Xi}{\rho c_p}\boldsymbol{\mathcal{L}}^{th} \\ \frac{1}{T_m}\langle\Phi(\frac{f}{\kappa}-1)\rangle\left[r_1(\xi,\vartheta)\mathbf{P} + r_2(\xi,\vartheta)\frac{\mathbf{P}:\mathbf{Q}}{\tilde{\boldsymbol{\tau}}:\mathbf{Q}}\tilde{\boldsymbol{\tau}}\right] \\ \Xi \\ \frac{1}{T_m}\langle\Phi(\frac{f}{\kappa}-1)\rangle\frac{\chi^*}{\rho c_p}\boldsymbol{\tau} : \mathbf{P} + \frac{\chi^{**}}{\rho c_p}\Xi \end{bmatrix},$$

[7]We shall follow here some fundamental results which have been discussed in Richtmyer and Morton (1967), Strang and Fix (1973), Richtmyer (1978), Dautray and Lions (1993) and Gustafsson, Kreiss and Oliger (1995).

$$
\mathcal{A} = \begin{bmatrix}
0 & \frac{\boldsymbol{\tau}}{\rho_{Ref}\rho}\mathrm{grad} & \frac{1}{\rho_{Ref}}\mathrm{div} & 0 & 0 & 0 \\
-\rho\mathrm{div} & 0 & 0 & 0 & 0 & 0 \\
I\!E : \mathrm{sym}\frac{\partial}{\partial \mathbf{x}} + 2\mathrm{sym}(\boldsymbol{\tau} : \frac{\partial}{\partial \mathbf{x}}) & 0 & 0 & 0 & 0 & 0 \\
2\mathrm{sym}(\boldsymbol{\alpha} : \frac{\partial}{\partial \mathbf{x}}) & 0 & 0 & 0 & 0 & 0 \\
0 & 0 & 0 & 0 & 0 & 0 \\
\frac{\vartheta}{c_p \rho_{Ref}} \frac{\partial \boldsymbol{\tau}}{\partial \vartheta} : \mathrm{sym}\frac{\partial}{\partial \mathbf{x}} & 0 & 0 & 0 & 0 & 0
\end{bmatrix},
\tag{4.2}
$$

where

$$
I\!E = \mathcal{L}^e - \frac{\vartheta}{c_p \rho_{Ref}} \mathcal{L}^{th} \frac{\partial \boldsymbol{\tau}}{\partial \vartheta}.
$$

It is noteworthy that the spatial operator $\mathcal{A}$ has the same form as in thermo–elastodynamics while all dissipative effects generated by viscoplastic flow phenomena influence the process through the nonlinear function $\mathbf{f}$. When an isothermal inelastic flow process is considered then

$$
\boldsymbol{\varphi} = \begin{bmatrix} \boldsymbol{\upsilon} \\ \rho \\ \boldsymbol{\tau} \\ \boldsymbol{\alpha} \\ \xi \end{bmatrix}, \qquad
\mathbf{f} = \begin{bmatrix}
0 \\
-\frac{\rho}{1-\xi}\Xi \\
-\frac{1}{T_m}\langle \Phi(\frac{f}{\kappa} - 1)\rangle \left(\mathcal{L}^e + \mathbf{g}\boldsymbol{\tau} + \boldsymbol{\tau}\mathbf{g}\right) : \mathbf{P} \\
\frac{1}{T_m}\langle \Phi(\frac{f}{\kappa} - 1)\rangle \left[\xi_1(\xi)\mathbf{P} - \xi_2(\xi)\boldsymbol{\alpha}\right] \\
\Xi
\end{bmatrix},
$$

$$
\mathcal{A} = \begin{bmatrix}
0 & \frac{\boldsymbol{\tau}}{\rho_0\rho}\mathrm{grad} & \frac{1}{\rho_0}\mathrm{div} & 0 & 0 \\
0 & 0 & 0 & 0 & 0 \\
\mathcal{L}^e : \mathrm{sym}\frac{\partial}{\partial \mathbf{x}} + 2\mathrm{sym}(\boldsymbol{\tau} : \frac{\partial}{\partial \mathbf{x}}) & 0 & 0 & 0 & 0 \\
2\mathrm{sym}(\boldsymbol{\alpha} : \frac{\partial}{\partial \mathbf{x}}) & 0 & 0 & 0 & 0 \\
0 & 0 & 0 & 0 & 0
\end{bmatrix}.
\tag{4.3}
$$

4.2 Application of finite difference method

Let us consider the evolution problem in the form of (4.1). Let us introduce in the Euclidean space E^3 a regular difference net of nodes (i, j, k) with convective coordinates $\chi^1_i = i\Delta\chi^1$, $\chi^2_j = j\Delta\chi^2$ and $\chi^3_k = k\Delta\chi^3$, $i, j, k \in N$, where N is a set of natural numbers, cf. Dornowski (1999). Of course, some of the nodes belong to the edge of the body and are used to approximate the boundary conditions. Time is approximated by a discrete sequence of moments $t_n = n\Delta t$, where Δt is time step, $n \in N$.

For all functions $\varphi = \hat{\varphi}(\mathbf{x}, t)$ of the analysed problem we postulate the following approximation in the domain $\Delta \mathrm{E} = \Delta\chi^1 \times \Delta\chi^2 \times \Delta\chi^3$ of a convective difference mesh (cf. Fig. 4.1):

$$\begin{aligned}\varphi(\mathbf{x}, t) \cong \varphi_h(\mathbf{x}, t) \;=\;& \mathbf{a}_1(t) + \mathbf{a}_2(t)\chi^1 + \mathbf{a}_3(t)\chi^2 + \mathbf{a}_4(t)\chi^3 \\ & + \mathbf{a}_5(t)\chi^1\chi^2 + \mathbf{a}_6(t)\chi^1\chi^3 + \mathbf{a}_7(t)\chi^2\chi^3 \\ & + \mathbf{a}_8(t)\chi^1\chi^2\chi^3, \\ & \quad \mathbf{x} \in \Delta\mathcal{S}. \end{aligned} \tag{4.4}$$

The functions $\mathbf{a}_1(t), \ldots, \mathbf{a}_8(t)$ depend only on time, are determined by the value of the function $\varphi_w(t) = [\varphi_1(t), \ldots, \varphi_8(t)]^T$ in the node points of difference mesh, (cf. Fig. 4.1). Hence the approximation functions (4.4) can be written in the form

$$\varphi_h(\mathbf{x}, t) = \mathbf{N}(\mathbf{x})\varphi_w(t), \qquad \mathbf{x} \in \Delta\mathcal{S}, \tag{4.5}$$

where

$$\begin{aligned}
N_1(\mathbf{x}) &= q\left(-\Delta\chi^1 + 2\chi^1\right)\left(-\Delta\chi^2 + 2\chi^2\right)\left(\Delta\chi^3 - 2\chi^3\right), \\
N_2(\mathbf{x}) &= q\left(\Delta\chi^1 + 2\chi^1\right)\left(-\Delta\chi^2 + 2\chi^2\right)\left(-\Delta\chi^3 + 2\chi^3\right), \\
N_3(\mathbf{x}) &= q\left(-\Delta\chi^1 + 2\chi^1\right)\left(\Delta\chi^2 + 2\chi^2\right)\left(-\Delta\chi^3 + 2\chi^3\right), \\
N_4(\mathbf{x}) &= q\left(\Delta\chi^1 + 2\chi^1\right)\left(\Delta\chi^2 + 2\chi^2\right)\left(\Delta\chi^3 - 2\chi^3\right), \\
N_5(\mathbf{x}) &= q\left(-\Delta\chi^1 + 2\chi^1\right)\left(-\Delta\chi^2 + 2\chi^2\right)\left(\Delta\chi^3 + 2\chi^3\right), \\
N_6(\mathbf{x}) &= q\left(\Delta\chi^1 + 2\chi^1\right)\left(\Delta\chi^2 - 2\chi^2\right)\left(\Delta\chi^3 + 2\chi^3\right), \\
N_7(\mathbf{x}) &= q\left(\Delta\chi^1 - 2\chi^1\right)\left(\Delta\chi^2 + 2\chi^2\right)\left(\Delta\chi^3 + 2\chi^3\right), \\
N_8(\mathbf{x}) &= q\left(\Delta\chi^1 + 2\chi^1\right)\left(\Delta\chi^2 + 2\chi^2\right)\left(\Delta\chi^3 + 2\chi^3\right), \\
& q = \frac{1}{8\Delta\chi^1\Delta\chi^2\Delta\chi^3}.
\end{aligned} \tag{4.6}$$

Equations (4.5) allow to determine values of the function $\varphi_h(\mathbf{x}, t)$ in any point of the difference mesh, $\mathbf{x} \in \Delta\mathcal{S}$. For the central point $\mathbf{x} = \mathbf{x}_0$, $N_1 = \cdots = N_8 = \frac{1}{8}$ and $\varphi_h(t) = [\varphi_1(t) + \ldots + \varphi_8(t)]\frac{1}{8}$.

By using (4.5) we can determine the matrix of the difference operators which approximate the first partial derivatives of the function $\varphi(\mathbf{x}, t)$ for $\mathbf{x} \in \Delta\mathcal{S}$,

$$\frac{\partial}{\partial \mathbf{x}}\varphi(\mathbf{x}, t) \cong \frac{\partial}{\partial \mathbf{x}}\varphi_h(\mathbf{x}, t) = \frac{\partial}{\partial \mathbf{x}}\mathbf{N}(\mathbf{x})\varphi_w(t) = \mathbf{R}(\mathbf{x})\varphi_w(t). \tag{4.7}$$

The matrix of the difference operator $\mathbf{R}(\mathbf{x})$ for the central point takes the form

$$\mathbf{R}(\mathbf{x}=\mathbf{x}_0) = \frac{\partial}{\partial \mathbf{x}}\mathbf{N}(\mathbf{x})\bigg|_{\mathbf{x}=\mathbf{x}_0} =$$

$$\begin{bmatrix} \frac{-1}{\Delta\chi^1} & \frac{1}{\Delta\chi^1} & \frac{-1}{\Delta\chi^1} & \frac{1}{\Delta\chi^1} & \frac{-1}{\Delta\chi^1} & \frac{1}{\Delta\chi^1} & \frac{-1}{\Delta\chi^1} & \frac{1}{\Delta\chi^1} \\ \frac{-1}{\Delta\chi^2} & \frac{-1}{\Delta\chi^2} & \frac{1}{\Delta\chi^2} & \frac{1}{\Delta\chi^2} & \frac{-1}{\Delta\chi^2} & \frac{-1}{\Delta\chi^2} & \frac{1}{\Delta\chi^2} & \frac{1}{\Delta\chi^2} \\ \frac{-1}{\Delta\chi^3} & \frac{-1}{\Delta\chi^3} & \frac{-1}{\Delta\chi^3} & \frac{-1}{\Delta\chi^3} & \frac{1}{\Delta\chi^3} & \frac{1}{\Delta\chi^3} & \frac{1}{\Delta\chi^3} & \frac{1}{\Delta\chi^3} \end{bmatrix}. \tag{4.8}$$

In similar way we can find the difference form of the spatial difference operator $\mathcal{A}(t,\varphi)$ of the considered evolution problem (4.1)

$$\mathcal{A}(t,\varphi)\varphi \cong \mathcal{A}(t,\varphi_h)\mathbf{N}(\mathbf{x})\varphi_w(t) = \mathcal{A}_h(t,\varphi_h)\varphi_w(t), \tag{4.9}$$

hence

$$\mathcal{A}_h(t,\varphi_h) = \mathcal{A}(t,\varphi_h)\mathbf{N}(\mathbf{x}) \quad \text{for} \quad \mathbf{x} \in \Delta\mathcal{S}. \tag{4.10}$$

For the central node, $\mathbf{x} = \mathbf{x}_0$ the difference operator (4.10) depends only on time.

As a result of the proposed approximation of the evolution problem (4.1) with respect to the spatial variables we obtain a set of differential equations with respect to time and difference equations with respect to spatial variables

$$\frac{\mathrm{d}\,\varphi_h(t)}{\mathrm{d}t} = \mathcal{A}_h\varphi_w(t) + \mathbf{f}_h(t). \tag{4.11}$$

For the approximation of (4.11) with respect to time we use the evident scheme of the first order in the form

$$\frac{\mathrm{d}\,\varphi_h(t)}{\mathrm{d}t} \cong \frac{\varphi_h^{n+1} - \varphi_h^n}{\Delta t} = \mathcal{A}_h\varphi_w^n + \mathbf{f}_h^n. \tag{4.12}$$

The solution of (4.12) is reduced to the realization of the recurrence relation

$$\varphi_h^{n+1} = \mathbf{C}_h(\Delta t)\varphi_w^n + \Delta t\mathbf{f}_h^n. \tag{4.13}$$

The difference operator

$$\mathbf{C}_h(\Delta t) = \Delta t\mathcal{A}_h + \mathbf{N} \tag{4.14}$$

couple dependent variables and various points of difference mesh.

For the finite difference approximation it can be proved that the introduced scheme (4.13) is consistent with equation (4.1).

4.3 Stability criterion and the Lax–Richtmyer equivalence theorem

In explicit finite difference scheme for a set of the partial differential equations (4.1)(i) of the hiperbolic type the condition of stability is the criterion of Courant–Friedrichs–Lewy, cf. Courant et al. (1928)

$$\Delta t_{n,n+1} \leq \min\left(\frac{\Delta L^n_{p,q,r}}{\left|c^n_{p,q,r}\right|}\right), \quad p = 1, 2, 3, \ldots, P;$$
$$q = 1, 2, 3, \ldots, Q; \quad r = 1, 2, 3, \ldots, R, \tag{4.15}$$

where $\Delta t_{n,n+1}$ denotes time step, $c^n_{p,q,r}$ denotes the velocity of the propagation of the disturbances in the vicinity of the central node (p, q, r), $\Delta L^n_{p,q,r}$ is the minimum distance between the mesh nodes which are in the vicinity of the node (cf. Fig. 4.1).

The Courant–Friedrichs–Lewy condition requires that the numerical domain of dependence of a finite–difference scheme include the domain of dependence of the associated partial differential equations, cf. Durran (1999).

Theorem Suppose that the evolution problem (4.1) is well–posed for $t \in [0, t_0]$ and that it is approximated by the scheme (4.13), which we assume consistent. Then the scheme is convergent if and only if it is stable.

The proof of the Lax equivalence theorem for the case when the partial differential operator $\mathcal{A}$ in (4.1) is independent of φ can be found in Dautray and Lions (1993).

4.4 Identification procedure

4.4.1 Assumptions of the material functions

The plastic potential function f is assumed in the form (cf. Perzyna (1984) and Shima and Oyane (1976))

$$f = \left\{\tilde{J}_2' + \left[n_1(\vartheta) + n_2(\vartheta)\xi\right]\tilde{J}_1^2\right\}^{\frac{1}{2}} \tag{4.16}$$

where

$$n_1(\vartheta) = 0, \qquad n_2(\vartheta) = \text{const.} \tag{4.17}$$

The isotropic work–hardening–softening function κ is postulated as (cf. Perzyna (1986) and Nemes and Eftis (1993))

$$\kappa = \hat{\kappa}(\in^p, \vartheta, \xi) = \{\kappa_s(\vartheta) - [\kappa_s(\vartheta) - \kappa_0(\vartheta)] \exp[-\delta(\vartheta) \in^p]\} \left[1 - \left(\frac{\xi}{\xi_F}\right)^{\beta(\vartheta)}\right], \tag{4.18}$$

where

$$\kappa_s(\vartheta) = \kappa_s^* - \kappa_s^{**}\overline{\vartheta}, \qquad \kappa_0(\vartheta) = \kappa_0^* - \kappa_0^{**}\overline{\vartheta}, \tag{4.19}$$
$$\delta(\vartheta) = \delta^* - \delta^{**}\overline{\vartheta}, \qquad \beta(\vartheta) = \beta^* - \beta^{**}\overline{\vartheta}, \qquad \overline{\vartheta} = \frac{\vartheta - \vartheta_0}{\vartheta_0}.$$

The overstress function $\Phi\left(\frac{f}{\kappa} - 1\right)$ is assumed in the form

$$\Phi\left(\frac{f}{\kappa} - 1\right) = \left(\frac{f}{\kappa} - 1\right)^m. \tag{4.20}$$

The evolution equation for the kinematic hardening parameter $\boldsymbol{\alpha}$ is assumed in the form (3.17) with

$$\zeta_1(\xi, \vartheta) = \zeta_1^* - \zeta_1^{**}\overline{\vartheta}, \qquad \zeta_2(\xi, \vartheta) = \zeta_2^* - \zeta_2^{**}\overline{\vartheta}. \tag{4.21}$$

The evolution equation for the porosity ξ is postulated as

$$\dot{\xi} = \dot{\xi}_{grow} = \frac{g^*(\xi, \vartheta)}{T_m \kappa_0(\vartheta)} \left[\tilde{I}_g - \tau_{eq}(\xi, \vartheta, \in^p)\right] \tag{4.22}$$

where (cf. Dornowski (1999))

$$\begin{aligned}
g^*(\xi, \vartheta) &= c_1(\vartheta)\frac{\xi}{1-\xi}, \\
\tilde{I}_g &= b_1 \tilde{J}_1 + b_2\sqrt{\tilde{J}_2}, \\
\tau_{eq}(\xi, \vartheta, \in^p) &= c_2(\vartheta)(1-\xi)\ln\frac{1}{\xi}\left\{2\kappa_s(\vartheta) - [\kappa_s(\vartheta) - \kappa_0(\vartheta)]\, F(\xi_0, \xi, \vartheta)\right\}, \\
& \quad c_1(\vartheta) = \text{const}, \qquad c_2(\vartheta) = \text{const}, \\
F(\xi_0, \xi, \vartheta) &= \left(\frac{\xi_0}{1-\xi_0}\frac{1-\xi}{\xi}\right)^{\frac{2}{3}\delta} + \left(\frac{1-\xi}{1-\xi_0}\right)^{\frac{2}{3}\delta}.
\end{aligned} \tag{4.23}$$

As in the infinitesimal theory of elasticity we assume linear properties of the material, i.e.

$$\mathcal{L}^e = 2\mu\mathbf{I} + \lambda(\mathbf{g} \otimes \mathbf{g}) \tag{4.24}$$

where μ and λ denote the Lamé constants, and the thermal expansion matrix is postulated as

$$\mathcal{L}^{th} = (2\mu + 3\lambda)\theta \mathbf{g}, \tag{4.25}$$

where θ is the thermal expansion constant.

To identify the material function η^* in the evolution equation

$$\boldsymbol{\omega}^P = \eta^* \Lambda(\boldsymbol{\alpha} \cdot \mathbf{P} - \mathbf{P} \cdot \boldsymbol{\alpha}) \tag{4.26}$$

we assume that the value of plastic spin is the same order of magnitude as the value of the rate of plastic deformation. This proposition was first introduced by Khan and Cheng (1996). It gives

$$\boldsymbol{\omega}^P = \frac{\Lambda(\boldsymbol{\alpha} \cdot \mathbf{P} - \mathbf{P} \cdot \boldsymbol{\alpha})}{[(\boldsymbol{\alpha} \cdot \mathbf{P} - \mathbf{P} \cdot \boldsymbol{\alpha}) : (\boldsymbol{\alpha} \cdot \mathbf{P} - \mathbf{P} \cdot \boldsymbol{\alpha})]^{1/2}}. \tag{4.27}$$

4.4.2 Identification of the material constants

To determine the material constants assumed in Section 11.4.1 we take advantage of the experimental observations presented by Chakrabarti and Spretnak (1975). They investigated the localized fracture mode for tensile steel sheet specimens simulating both plane stress and plane strain processes. The material used in their study was AISI 4340 steel. The principal variable in this flat specimen test was the width – to – thickness ratio. Variation in specimen geometry produces significant changes in stress state, directions of shear bands, and ductility. They found that fracture propagated consistently along the shear band localized region.

Let us now consider the adiabatic dynamic process for a thin steel plate under condition of plane stress state. In fact we idealize the initial–boundary value problem investigated by Chakrabarti and Spretnak (1975) by assuming the velocity driven adiabatic process for a thin steel plate. Dimensions of the plate and the variation in time of the kinematic constraints are presented in Fig. 4.2. The problem has been solved by using the finite difference method. A thin sheet is modelled via $N \times M$ elements, cf. Fig. 4.2.

In numerical calculations it is assumed:

$$\dot{V}_0 = 1.5 \text{ m/s}, \qquad t_0 = 50\ \mu\text{s}, \qquad t_f = 800\ \mu\text{s}.$$

The material of a plate is AISI 4340 steel.

Based on the best curve fitting of the experimental results obtained by Chakrabarti and Spretnak (1975) for the stress-strain relation (cf. Fig. 4.3) the identification of the material constants has been done, cf. Table 4.1.

Table 4.1: Material constants for AISI 4340 steel

$\kappa_s^* = 809$ MPa,	$\kappa_s^{**} = 228$ MPa,	$\kappa_0^* = 598$ MPa,
$\kappa_0^{**} = 168$ MPa,	$\delta^* = 14.00$,	$\delta^{**} = 3.94$,
$\vartheta_0 = 293$ K,	$\xi_F = 0.20$,	$\rho_{Ref} = 7850$ kg/m^3
$\lambda = 115.38$ GPa,	$\theta = 12 \cdot 10^{-6}$ K^{-1},	$T_m = 2.5$ ms,
$\zeta_1^* = 15.00$ GPa,	$\zeta_1^{**} = 4.22$ GPa	$\zeta_2^* = 69.60$
$c_1 = 0.202$,	$c_2 = 6.7 \cdot 10^{-2}$,	$m = 1$,
$\xi_0 = 6 \cdot 10^{-4}$,	$\overline{\chi} = 0.85$,	$\overline{\overline{\chi}} = 0$
$\beta^* = 9.00$,	$\beta^{**} = 2.53$,	$\mu = 76.92$ GPa,
$b_1 = 1.00$,	$b_2 = 1.30$	$\zeta_2^{**} = 19.60$,
$c_p = 455$ J/kg K.		

4.4.3 Investigation of stability and convergence

Using the same initial–boundary value problem we can investigate the convergence of the numerical method based on the finite difference discretization. In this analysis we assume that the micro–damage evolution is investigated. For a thin steel plate five different meshes have been assumed. In Fig. 4.4 distributions of the equivalent plastic deformation along the middle cross–section ($X = a_0/2$) for various meshes are presented. Similar results for temperature are showed in Fig. 4.5. All curves are plotted in the initial configuration of the specimen and for finite time of the process, $t_f = 800$ μs. The deformed meshes with various densities of nodes are presented in Fig. 4.6. Both analysed values shown in Fig. 4.4 and 4.5 have strong increase in the middle region of the plate. This suggests that in that region of the specimen the localization of plastic deformation takes place. Based on the results for distributions of the equivalent plastic deformation plotted in Fig. 4.4 we can observe that the finite width of the localized region needs appropriate density for the finite difference mesh (in considered case higher than 864 nodes).

The numerical procedure for the explicit finite difference scheme (4.13) has been accomplished with various time step Δt. The value of the time step has been determined by using the stability criterion (4.15). In Fig. 4.7 the evolution of the time step Δt in the deformation process for the mesh $N \times M = 3456$ has been shown. For the considered deformation process the time step Δt is decreasing function of time.

The stability of the numerical procedure with constant time step can be keeping in entire interval $[0, t_f]$ provided we assume $\Delta t = 0.025$ μs. Then as a result for the considered deformation process we have 32 000 time steps.

While the analogical process with various time step Δt is accomplished for 26 603 time steps. This decreases very much time of computation. Very intense change of time step Δt observed from $t = 0.6$ ms in Fig. 4.7 is caused by the formation of the shear band localization.

The results obtained show that the satisfaction of the stability criterion (4.15) leads to the convergence of the numerical procedure. We can conclude that the assumed finite difference scheme (4.13) approximates the well posed initial–boundary value problem (4.1). These results can be treated as the numerical proof of the Lax–Richtmyer equivalence theorem formulated in Section 4.3.

4.4.4 Investigation of localization and fracture phenomena

Let us consider again the adiabatic dynamic process for a thin steel plate under condition of plane stress state which idealized the experimental observations of the localized fracture mode for tensile steel sheet specimens performed by Chakrabarti and Spretnak (1975), cf. Section 4.4.2.

In Fig. 4.8 distributions of the equivalent plastic deformation in a thin plate for chosen instants of the deformation process are presented.

The development of macrocracks in a thin plate has been shown in Fig. 4.9 and the deformed configuration for final fracture of a plate is presented in Fig. 4.10.

The results presented permit to draw some conclusions concerning the adiabatic dynamic deformation process for a thin steel plate considered. We observe that two cross shear bands are developed in the center of the specimen. With continued loading process, one of the instability bands becomes more predominant and most of the plastic deformation is confirmed to this particular adiabatic shear band. Fracture occurs along the boundary of this active instability band. The symmetry breaking in the deformation process observed is only due to very small imperfections generated by numerical procedure. The obtained results are in accord with the experimental observations of Chakrabarti and Spretnak (1975).

4.5 Particular example

4.5.1 Dynamic, adiabatic, cyclic loading processes for a thin plate with sharp notch

To investigate more precisely the development of fatigue damage and the propagation of the macrocrack let us consider a dynamic adiabatic cyclic loading process for a thin steel plate with sharp notch, as shown in Fig. 4.11. To the upper edge of the plate cyclic constraint realized by rigid rotation of the edge of the plate is applied while the lower edge is supported rigidly. It is considered the cyclic rotation constraint in the form of the loading characteristic (adiabatic) in time, namely slow–fast. This type of constraint is represented by positive cycle, pulsating from zero and having the amplitude $\alpha_{max} = 1.2$ deg and the period $T = 4.0$ ms. It has different time for positive rotation $T_p = 3.6$ ms and negative rotation $T_n = 0.4$ ms. The positive and negative rotations are described by the sine functions. The material of the plate is AISI steel (cf. Table 4.1).

Utilizing the finite difference method for regularized thermo–elasto–visco-plastic model, the numerical investigation of the three–dimensional dynamic adiabatic deformation in a thin plate with sharp notch is presented.

4.5.2 Discussion of localization and localized fracture phenomena

The distribution of the norm of the Kirchhoff stress tensor $\|\boldsymbol{\tau}\|$ for a thin plate with sharp notch for chosen instants during two deformation cycles has been shown in Fig. 4.12. The results illustrate the formation of the greatest stress zones in the vicinity of the notch. The characteristic unsymmetrical distribution of these zones is a result of the assumed boundary conditions. In Fig. 4.13 the evolution of the plastic equivalent deformation in the vicinity of the developed fatigue macrocrack is presented. In the initial part of the cyclic deformation process (several cycles), the plastic zone has a characteristic shape (seed of maple). Such a form of the plastic zone has been observed experimentally. For the advanced cyclic deformation process (i.e. when the number of cycles is increased) the plastic zone is very much restricted to the vicinity of the macrocrack. The macrocrack direction is consistent with the least radius direction of the initial plastic zone. The strong concentration of plastic deformation has been seen on the front of the macrocrack. The evolution of temperature is shown in Fig. 4.14. Zones of increased temperature correspond to the plastic zones. The maximum value of temperature is $\vartheta_{max} = 879$ K. The effect of such a strong heating of the material results from its mechanical properties, i.e. the high strength steel, $R_m = 1400$ MPa.

In Fig. 4.15 the evolution of microdamage is presented. The domain of microdamage is limited to the vicinity of the macrocrack. This effect is a result of the strong concentration of microdamage process on the front of the macrocrack. The complex evolution of the plastic rotation is shown in Fig. 4.16. In the domain lying above the macrocrack the plastic rotation has negative value, i.e. the rotation in the left direction in relation to the assumed coordinate system, while in the domain lying below the macrocrack it has positive value. It is noteworthy that the border between these two domains is consistent with the macrocrack direction. The results for the length of macroscopic fatigue damage crack versus number of cycles for various cyclic loading processes (slow–fast (T_p = 3.6 ms, T_n = 0.4 ms), equal–equal (T_p = 2.0 ms, T_n = 2.0 ms) and fast–slow (T_p = 0.4 ms, T_n = 3.6 ms)) have been plotted in Fig. 4.17. These results have clearly shown that the length of the macroscopic fatigue damage crack distinctly depends on the wave shape of the assumed loading cycle. It has been proved that the most dangerous situation we have for the slow–fast cyclic loading process. This result is in accord with the experimental observations performed by Sidey and Coffin (1979). This conclusion is approved by the comparison of the speed of macroscopic crack propagation for various loading processes, cf. Fig. 4.18. From this comparison it is clear that the speed value of macroscopic crack propagation for the slow–fast cyclic loading has dominating character almost for the entire process considered.

To show how the modification of the evolution equation for the porosity parameter ξ helps to describe the accumulation of damage during dynamic cyclic loading process the evolution of the microdamage ξ_{grow} in a node near the tip of the hole (black quadrangle) for different forms of the stress intensity invariant for adiabatic process and various wave shape loadings is presented in Fig. 4.19.

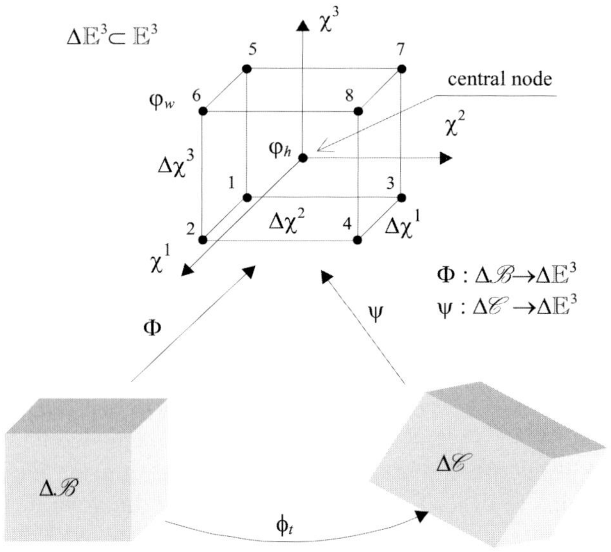

Figure 4.1: Convective finite difference mesh of nodes

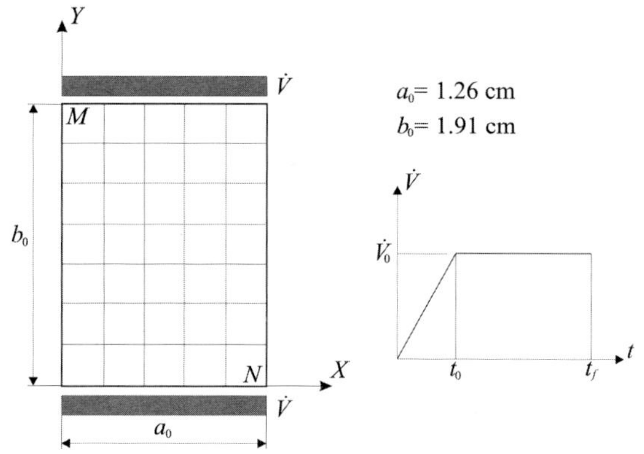

Figure 4.2: Dimension of the plate and the variation in time of the kinematic constrains (After Chakrabarti and Sprentak (1975))

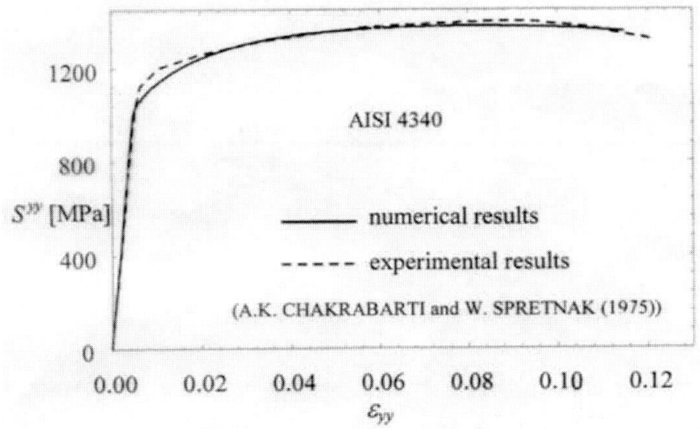

Figure 4.3: Component S^{YY} of the second Piola–Kirchhoff stress tensor as a function of the logarithmic strain ε_{YY}

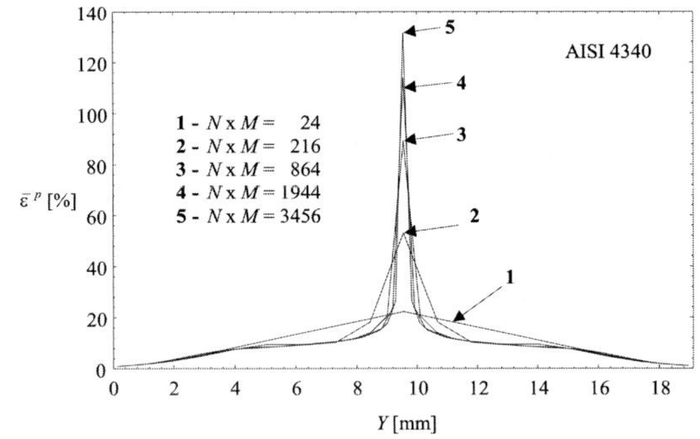

Figure 4.4: Distributions of the equivalent plastic deformation along the middle cross–section ($X = a_0/2$) for various meshes

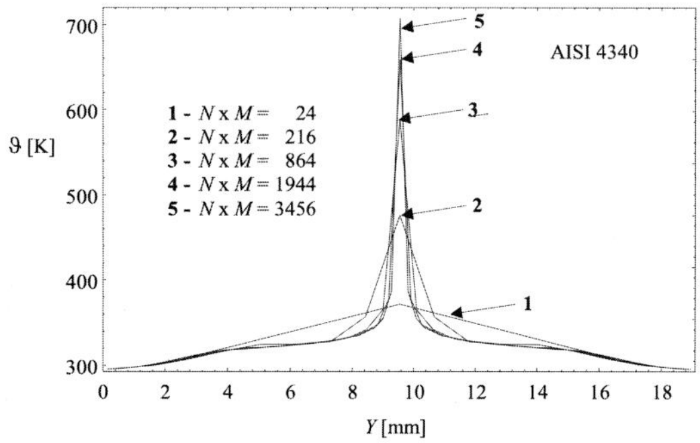

Figure 4.5: Distributions of the temperature along the the middle cross-section ($X = a_0/2$) for various meshes

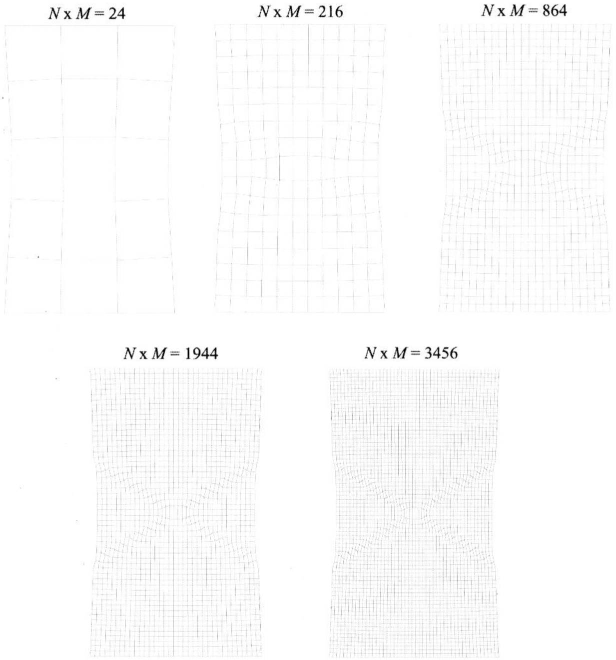

Figure 4.6: Deformed meshes with various densities of nodes

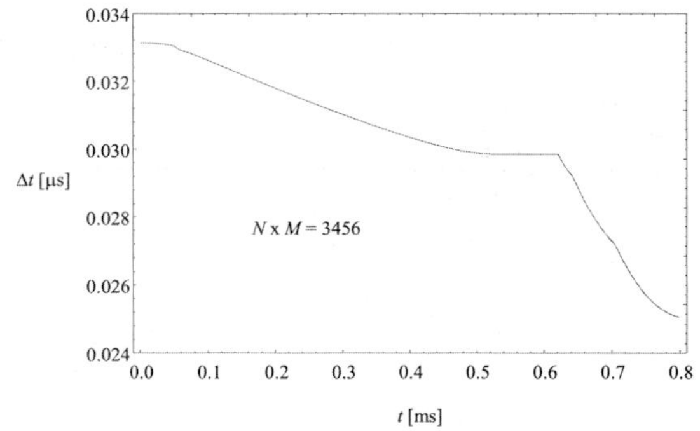

Figure 4.7: Evolution of the time step Δt for a mesh with 3456 nodes

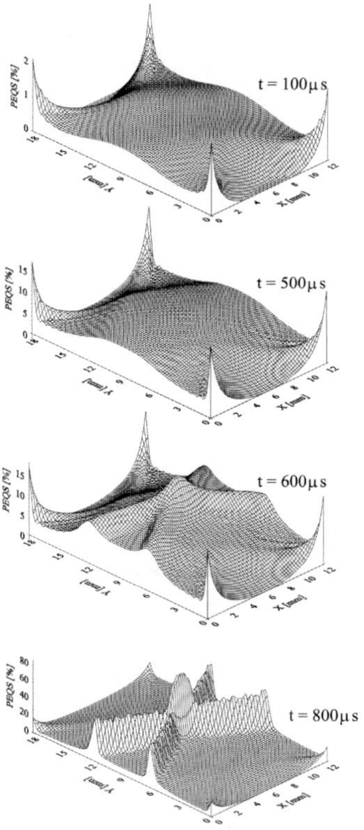

Figure 4.8: Distributions of the equivalent plastic strain in a thin plate for chosen instants of the deformation process

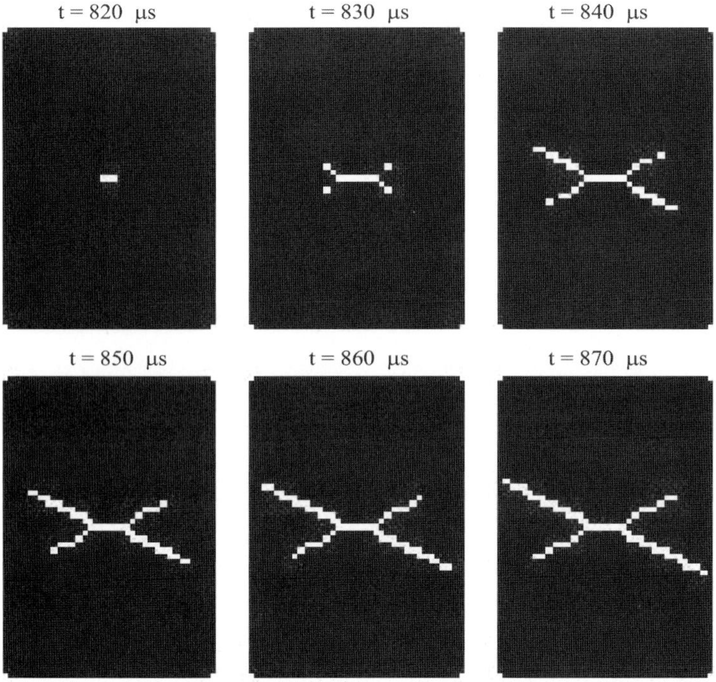

Figure 4.9: Development of macrocracks in a thin plate

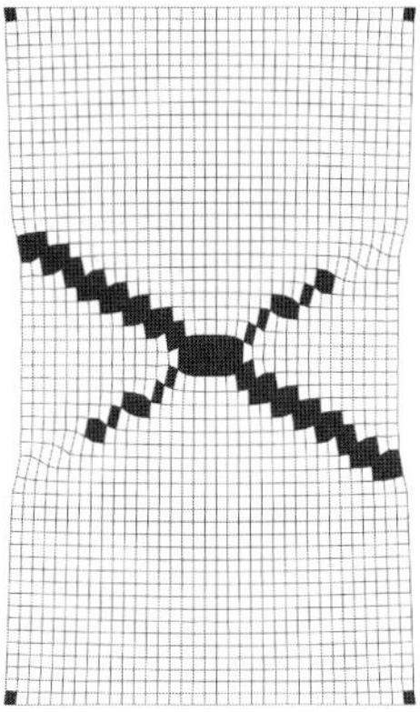

Figure 4.10: Deformed configuration for final fracture of the plate

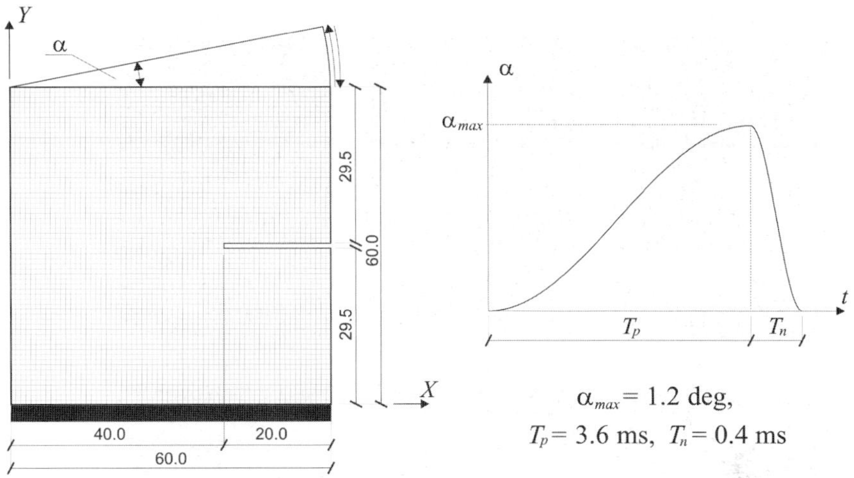

Figure 4.11: Geometry and kinematic constrains of the thin steel plate with sharp notch

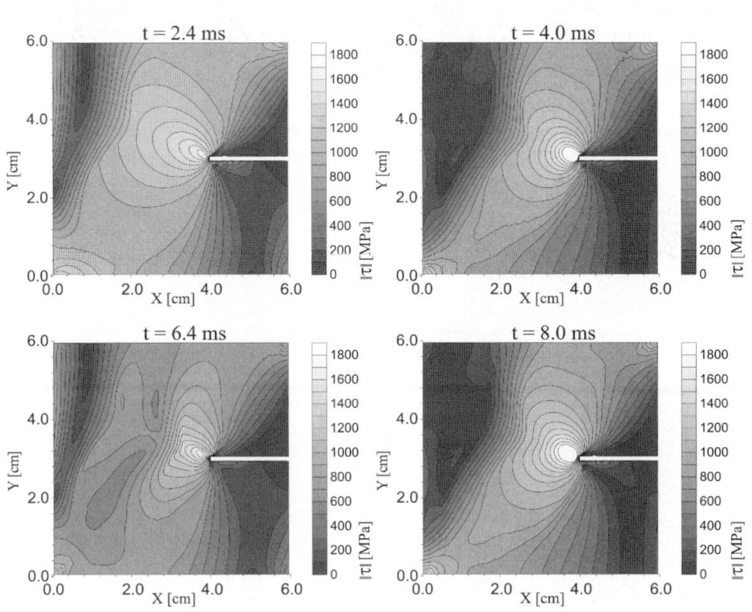

Figure 4.12: Distribution of the norm of the Kirchhoff stress for chosen instants during two deformation cycles

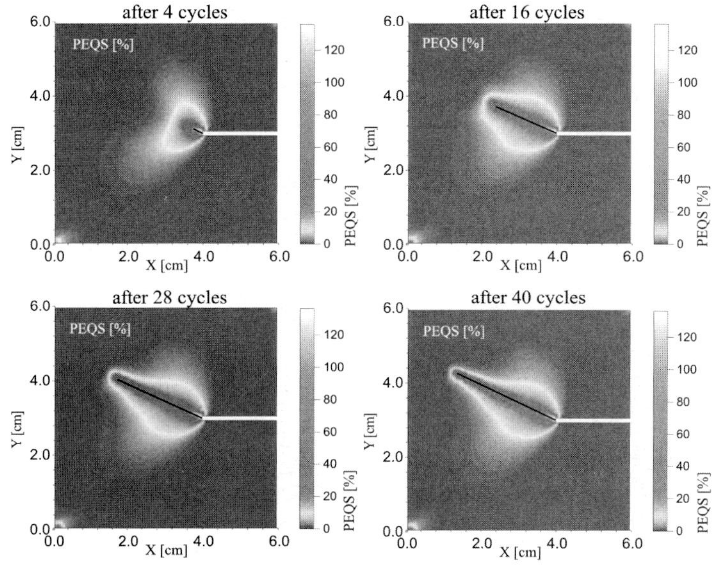

Figure 4.13: Evolution of the equivalent plastic deformation in the vicinity of the developed fatigue damage

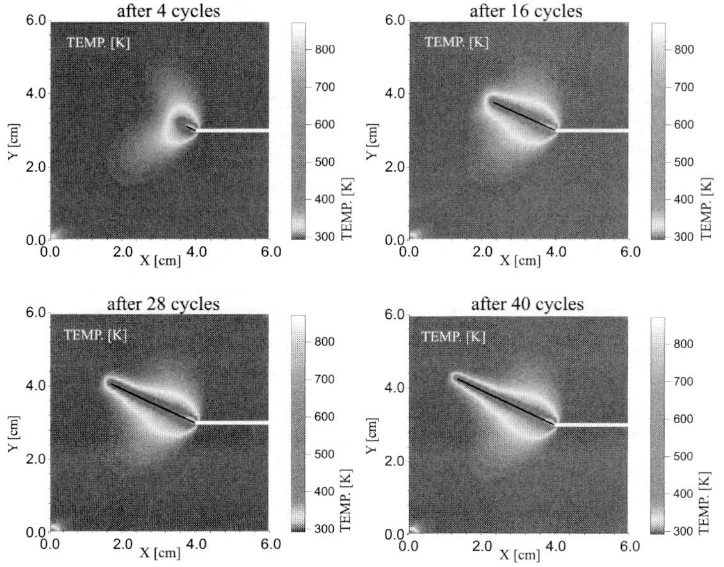

Figure 4.14: Evolution of temperature in the vicinity of the developed fatigue damage

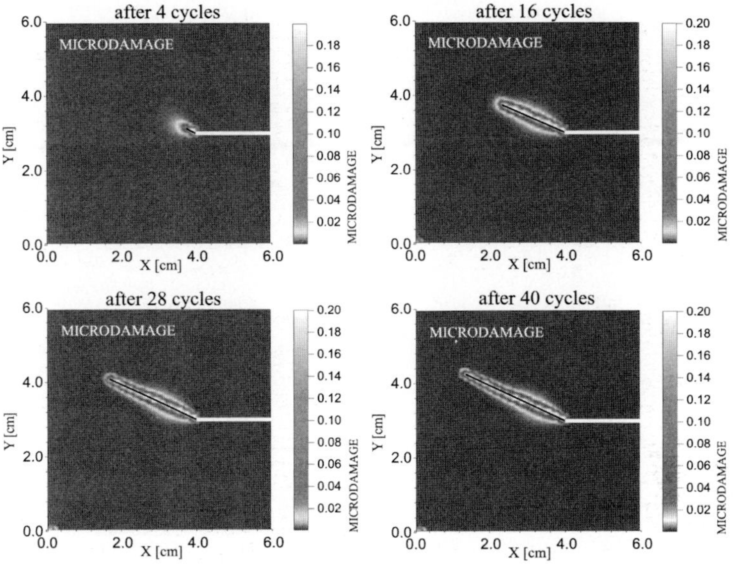

Figure 4.15: Evolution of the microdamage in the vicinity of the developed fatigue macrocrack

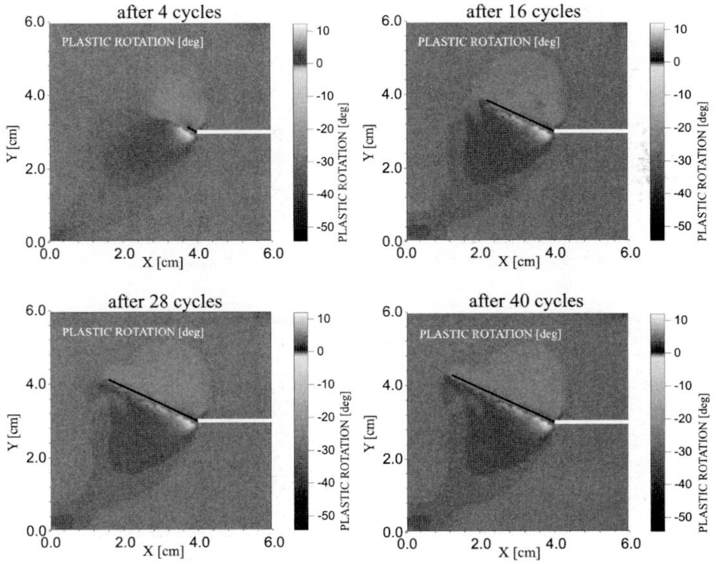

Figure 4.16: Evolution of the plastic rotation in the vicinity of the developed fatigue macrocrack

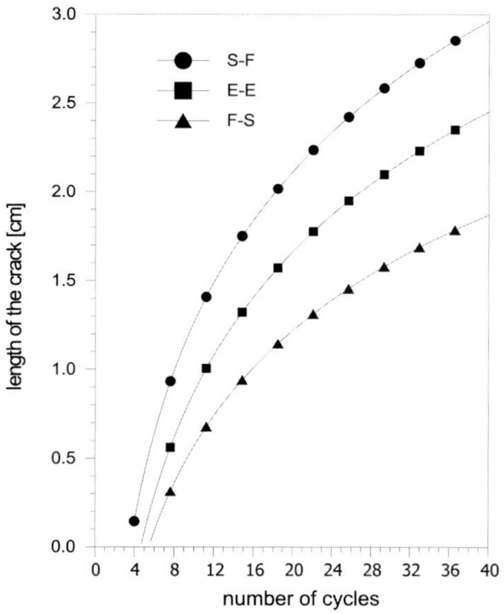

Figure 4.17: The length of macroscopic fatigue damage crack versus number of cycles for various loading processes

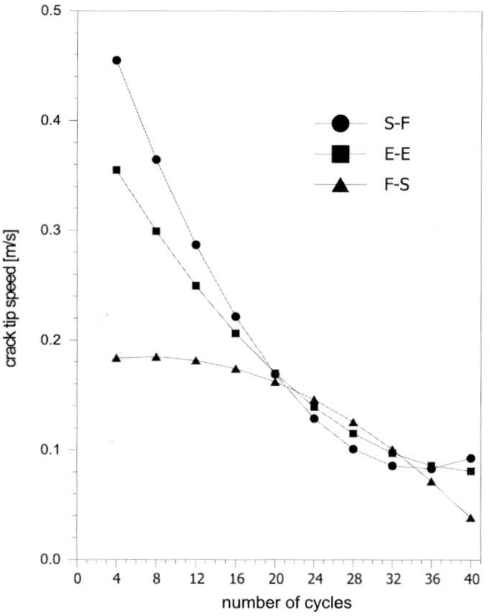

Figure 4.18: The speed of macroscopic crack propagation versus number of cycles for various loading processes

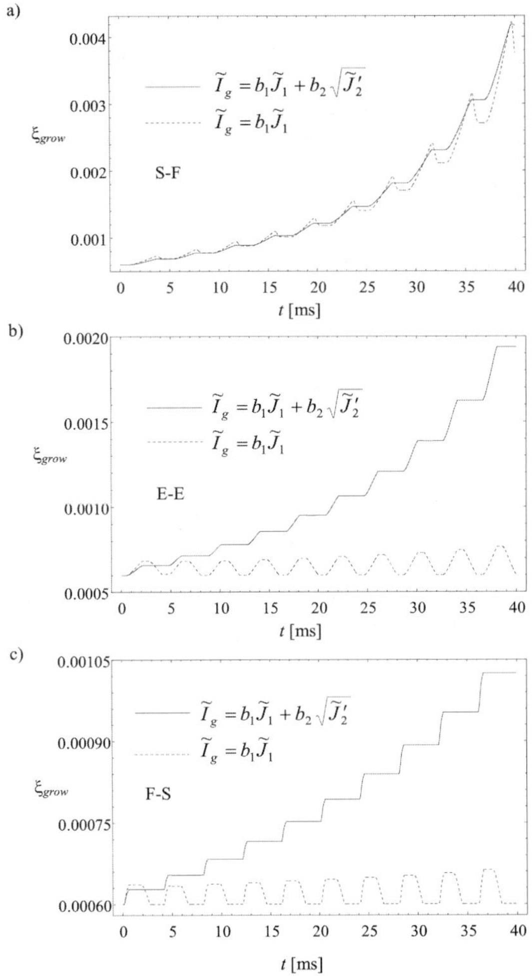

Figure 4.19: Evolution of the damage ξ_{grow} for different forms of the stress intensity invariant (adiabatic process), (a) slow–fast, (b) equal–equal, (c) fast–slow

5. Numerical Analysis of Macrocrack Propagation along a Bimaterial Interface under Dynamic Loading Processes

5.1 Experimental motivation

Classical dynamic fracture theories predict the surface wave speed to be the limiting speed for propagation of in-plane cracks in homogeneous, linear elastic materials subjected to remote loading. Cracks or fractures are displacement discontinuities in an otherwise intact material. On the basis of the nature the displacement discontinuity near the crack tip, three distinct fracture modes can be defined (cf. Rosakis, Samudrala and Coker (1999)): mode I, the in-plane opening mode resulting from normal separation of the crack faces (opening displacement discontinuity); mode II, the in-plane mode resulting from relative sliding of crack faces perpendicular to the crack edge (sliding displacement discontinuity); and mode III, the anti-plane shearing mode resulting from relative out-of-plane sliding of the crack faces (tearing displacement discontinuity).

Rosakis, Samudrala and Coker (1999) sought to determine experimentally whether in-plane intersonic crack growth could be obtained in laboratory specimens under remote shear loading conditions. In monolithic, prenotched laboratory specimens subjected to shear loading, after initiation from the notch tip the crack does not follow a straight path in line with the notch; it invariably kinks in the local symmetric opening direction. To make shear crack growth possible by suppressing kinking, they introduced a weak plane ahead of notch tip in the form of a bond between two identical pieces of isotropic material. The bonding process was carefully chosen that the constitutive properties of the bond were close to those of the bulk material. They thus constructed a material system that, although not monolithic, can be considered homogeneous with regard to its linear elastic constitutive description. However, fracture toughness along the bond line is lower, so that the material is inhomogeneous with regard to its fracture properties.

In the experimental investigation performed by Rosakis, Samudrala and Coker (1999) dynamic photoelasticity was chosen for capturing the stress field near the propagation crack tip because of its ability to visualize shear waves anticipated by the intersonic crack solutions.

In Fig. 5.1 the dynamic photoelasticity setup is shown. A Homalite-100 (Rosakis, Samudrala and Coker (1998)) specimen is subjected to asymmetric impact by a projectile fired from a high-speed gas gun. The coordinate system (x_1, x_2, x_3) is centered at the crack tip. Dimensions are in millime-

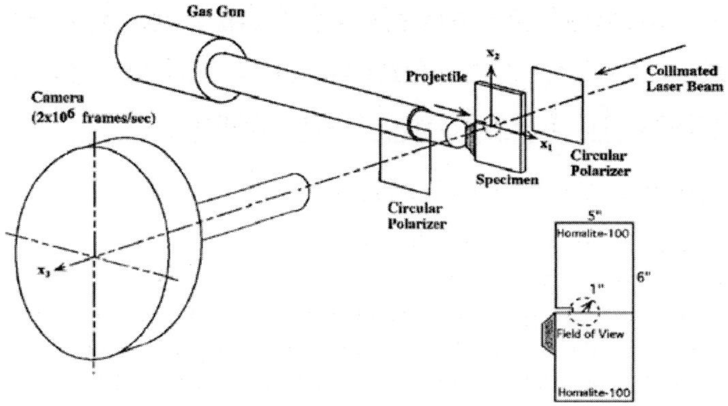

Figure 5.1: The dynamic photoelasticity setup (After Rosakis, Samudrala and Coker (1998))

ters. The specimen is 4 mm thick and the bond thickness is about 20 to 30 μm. The initial notch is 25 mm long and 2.3 mm wide. For Homalite-100, $c_L = 2200$ m/s and $c_S = 1255$ m/s. The steel projectile (length 75 mm, diameter 50 mm) impacts a steel piece, which was bonded to the specimen at the impact site to prevent shattering and to induce a planar loading wave front. The compressive longitudinal wave loads the notch tip in a predominantly shear mode. The dynamic stress field produced by the loading was recorded using photoelasticity in conjunction with high-speed photography. A coherent, monochromatic, plane-polarized, collimated laser beam (diameter $\phi = 50$ mm) was transmitted through the specimen. The specimen was placed in a circular polariscope, and the resulting isochromatic fringe pattern was recorded by a rotating mirror-type high-speed camera capable of recording 80 frames at framing rates up to 2 million frames per second.

The specimen was subjected to asymmetric impact loading with a projectile at 25 m/s.

Sequences of isochromatic fringe patterns were recorded around a shear crack as it initiated and propagated along the interface between two Homalite halves, cf. Fig. 5.2.

The frames included in the sequence are selected from two different experiments performed under identical conditions, except for the position of the field of view. Time after impact and crack tip speed are shown in each frame. The fringe pattern around pattern the propagating crack in the last frame is similar to that in the previous frame, indicating that the propagating crack has reached a steady state.

Evolution of a crack tip speed as the mode-II crack propagates along a

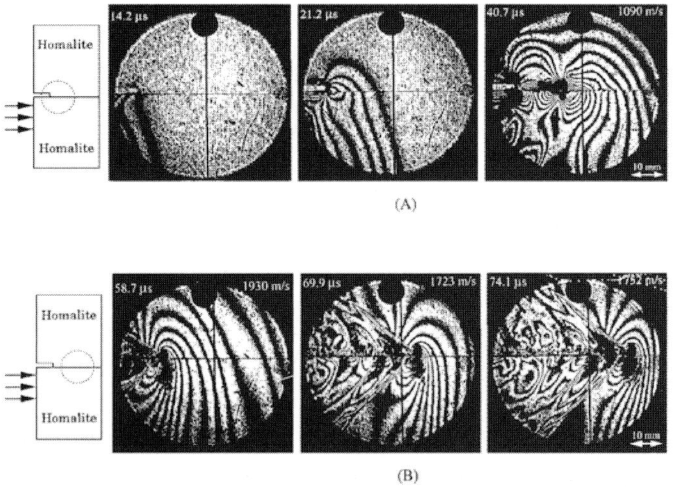

Figure 5.2: Selected sequence of high-speed images showing the isochromatic fringe pattern around a propagating shear crack along a weak plane in Homalite-100. (A) Field of view enclosing the notch tip. (B) Field of view ahead of the notch tip (After Rosakis, Samudrala and Coker (1999))

weak plane in Homalite-100 is presented in Fig. 5.3.

From the results of Fig. 5.3 we see that the initially recorded crack tip is close to the shear wave speed of Homalite beyond it accelerates, thus becoming intersonic. Thereafter, it continues to accelerate up to the plane stress dilatational wave speed of Homalite, then decelerates and ultimately reaches a steady-state value of about $\sqrt{2}$ times the shear wave speed. The experiments performed by Rosakis, Samudrala and Coker (1999) have provided evidence that shearing mode conditions near a propagating crack tip can drive the crack intersonic speeds, which are found to be possible even in purely homogeneous systems with only one distinct set of wave speeds.

5.2 Thermo–elasto–viscoplastic model of a material

For our practical purposes it is sufficient to assume that the internal state vector $\boldsymbol{\mu}$ has the form

$$\boldsymbol{\mu} = (\in^p, \xi), \tag{5.1}$$

where $\in^p$ is the equivalent viscoplastic deformation, and ξ is volume fraction porosity and takes account for micro-damaged effects.

Let us introduce the plastic potential function $f = f(J_1, J_2, \vartheta, \boldsymbol{\mu})$, where J_1, J_2 denote the first two invariants of the Kirchhoff stress tensor $\boldsymbol{\tau}$.

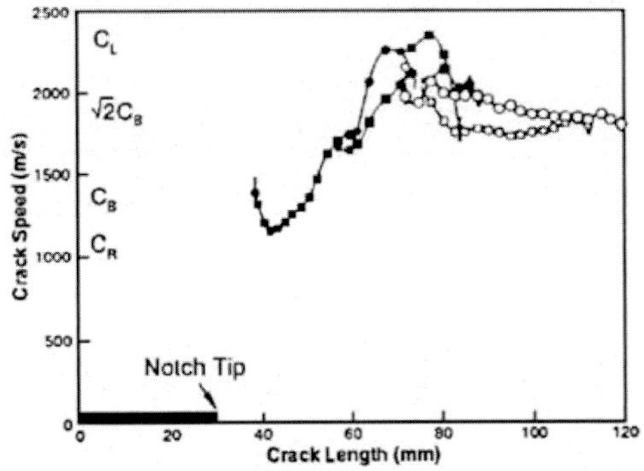

Figure 5.3: Evolution of crack speed as the shear crack propagates along a weak plane in Homalite-100. Crack tip speed was obtained from crack length history (squares) and from shock wave angles (circles) for a field of view around the notch tip (solid symbols) and for a field of view ahead of the notch (open symbols) (After Rosakis, Samudrala and Coker (1999))

Let us postulate the evolution equations as follows

$$\mathbf{d}^p = \Lambda \mathbf{P}, \qquad \dot{\xi} = \Xi, \tag{5.2}$$

where for elasto–viscoplastic model of a material we assume (cf. Perzyna (1963, 1966, 1971, 1995))

$$\Lambda = \frac{1}{T_m} \langle \Phi(\frac{f}{\kappa} - 1) \rangle, \tag{5.3}$$

T_m denotes the relaxation time for mechanical disturbances, the isotropic work-hardening-softening function κ is

$$\kappa = \hat{\kappa}(\in^p, \vartheta, \xi), \tag{5.4}$$

Φ is the empirical overstress function, the bracket $\langle \cdot \rangle$ defines the ramp function,

$$\mathbf{P} = \left. \frac{\partial f}{\partial \boldsymbol{\tau}} \right|_{\xi = const} \left(\left\| \frac{\partial f}{\partial \boldsymbol{\tau}} \right\| \right)^{-1}, \tag{5.5}$$

Ξ denotes the evolution function which has to be determined.

To determine the evolution function Ξ we can take advantage of the Section 3.3.

So, a set of the constitutive equations of the rate type has the form as follows

$$
\begin{aligned}
\mathrm{L}_{\boldsymbol{\upsilon}}\boldsymbol{\tau} &= \mathcal{L}^e : \mathbf{d} - \mathcal{L}^{th}\dot{\vartheta} - \left[(\mathcal{L}^e + \mathbf{g}\boldsymbol{\tau} + \boldsymbol{\tau}\mathbf{g}) : \mathbf{P}\right] \frac{1}{T_m} \langle \Phi(\frac{f}{\kappa} - 1)\rangle, \\
\rho c_p \dot{\vartheta} &= -\mathrm{div}\mathbf{q} + \vartheta \frac{\rho}{\rho_{Ref}} \frac{\partial \boldsymbol{\tau}}{\partial \vartheta} : \mathbf{d} + \rho\chi^* \frac{1}{T_m} \langle \Phi(\frac{f}{\kappa} - 1)\rangle \boldsymbol{\tau} : \mathbf{P} + \rho\chi^{**}\dot{\xi}, \\
\dot{\xi} &= \frac{h^*(\xi, \vartheta)}{T_m} \left[\exp \frac{m^*(\vartheta) \mid I_n - \tau_n(\xi, \vartheta, \in^p) \mid}{k\vartheta} - 1 \right] \\
&\quad + \frac{g^*(\xi, \vartheta)}{T_m \kappa_0} \left[I_g - \tau_{eq}(\xi, \vartheta, \in^p) \right].
\end{aligned}
\tag{5.6}
$$

All the material functions and the material constants should be identified based on available experimental data.

5.3 Identification procedure

5.3.1 Assumption of the material functions for an adiabatic process

To do the proper identification procedure we first make assumption of the material functions (cf. Dornowski and Perzyna (2000)).

The plastic potential function f is assumed in the form (cf. Perzyna (1984) and Shima and Oyane (1976))

$$
f = \left\{ J_2^{'} + \left[n_1(\vartheta) + n_2(\vartheta)\xi \right] J_1^2 \right\}^{\frac{1}{2}} \tag{5.7}
$$

where

$$
n_1(\vartheta) = 0, \qquad n_2(\vartheta) = n = \text{const.} \tag{5.8}
$$

The isotropic work-hardening-softening function κ is postulated as (cf. Perzyna (1986a) and Nemes and Eftis (1993))

$$
\kappa = \hat{\kappa}(\in^p, \vartheta, \xi) = \left\{ \kappa_s(\vartheta) - \left[\kappa_s(\vartheta) - \kappa_0(\vartheta) \right] \exp\left[-\delta(\vartheta) \in^p \right] \right\} \left[1 - \left(\frac{\xi}{\xi_F} \right)^{\beta(\vartheta)} \right], \tag{5.9}
$$

where

$$
\begin{aligned}
\kappa_s(\vartheta) &= \kappa_s^* - \kappa_s^{**}\overline{\vartheta}, \qquad \kappa_0(\vartheta) = \kappa_0^* - \kappa_0^{**}\overline{\vartheta}, \\
\delta(\vartheta) &= \delta^* - \delta^{**}\overline{\vartheta}, \qquad \beta(\vartheta) = \beta^* - \beta^{**}\overline{\vartheta}, \qquad \overline{\vartheta} = \frac{\vartheta - \vartheta_0}{\vartheta_0}.
\end{aligned}
\tag{5.10}
$$

The overstress function $\Phi\left(\frac{f}{\kappa}-1\right)$ is assumed in the form

$$\Phi\left(\frac{f}{\kappa}-1\right)=\left(\frac{f}{\kappa}-1\right)^{m}. \tag{5.11}$$

The evolution equation for the porosity ξ is postulated as

$$\dot{\xi}=\dot{\xi}_{grow}=\frac{g^{*}(\xi,\vartheta)}{T_m\kappa_0(\vartheta)}\left[I_g-\tau_{eq}(\xi,\vartheta,\in^p)\right] \tag{5.12}$$

where (cf. Dornowski (1999))

$$\begin{aligned}
g^{*}(\xi,\vartheta) &= c_1(\vartheta)\frac{\xi}{1-\xi},\\
I_g &= b_1 J_1 + b_2\sqrt{J_2},\\
\tau_{eq}(\xi,\vartheta,\in^p) &= c_2(\vartheta)(1-\xi)\ln\frac{1}{\xi}\left\{2\kappa_s(\vartheta)-\left[\kappa_s(\vartheta)-\kappa_0(\vartheta)\right]\right.\\
&\quad \left. F(\xi_0,\xi,\vartheta)\right\}, \qquad (5.13)\\
c_1(\vartheta) &= \text{const}, \quad c_2(\vartheta)=\text{const},\\
F(\xi_0,\xi,\vartheta) &= \left(\frac{\xi_0}{1-\xi_0}\frac{1-\xi}{\xi}\right)^{\frac{2}{3}\delta}+\left(\frac{1-\xi}{1-\xi_0}\right)^{\frac{2}{3}\delta}.
\end{aligned}$$

As in the infinitesimal theory of elasticity we assume linear properties of the material, cf. Section 4.1.1.

5.3.2 Assumption of the material functions for an isothermal process

Let us assume the plastic potential function in the form

$$f=\left(J_2^{'}+n\xi J_1^2\right)^{\frac{1}{2}}, \tag{5.14}$$

the isotropic hardening-softening function

$$\kappa=\left[\kappa_s-(\kappa_s-\kappa_0)\exp\left(-\delta\in^p\right)\right]\left[1-\left(\frac{\xi}{\xi_F}\right)^{\beta}\right], \tag{5.15}$$

and the functions in the evolution equation for porosity

$$\begin{aligned}
g^{*} &= c_1\frac{\xi}{1-\xi}, \qquad I_g=b_1J_1+b_2\sqrt{J_2^{'}},\\
\tau_{eq}(\xi,\in^p) &= c_2(1-\xi)\ln\frac{1}{\xi}\left[2\kappa_s-(\kappa_s-\kappa_0)F(\xi_0,\xi)\right], \qquad (5.16)\\
F(\xi_0,\xi) &= \left(\frac{\xi_0}{1-\xi_0}\frac{1-\xi}{\xi}\right)^{\frac{2}{3}\delta}+\left(\frac{1-\xi}{1-\xi_0}\right)^{\frac{2}{3}\delta}.
\end{aligned}$$

Table 5.1: Material A (plate)

$\rho_{Ref} = 7830$ kg/m^3	$E = 208$ GPa	$v = 0.3$	$\kappa_0 = 635$ MPa
$\kappa_s = 808$ MPa	$\delta = 28$	$\beta = 2.2$	$T_m = 0.01\ \mu$s
$m = 1$	$c_1 = 0.202$	$c_2 = 0.067$	$b_1 = 1.0$
$b_2 = 1.3$	$\xi_0 = 6 \cdot 10^{-4}$	$\xi_F = 0.2$	$n = 0.25$

We keep the form of the overstress function as in (5.11).

5.3.3 Determination of the material constants

To determine the material constants assumed we take advantage of the experimental observations presented by Chakrabarti and Spretnak (1975). They investigated the localized fracture mode for tensile steel sheet specimens simulating both plane stress and plane strain processes. The material used in their study was AISI 4340 steel. The principal variable in this flat specimen test was the width – to – thickness ratio. Variation in specimen geometry produces significant changes in stress state, directions of shear bands, and ductility. They found that fracture propagated consistently along the shear band localized region.

Let us now consider the isothermal dynamic process for a thin steel plate under condition of plane stress state. In fact we idealize the initial-boundary value problem investigated by Chakrabarti and Spretnak (1975) by assuming the velocity driven isothermal process for a thin steel plate. The problem has been solved by using the finite difference method.

In numerical calculations it is assumed:

$$\dot{V}_0 = 1.5 \text{ m/s}, \quad t_0 = 50\ \mu\text{s}, \quad t_f = 800\ \mu\text{s}.$$

The material of a plate is AISI 4340 steel.

Based on the best curve fitting of the experimental results obtained by Chakrabarti and Spretnak (1975) for the stress-strain relation the identification of the material constants has been done for both materials A and B, cf. Table 5.1 and 5.2.

Let us now assume the velocity driven adiabatic process for a thin steel plate and proceed similarly as in the case of an isothermal process. Then the material constants for an adiabatic process for both materials A and B can be determined, cf. Table 5.3 and 5.4.

Table 5.2: Material B (cohesive band)

$\rho_{Ref} = 1190$ kg/m^3	$E = 3.24$ GPa	$v = 0.35$	$\kappa_0 = 5$ MPa
$\kappa_s = 92$ MPa	$\delta = 80$	$\beta = 2.0$	$T_m = 0.001$ μs
$m = 1$	$c_1 = 0.202$	$c_2 = 0.067$	$b_1 = 1.0$
$b_2 = 1.3$	$\xi_0 = 6 \cdot 10^{-4}$	$\xi_F = 0.09$	$n = 0.25$

Table 5.3: Material A (plate)

$\kappa_s^* = 808$ MPa	$\kappa_s^{**} = 230$ MPa	$\kappa_0^* = 635$ MPa
$\kappa_0^{**} = 181$ MPa	$\delta^* = 28$	$\delta^{**} = 8$
$\beta^* = 2.2$	$\beta^{**} = 0.63$	$E = 208$ GPa
$\vartheta_0 = 293$ K	$\xi_F = 0.2$	$\rho_{Ref} = 7850$ kg/m^3
$\nu = 0.3$	$\theta = 12 \cdot 10^{-6}$ K^{-1}	$T_m = 0.01$ μs
$c_1 = 0.202$	$c_2 = 0.067$	$b_1 = 1.0$
$\xi_0 = 6 \cdot 10^{-4}$	$n = 0.25$	$\chi^* = 0.9$
$m = 1$	$b_2 = 1.3$	$c_p = 455$ J/kg K

Table 5.4: Material B (plate)

$\kappa_s^* = 92$ MPa	$\kappa_s^{**} = 26$ MPa	$\kappa_0^* = 5$ MPa
$\kappa_0^{**} = 1.5$ MPa	$\delta^* = 80$	$\delta^{**} = 22$
$\beta^* = 2.0$	$\beta^{**} = 0.57$	$E = 3.24$ GPa
$\vartheta_0 = 293$ K	$\xi_F = 0.09$	$\rho_{Ref} = 1190$ kg/m^3
$\nu = 0.35$	$\theta = 12 \cdot 10^{-6}$ K^{-1}	$T_m = 0.01$ μs
$c_1 = 0.202$	$c_2 = 0.067$	$b_1 = 1.0$
$\xi_0 = 6 \cdot 10^{-4}$	$n = 0.25$	$\chi^* = 0.9$
$m = 1$	$b_2 = 1.3$	$c_p = 455$ J/kg K

5.4 Numerical solution of the initial boundary–value problems

5.4.1 Formulation of the initial boundary–value problem

We analyse the edge-cracked bimaterial thin specimen performed from two identical elements joined by the cohesive band, (Fig. 5.4). The material of joint elements (material A) and the cohesive band (material B) are modelled as elasto-viscoplastic with isotropic hardening-softening effects. We assume that the material softening is caused by intrinsic microdamage mechanism. The height of the specimen is equal to 30 cm, width is 12.5 and length of the initial crack (notch) is equal to 2.5 cm. The cohesive band width of 1 mm. We intend to consider both plane states, i.e. the plane stress state, and the plane strain state. As it is shown in Fig. 5.4, this specimen is loaded symmetrically (mode-I) or asymmetrically (mode-II) to the cohesive axis.

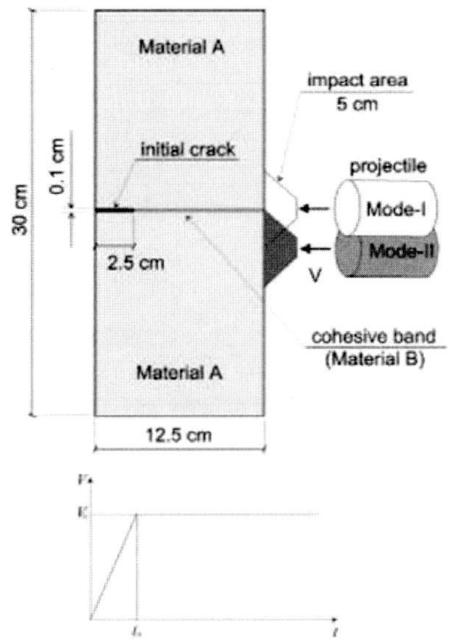

Figure 5.4: Specimen geometry

To obtain the solution of the initial-boundary value problem formulated the finite difference method for regularized elasto-viscoplastic model is used, cf. Dornowski and Perzyna (2000).

The loading condition is modelled by the velocity of nodes lying on the edge section with the length equals to 5 cm, according to the relation

$$V(t) = V_0 t/t_0 \quad \text{for} \quad t \leq t_0 \quad \text{and} \quad V(t) = V_0 \quad \text{for} \quad t > t_0 \tag{5.17}$$

The rise time t_0 is fixed at 0.1 μs. Initial conditions of the problem are homogeneous. In the discussion of the numerical results attention is focused mostly on differences resulting from the formulation of the plane problem (plane stress or plane strain conditions) and from the spatial location of loading (mode-I or II). Both, isothermal and adiabatic processes are considered.

The finite difference method with the explicit time integration scheme (conditionally stable) is used. The stress state in a nodal environment is determined by the iterative procedure of solving the dynamical yield condition with respect to the norm of the plastic deformation rate tensor.

The elaborated algorithm satisfies the material objectivity principle with respect to diffeomorphism (any motion). The discretization parameters are assumed as follows: $\Delta x = 1$ mm, $\Delta y = 1$ mm, $\Delta t_{cr} = 0.1331$ μs, and number of nodes is $N \times M = 126 \times 300 = 37800$.

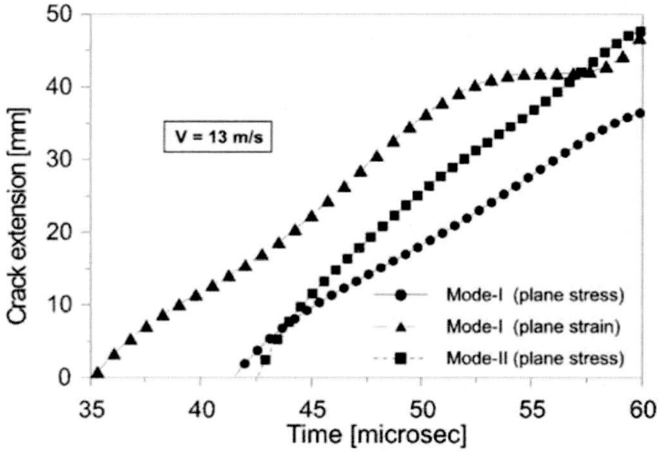

Figure 5.5: Crack tip position history

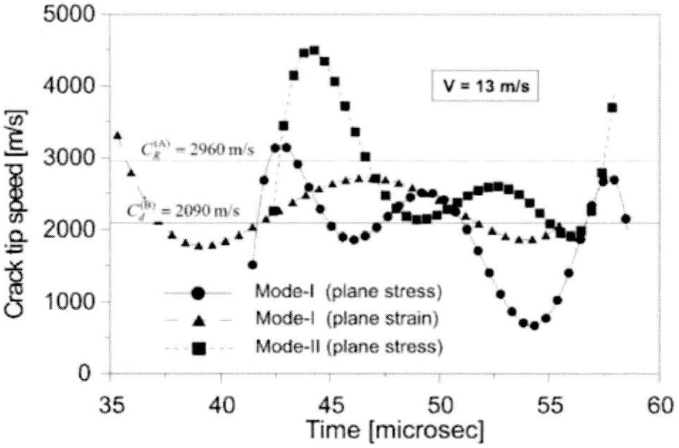

Figure 5.6: Crack tip speed history

5.4.2 Investigation of the macrocrack propagation along a bimaterial interface

Figure 5.5 shows the crack tip position history for both the mode-I and mode-II cases. For the asymmetrical loading and for plane strain conditions the crack starts simultaneously in the notch tip as well as in some distance from it. From this reason this case is not illustrated in Fig. 5.5.

We can see the distinct difference between instants at which the crack is initiated by mode-I loading. For plane strain conditions the crack is initiated at $t = 35.33\ \mu$s, however for plane stress conditions at $t = 41.45\ \mu$s, i.e. later about 6.12 μs. This difference determines 25 percent of total time of the crack propagation. The result clearly indicates on the importance of the choice of the plane state used in numerical calculations.

Figure 5.5 shows also that the asymmetrical loading (mode-II), for plane stress conditions generates the crack with a greater length than the same loading applied symmetrically (mode-I).

The crack tip speed history is shown in Fig. 5.6. It can be seen that the speed of crack propagation generated by the asymmetrical impact reaches some greatest values for plane stress condition. In the process of advanced cracking for all types of loading the analyzed speeds are changing more or less in the range determined by the speed $C_d^{(B)} = 2090$ m/s of the dilatational wave propagating in the cohesive material (material B) and the speed $C_R^{(A)} = 2960$ m/s of the Rayleigh wave propagating in the specimen material (material A). Some visible crack speed oscillations result from interaction of mentioned waves.

An influence of the loading mode on the deformability of the cohesive

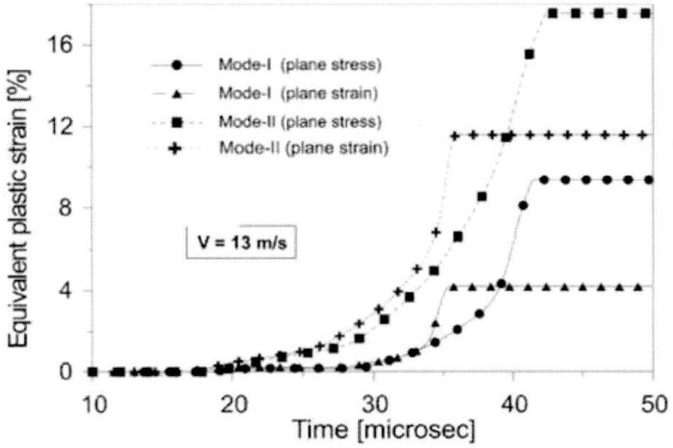

Figure 5.7: Equivalent plastic strain at the initial crack tip as a function of time

band for different formulations of a plane problem is illustrated in Fig. 5.7. It is assumed that the equivalent plastic deformation, calculated at the initial crack tip, is a measure of this deformability. It reaches some greatest values (about 18% at local fracture) for the asymmetrical impact (mode-II), which causes a most complex strain state in the cohesive band. The cohesive band deformability decreases more intensely when the strain tensor component, normal to the surface, is reduced (plane strain state).

In Fig. 5.8 the Mises stress ($\tau_M = \sqrt{\frac{2}{3}\tau' : \tau'}$) as a function of time in the initial crack environment is shown. This figure may be treated as an illustration of the cohesive band strength evolution to rupture, in processes generated by different loading modes. The mode-II creates the most complex stress states in the cohesive band. In the case of symmetrical loading (mode-I) the shear stress vanishes in the cohesive band – it is subjected to many-side extension or to compression only. In the analyzed node we can see also the transient unloading of a material (effect of passing the reflected wave). For plane strain conditions the cohesive band material loses its carrying capacity more quickly than for plane stress conditions – crack initiation follows more quickly too.

Figure 5.9 represents the evolution of the micro-damage parameter in the initial crack environment. It is assumed that the cohesive band material undergoes to local fracture at $\xi = \xi_F = 0.09$. For the plane strain state the micro-damage growth starts sooner and it is characterized by greater intensity.

It can be seen that the symmetrical loading initiates the crack prop-

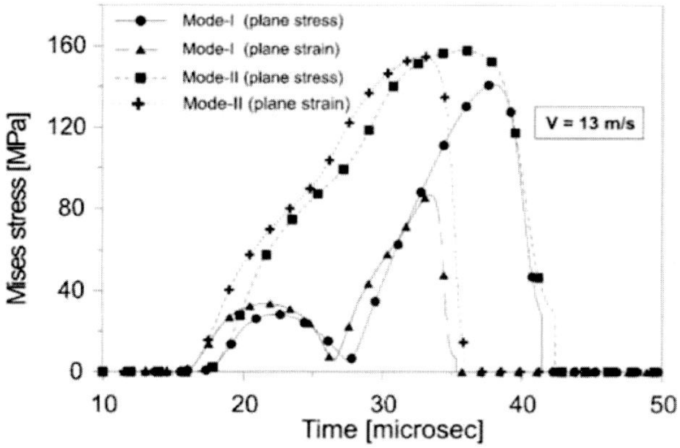

Figure 5.8: Mises stress at the initial crack tip as a function of time

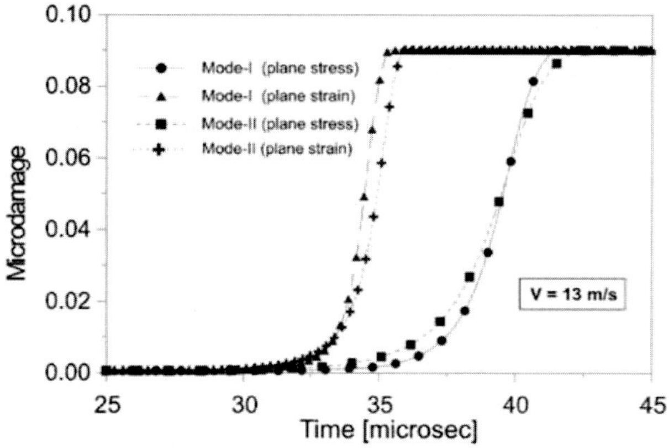

Figure 5.9: Microdamage at the initial crack tip as a function of time

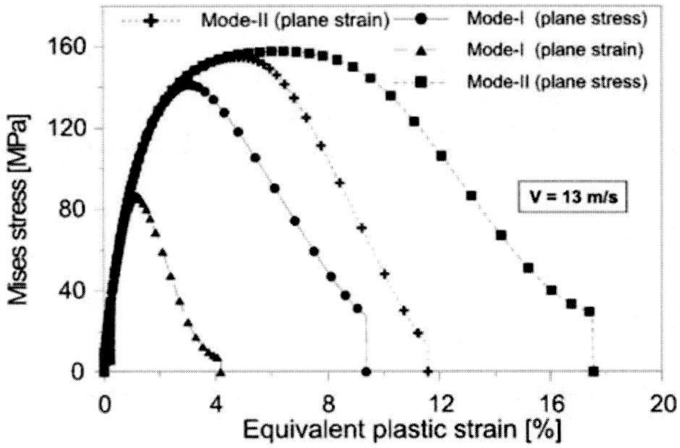

Figure 5.10: Mises stress at the initial crack tip as a function of the equivalent plastic strain

agation more quickly than the asymmetrical one. However, the mode-II loading leads to the greater speed of crack propagation and as a result the final crack is longer.

A dependence of the Mises stress on the equivalent plastic deformation at the considered node is presented in Fig. 5.10. The different types of loading and different kinds of plane state formulations are considered.

Figure illustrates mainly the nonelastic stress-strain characteristic of the cohesive band material. At the beginning of the deformation process the cohesive band undergoes to isotropic hardening, after the material softening takes place. The evolutional approach to softening effects is very well visible here. It is clearly shown that the beginning of softening and the softening intensity distinctly depend on the loading mode (spatial stress configuration). Comparison of the results obtained for different plane states confirms this conclusion. In such a case the spatial stress configuration is determined directly by the plane state indeed.

The Mises stress as a function of micro-damage is shown in Fig. 5.11. It can be noticed that the value of micro-damage, for which the Mises stress reaches the maximum, depends on the loading mode and on the plane state formulation. Besides all the conclusions and observations made sooner keep own validity.

Figure 5.12 shows an evolution of the spatial configuration of stress waves at the beginning of the loading process (mode-I) for plane stress conditions. It is obtained by shaded relief mapping some distributions of the Kirchhoff stress component τ^{yy}, for selected instants.

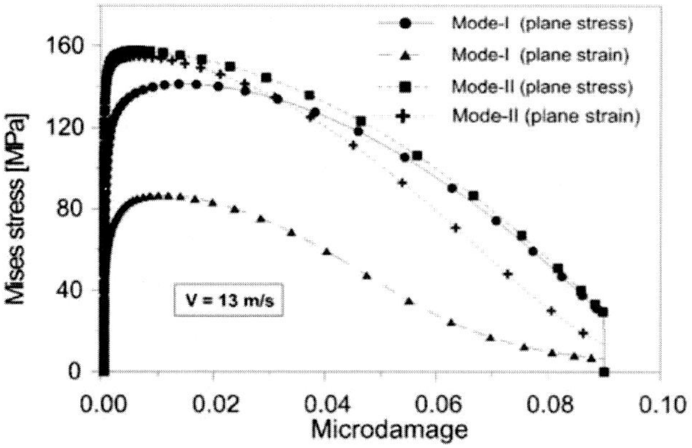

Figure 5.11: Mises stress at the initial crack tip as a function of the microdamage

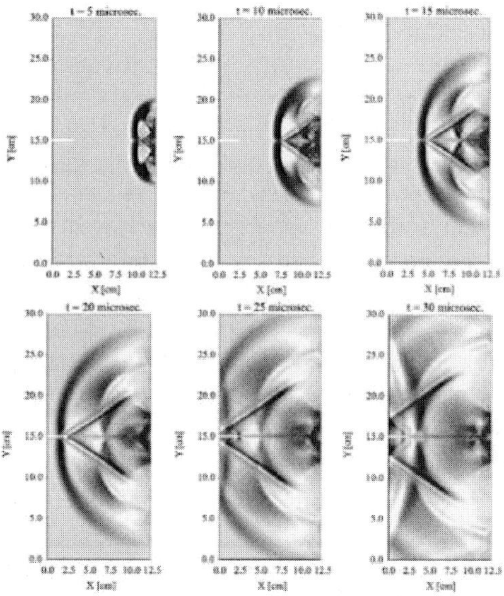

Figure 5.12: Evolution of spatial configuration of stress waves at the beginning of the process under symmetric mode-I loading (plane stress)

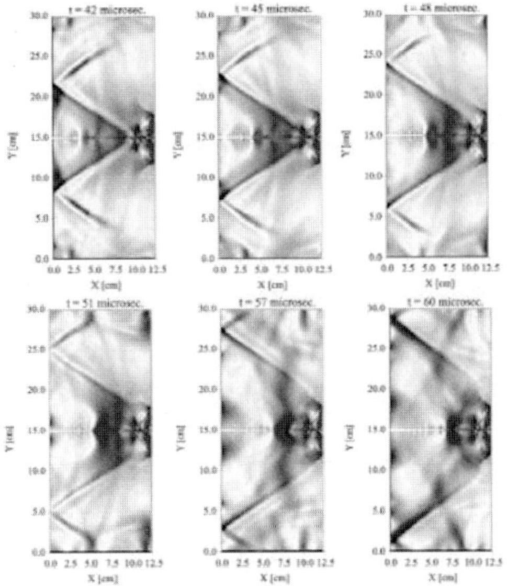

Figure 5.13: Evolution of spatial configuration of stress waves in the cracking phase under symmetric mode-I loading (plane stress)

Two circular fronts of a plastic wave are visible behind a front of the analyzed stress wave. These waves move symmetrically to the cohesive band axis. Waves move obliquely to the cohesive band are reflections of the longitudinal wave from the boundary of two different materials. We can see the main effect of impact loading which is a wave carrying a normal stress τ^{yy} that propagates across the specimen and reaches the crack tip at 16.2 μs. At 20.4 μs the loading wave reaches the free surface at $x = 0$ and at 27.5 μs the reflected wave reaches the initial crack tip. Later, the crack tip environment is subject to many-sided compression. An intensive increase of the tensile normal stress τ^{yy} at the crack tip precedes directly the cracking phase illustrated in Fig. 5.13. In this figure an evolution of the spatial configuration of stress waves in the cracking phase is shown. The crack growth begins at 41.45 μs in the initial crack tip and the crack line develops to the impact area. It can be seen that the overall stress pattern is fairly complex due to elastic mismatch across the crack line and due to reflection from the free surfaces.

There is the unloaded domain in the crack environment when the stress wave is reflected from the impact edge – crack propagation stopped ($t \approx 60$ μs).

Figure 5.14 shows distributions of τ^{yy} at various stages of crack propagation ($t = 42$ μs and $t = 60$ μs). The region of opening stress (positive stress) concentration can be seen emanating from the crack tip into the specimen. The large compressive stress is seen in the vicinity of the impact region, especially at corners of this region where the local plastic deformation can occur. It can be also seen that the edge is strongly stretching outside of the impact region. The region of intensive opening stresses decreases at the end of crack propagation.

Figure 5.15 shows the evolution of the spatial configuration of stress waves τ^{yy} generated by asymmetrical loading, for plane strain conditions. Also in this case, similarly as for mode-I loading, the exact analysis of propagation of the considered wave is possible. However, we limit our discussion to the indication some essential differences between these results. Herein the impact zone is situated at the bottom part of the specimen therefore the disturbance transfers to the upper part through the joint layer. Because of energy dissipation on plastic deformations of the cohesive band, this disturbance has smaller amplitudes.

The spatial configuration of waves is asymmetrical. The cohesive band material fracture is preceded by shear deformation mostly. The crack propagation is more complicated when it is generated in the shear way for plane strain conditions. Delamination of the cohesive band material can start simultaneously in different places. This phenomenon is illustrated more exactly in Fig. 5.17.

The evolution of the spatial configuration of shear stress τ^{xy} waves for the crack propagation under asymmetric mode-II loading (plane strain) is shown in Fig. 5.16. Selected instants of the deformation process correspond to those in Fig. 5.15.

Figure 5.17 shows the complex crack propagation under asymmetric mode-II loading (plane strain). Complexity of this phenomenon is generated by the fact that the cracking can start simultaneously in different places. In this case the cracks initiate in the initial crack tip and behind it in the distance equal to 2 cm. Both cracks come together at time 38 μs.

The crack tip speed versus crack length for the asymmetric mode-II loading (plane stress) is illustrated in Fig. 5.18. The impact speed is changed from 10 m/s to 25 m/s. We can see that the maximum of the crack tip speed distinctly depends on the impact speed and it can be greater than the speed of the dilatational wave propagating in the specimen material (supersonic cracking). It can be observe that the certain crack length exists for which the crack tip speed is equal to Rayleigh wave speed for every impact speed applied. In our case this crack length is 16.7 mm. The subsonic cracking takes place above this magnitude.

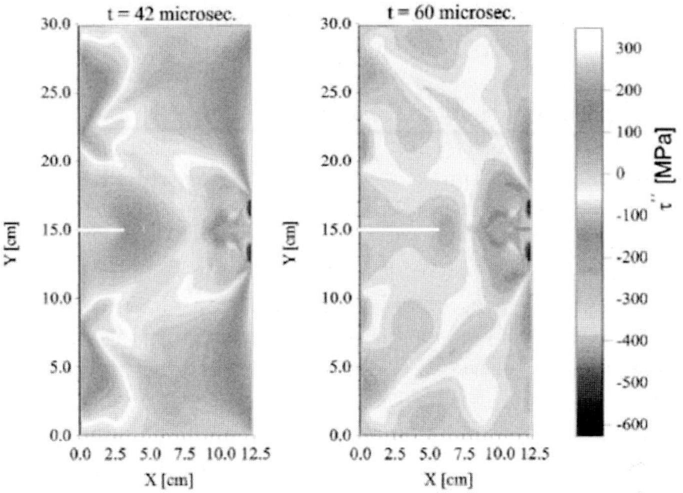

Figure 5.14: Distribution of the component τ^{yy} of the Kirchhoff stress for chosen instants of the crack growth under symmetric mode-I loading (plane stress)

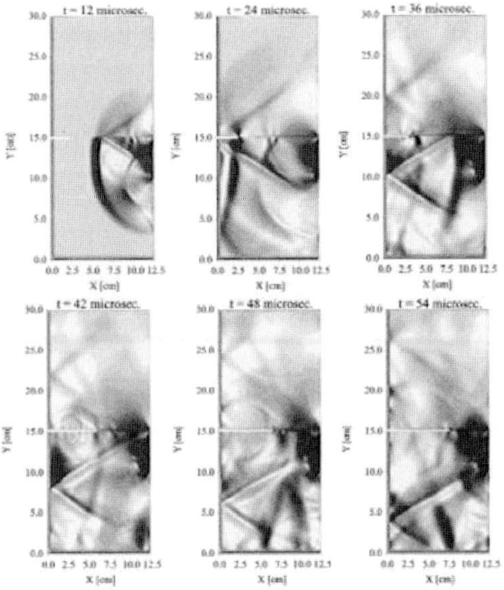

Figure 5.15: Evolution of spatial configuration of normal stress waves for the crack growth under asymmetric mode-II loading (plane strain)

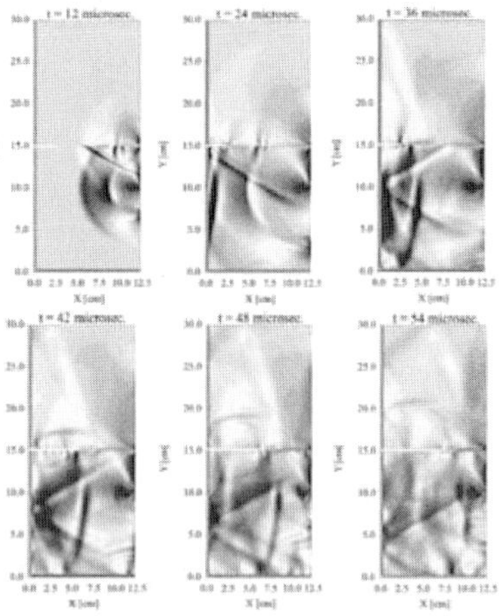

Figure 5.16: Evolution of spatial configuration of shear stress waves for the crack growth under asymmetric mode-II loading (plane strain)

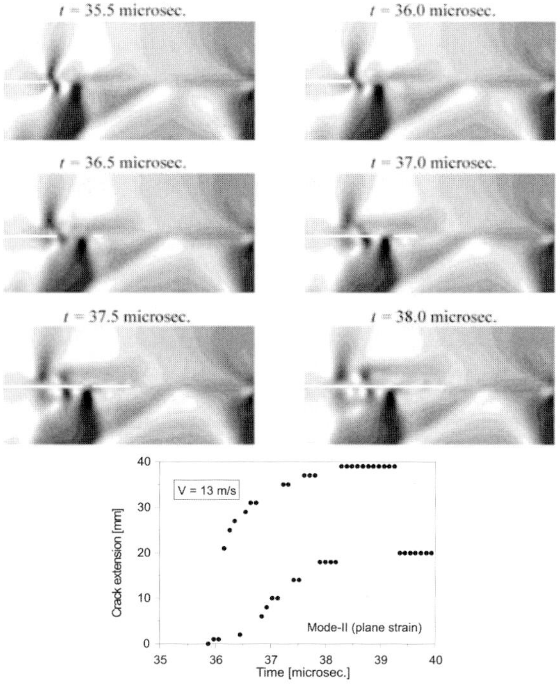

Figure 5.17: Complex crack propagation under asymmetric mode-II loading (plane strain)

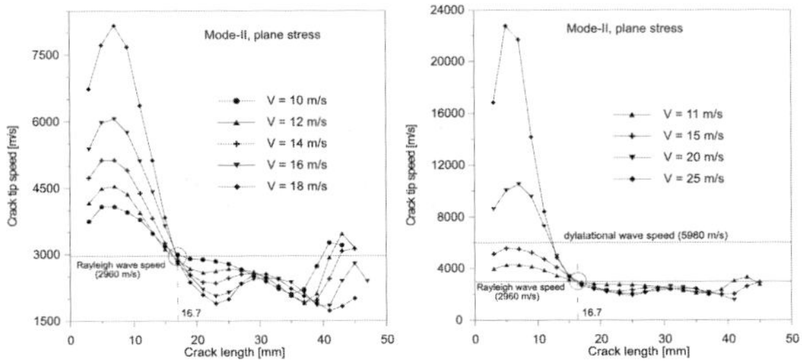

Figure 5.18: Crack tip speed versus length for the asymmetric mode-II loading (plane stress)

The results obtained are in accord with the experimental observations performed by Rosakis, Samudrala and Coker (1999).

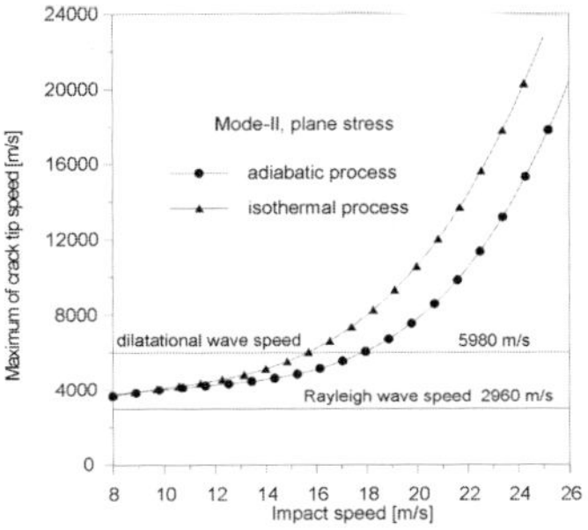

Figure 5.19: Maximal crack tip speed as a function of impact speed for the asymmetric mode-II loading (plane stress)

5.4.3 Influence of temperature effects for an adiabatic process

The maximal crack tip speed as a function of the impact speed for the asymmetric mode-II loading (plane stress) for isothermal and adiabatic process is shown in Fig. 5.19. Temperature at the initial crack tip as a function of time is plotted in Fig. 5.20. The influence of temperature for an adiabatic process for asymmetric mode-II (plane stress) on the evolution of the crack extension is shown in Fig. 5.21a and on the crack speed versus crack length curve in Fig. 5.21b.

Fig. 5.22 shows the evolution of equivalent plastic deformation along a weak cohesive band during an adiabatic processes for the plane stress

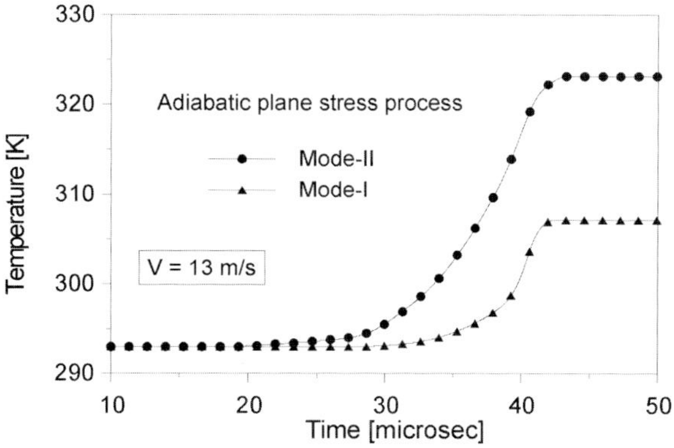

Figure 5.20: Temperature at the initial crack tip as a function of time

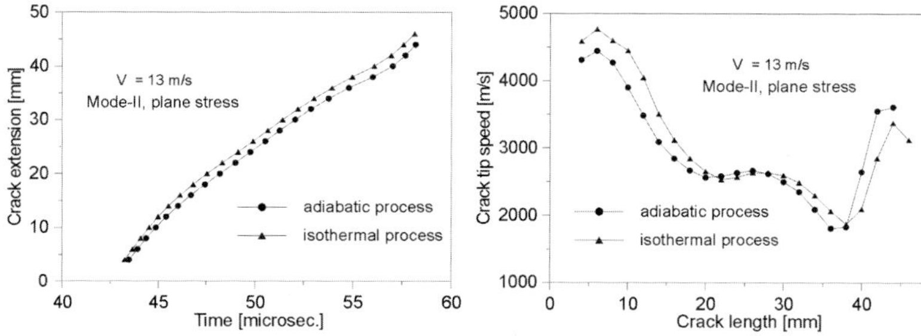

Figure 5.21: Effects of temperature for an adiabatic process (a) Crack extension versus time; (b) Crack speed versus crack length

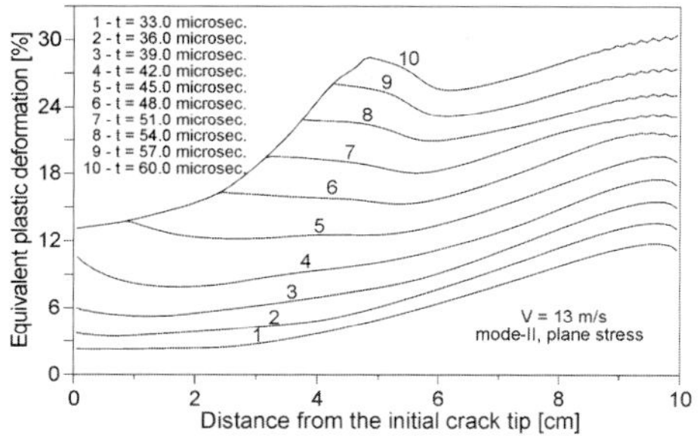

Figure 5.22: Evolution of equivalent plastic deformation along a weak cohesive band during an adiabatic process for the plane stress mode-II loading

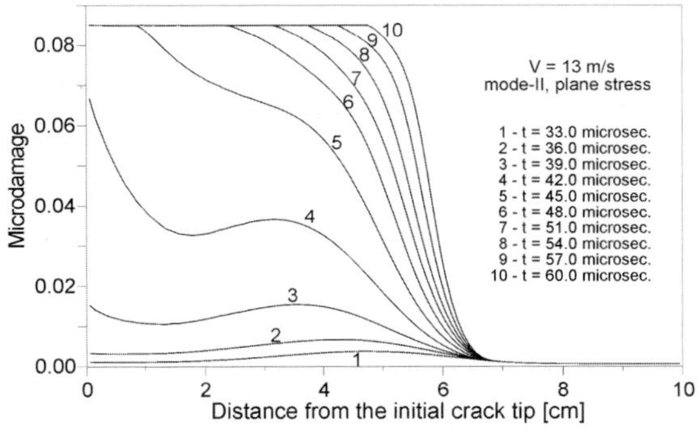

Figure 5.23: Evolution of microdamage along a weak cohesive band during an adiabatic process for the plane stress mode-II loading

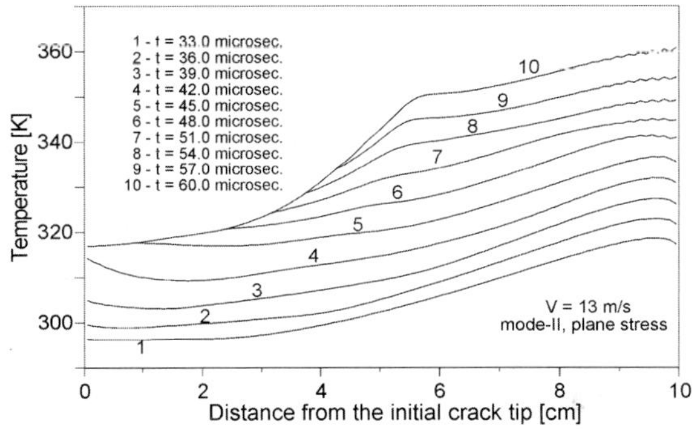

Figure 5.24: Evolution of temperature along a weak cohesive band during an adiabatic process for the plane stress mode-II loading

mode-II loading. Similar results for the evolution of microdamage and temperature have been presented in Figs. 5.23 and 5.24, respectively.

The results for an adiabatic plane stress process show that the increase of temperature on the evolution of the crack extension as well as on the crack speed is very small, cf. Figs. 5.20 and 5.21.

6. Numerical Investigation of Dynamic Shear Bands in Inelastic Solids as a Problem of Mesomechanics

6.1 Physical and experimental motivation

6.1.1 Analysis of mesomechanical problems

In modern technology we observe recently very important application of metals, ceramics and polymers at mesoscale. Micromachines are in this size range clearly will be of increasing technological significance. Processes that control the mechanical integrity of microelectronic devices take also place on this size scale, cf. Needleman (2000) and Hutchinson (2000).

It is considerable experimental evidence that plastic flow and particularly localization of plastic deformation and fracture phenomena in crystalline solids are inherently size dependent over mesoscale range. It is generally accepted that: "smaller is stronger" or "smaller is harder".

Plastic behaviour at mesoscale range can not be characterized by conventional plasticity theories because they incorporate no material length scale and predict no size effect.

In recent years a variety of theoretical frameworks is emerging to describe inelastic deformation at the mesoscale. Four such frameworks (constitutive structures), each involving a length scale, are as follows: (i) discrete dislocation plasticity; (ii) nonlocal plasticity; (iii) the coupling of matter diffusion and deformation; (iv) elasto-viscoplasticity.

The mesomechanical problems pose also numerical challenges. Computations on smaller size scale require smaller time steps. Since size dependent phenomena come into play when there are gradients of deformation and stress, hence numerical methods are usually needed to obtain solutions. Finite strains and rotations have to be taken into consideration.

At the mesoscale problems the dominant numerical methods are the finite element and finite difference methods.

It is noteworthy to add that the mesoscale continuum mechanics is in an early stage of development, both in terms of the theoretical framework as well as the computational methods.

In our consideration we shall use the thermo-elasto-viscoplasticity as a constitutive model of the material and apply the finite element method in numerical computations, cf. Glema, Łodygowski and Perzyna (2008).

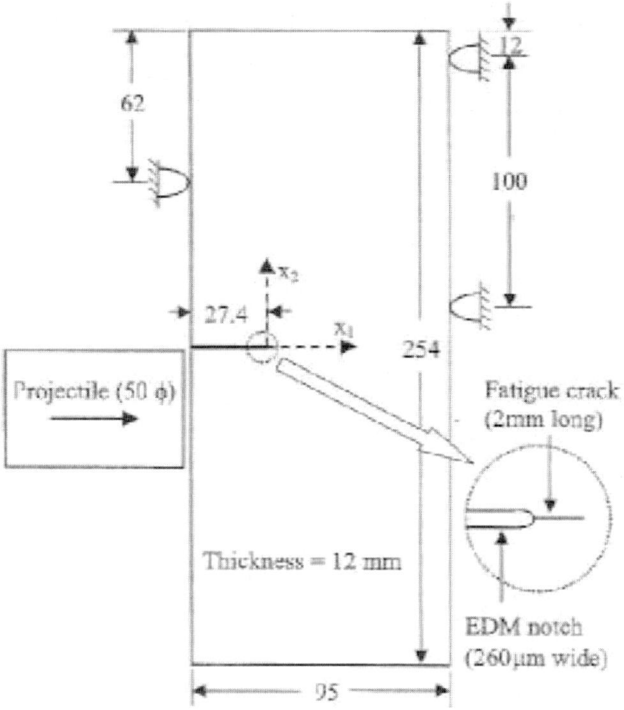

Figure 6.1: Specimen geometry and impact arrangement. The projectile is 127 mm long. All dimensions shown are in millimeters. (After Guduru, Rosakis and Ravichandran (2001)).

6.1.2 Experimental investigation of the initiation and propagation of shear bands

Guduru, Rosakis and Ravichandran (2001) presented an experimental investigation of the initiation and propagation characteristics of dynamic shear band in C300 maraging steel. An elastic discharge machining (EDM) notch (260 μm wide) was further extended by 2 mm by fatigue loading, cf. Fig. 6.1.

In experimental investigation of Guduru, Rosakis and Ravichandran (2001), two diagnostic techniques were used to observe the crack tip, the propagating shear band and the temperature field evolution during the initiation and propagation of the shear band. On one side of the specimen, the optical technique of coherent gradient sensing (CGS) in reflection was used to monitor the evolution of the stress intensity factors as a function

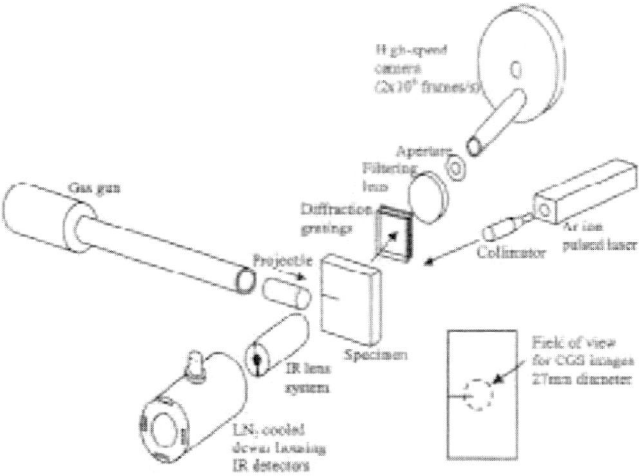

Figure 6.2: Schematic illustration of the experimental setup. The CGS setup measures the out plane displacement gradients on the rear side of the specimen. Simeltaneous thermal imaging is accomplished using the IR camera on the facing side of the specimen. (After Guduru, Rosakis and Ravichandran (2001)).

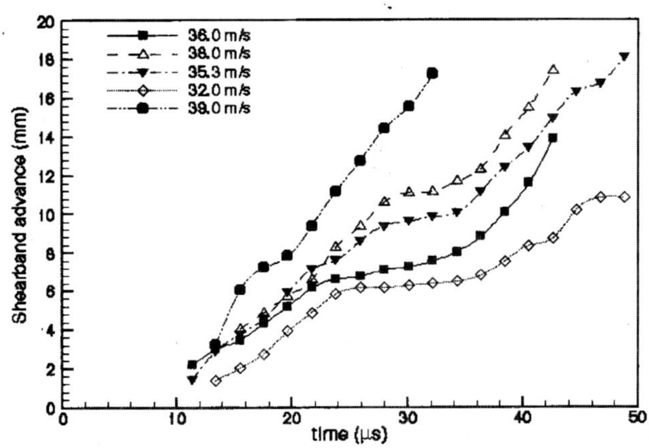

Figure 6.3: Shear band advance as a function of time. (After Guduru, Rosakis and Ravichandran (2001)).

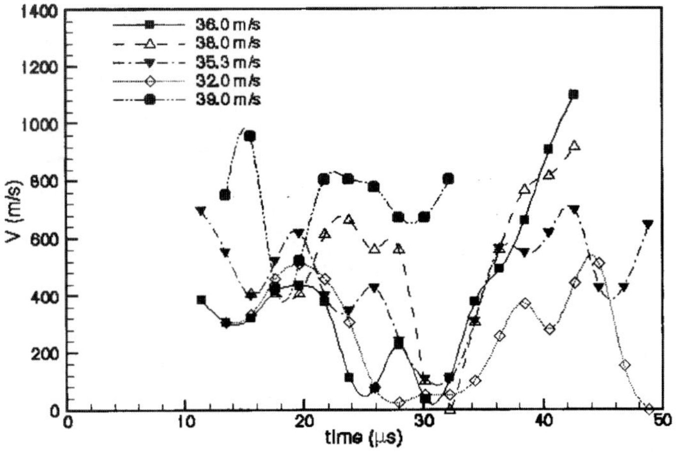

Figure 6.4: Shear band velocity as a function of time. (After Guduru, Rosakis and Ravichandran (2001)).

of time. On the other side of specimen, a newly developed full-field, high-speed infrared (IR) imaging system was employed to measure the evolution, 2-D temperature field, cf. Fig. 6.2. They measure the advance of the shear band and its velocity in five different experiments, cf. Fig. 6.3 and 6.4. The shear band velocity can be seen to be highly transient and a function of the impact speed. In all but one experiment, the band arrests momentarily at about 30 μs, before accelerating to high speeds. The maximum shear band velocity observed here is about 1100 m/s, cf. also Zhou et al. (1996).

6.1.3 Fracture phenomena along localized shear bands

Fractured specimens were examined using an optical microscope to study the features of the shear bands such as its width, trajectory, the fracture surface, etc. The shear band is revealed as a white stripe. The thickness of the band is about 40 μm, cf. Fig. 6.5. A scanning electron microscope (SEM) image of the specimen surface that failed by shear band propagation, shows elongated voids, with sheared edges that are characteristic of such a failure mode, cf. Fig. 6.6. The presence of voids reveals the development of triaxiality tensile stress state that led to void growth and eventual fracture.

Figure 6.5: An optical micrograph of an arrested shear band. (After Guduru, Rosakis and Ravichandran (2001)).

Figure 6.6: An SEM image of the fracture surface that failed by shear banding. Arrow shows the crack propagation direction. (After Guduru, Rosakis and Ravichandran (2001)).

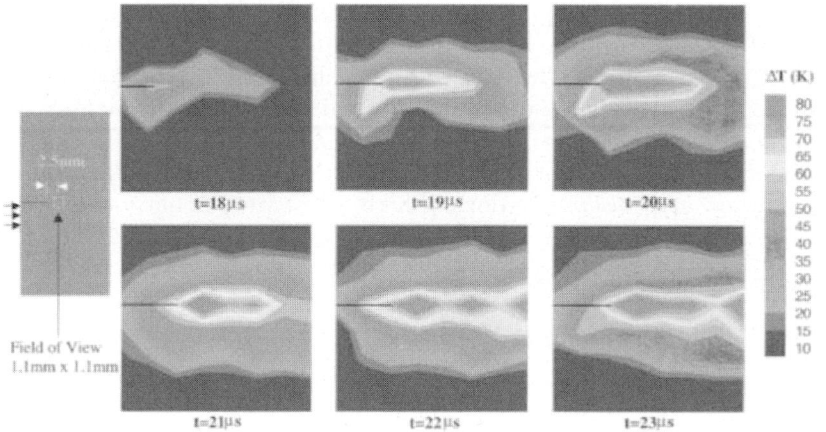

Figure 6.7: A sequence of thermal images showing the transition of crack tip plastic zone into a shear band. (After Guduru, Rosakis and Ravichandran (2001)).

6.1.4 Temperature measurement

Guduru, Rosakis and Ravichandran (2001) performed also broad experimental observations of the temperature field evolution during the initiation and propagation of the shear band. One of their objectives of imaging the temperature field was to visualize the development of the plastic zone at the tip of the initial crack and to observe its evolution, through further localization, into a shear band. The IR camera was focussed at the tip of the fatigue crack as illustrated on the left-hand side of Fig. 6.7. The impact speed was 35 m/s. They show a sequence of thermal images revelling the development of the temperature field as a function of time. Starting at about 21 μs, the central hot region extends to the right, as indicated by the contour lines, signifying the process of shear localization. The measured highest temperature rise within the plastic zone when this happened was at least 80 K, cf. Fig. 6.7.

Let us now focus our attention on the temperature field associated with the tip of a propagating shear band. The gradual nature of temperature rise at the front end of shear band supports the notion of a very diffuse shear band tip, as opposed to a crack tip which carries a strong singularity in the field quantities.

As the shear band propagates, the material within the band progressively accumulates large plastic shear strains within short times and the temperature can quickly reach very high value. Of special interest in the investigation of Guduru, Rosakis and Ravichandran (2001) has been the temperature distribution along a well-defined shear band. They have been consistently observed, in all experiments where a propagating shear band was imaged, that the temperature distribution along the band is highly non-uniform, with discrete regions of high temperature, that look like "hot spots". These hot spots are also seen to translate along the length of the band.

6.1.5 Possible shear band branching

Guduru, Rosakis and Ravinchandran (2001) discussed also the possibility of the shear band branching. Figure 6.8 shows their three micrographs that confirm observation about the possibility of shear band branching mode by Meyers (1994). In the first two, the left edges are the failure paths, caused by a propagating shear band. The third image is taken from an arrested shear band. The branching behaviuor is clearly seen in all three cases. They added currently there is no theoretical framework to explain such a phenomenon and they stressed that this interesting observation requires further investigation to understand the conditions under which such a bifurcation can take place.

6.2 Formulation of the evolution problem

6.2.1 Thermodynamic theory of elasto–viscoplasticity

We shall use a constitutive model of elasto–viscoplastic polycrystalline solids developed within the thermodynamic framework of the rate type covariance structure with a finite set of the internal state variables (cf. Perzyna (1963), (1966), (1971), (1995), (2001), (2005), Duszek–Perzyna and Perzyna (1991), (1994)). It is postulated that the internal state vector $\boldsymbol{\mu}$ has the form $\boldsymbol{\mu} = (\in^p)$, where $\in^p = \int_0^t \left(\frac{2}{3}\mathbf{d}^p : \mathbf{d}^p\right)^{\frac{1}{2}}$ defines the equivalent deformation and describes the dissipation effects generated by viscoplastic phenomena. The rate type constitutive equations for Kirchhoff stress tensor $\boldsymbol{\tau}$ and temperature ϑ have the form as follows

$$\begin{aligned} \mathrm{L}_{\boldsymbol{\upsilon}}\boldsymbol{\tau} &= \mathcal{L}^e : \mathbf{d} - \mathcal{L}^{th}\dot{\vartheta} - \left[(\mathcal{L}^e + \mathbf{g}\boldsymbol{\tau} + \boldsymbol{\tau}\mathbf{g}) : \mathbf{P}\right]\frac{1}{T_m}\langle\Phi(\frac{f}{\kappa} - 1)\rangle, \\ \rho c_p \dot{\vartheta} &= -\mathrm{div}\mathbf{q} + \vartheta\frac{\rho}{\rho_{Ref}}\frac{\partial \boldsymbol{\tau}_c}{\partial \vartheta} : \mathbf{d} + \rho\chi^*\boldsymbol{\tau} : \mathbf{d}^p, \end{aligned} \tag{6.1}$$

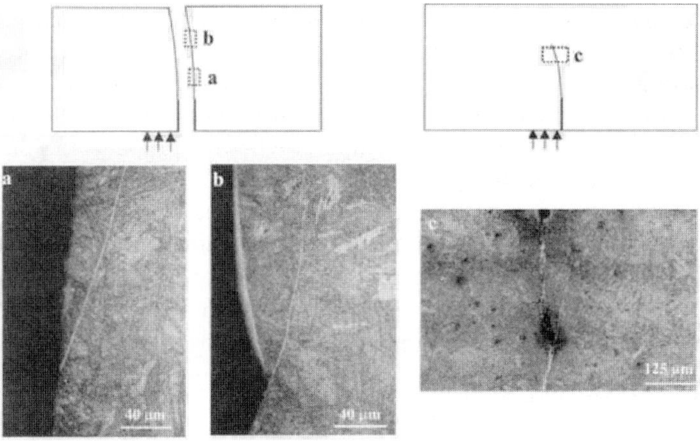

Figure 6.8: Optical micrograph of possible shear band bifurcation. (After Guduru, Rosakis and Ravichandran (2001)).

where $\mathrm{L}_{\boldsymbol{\upsilon}}$ denotes the Lie derivative and

$$\begin{aligned} \mathcal{L}^e &= \rho_{Ref} \frac{\partial^2 \hat{\psi}}{\partial \mathbf{e}^2}, \quad \mathcal{L}^{th} = -\rho_{Ref} \frac{\partial^2 \hat{\psi}}{\partial \mathbf{e} \partial \vartheta}, \\ \psi &= \hat{\psi}(\mathbf{e}, \mathbf{F}, \vartheta, \boldsymbol{\mu}), \quad \mathbf{P} = \frac{\partial f}{\partial \boldsymbol{\tau}} \left(\left\| \frac{\partial f}{\partial \boldsymbol{\tau}} \right\| \right)^{-1}, \\ f &= f(J_2, \vartheta, \boldsymbol{\mu}), \quad \kappa = \hat{\kappa}(\in^p, \vartheta). \end{aligned} \tag{6.2}$$

We interpret $\mathcal{L}^e$ as the elastic matrix, $\mathcal{L}^{th}$ as the thermal expansion matrix, ψ is the free energy function, f is the plastic potential function, J_2 denotes the second invariant of the Kirchhoff stress deviator $\boldsymbol{\tau}'$, T_m denotes the relaxation time for mechanical disturbances, κ is the isotropic work–hardening–softening function, $c_p = -\vartheta \frac{\partial^2 \hat{\psi}}{\partial \vartheta^2}$ defines the specific heat, $\mathbf{q}$ denotes the heat flow vector field, $\mathbf{d} = \mathbf{d}^e + \mathbf{d}^p$ is the total deformation rate tensor, $\mathbf{g}$ denotes the metric tensor in the current configuration, Φ is the empirical overstress viscoplastic function, ρ_{Ref} and ρ denote the mass density in the reference and current configuration respectively, and χ^* is the irreversibility coefficient.

6.2.2 Initial–boundary value problem (evolution problem)

Find φ as function of t and $\mathbf{x}$ satisfying

$$\left.\begin{array}{ll} \text{(i)} & \dot{\varphi} = \mathcal{A}(t, \varphi)\varphi + \mathbf{f}(t, \varphi); \\ \text{(ii)} & \varphi(0) = \varphi^0(\mathbf{x}); \\ \text{(iii)} & \text{The boundary conditions;} \end{array}\right\} \tag{6.3}$$

where the unknown φ takes values in a Banach space, $\mathcal{A}(t, \varphi)$ is a spatial linear differential operator (in general unbounded) depending on t and φ, $\mathbf{f}$ is a nonlinear function, and the dot denotes the material derivative.

The evolution problem (6.3) describes an adiabatic inelastic flow process provided

$$\varphi = \begin{bmatrix} \boldsymbol{\upsilon} \\ \rho \\ \boldsymbol{\tau} \\ \vartheta \end{bmatrix}, \tag{6.4}$$

$$\mathbf{f} = \begin{bmatrix} 0 \\ 0 \\ -\frac{\langle\Phi(\frac{f}{\kappa}-1)\rangle}{T_m}\left[\left(\mathcal{L}^e + \frac{\chi^* \boldsymbol{\tau}}{\rho_{Ref}}\mathcal{L}^{th} + \mathbf{g}\boldsymbol{\tau} + \boldsymbol{\tau}\mathbf{g}\right) : \mathbf{P}\right] \\ \frac{1}{T_m}\langle\Phi(\frac{f}{\kappa} - 1)\rangle\frac{\chi^*}{\rho_{Ref}}\boldsymbol{\tau} : \mathbf{P} \end{bmatrix},$$

$$\mathcal{A} = \begin{bmatrix} 0 & 0 & \frac{\boldsymbol{\tau}}{\rho_{Ref}\rho}\text{grad} & \frac{1}{\rho_{Ref}}\text{div} \\ 0 & -\rho\text{div} & 0 & 0 \\ 0 & I\!\!E : \text{sym}\frac{\partial}{\partial \mathbf{x}} + 2\text{sym}(\boldsymbol{\tau} : \frac{\partial}{\partial \mathbf{x}}) & 0 & 0 \\ 0 & \frac{\vartheta}{c_p\rho_{Ref}}\frac{\partial \boldsymbol{\tau}}{\partial \vartheta} : \text{sym}\frac{\partial}{\partial \mathbf{x}} & 0 & 0 \end{bmatrix}.$$

where

$$I\!\!E = \mathcal{L}^e - \frac{\vartheta}{c_p\rho_{Ref}}\mathcal{L}^{th}\frac{\partial \boldsymbol{\tau}}{\partial \vartheta} \tag{6.5}$$

denotes the thermo-elastodynamic matrix for adiabatic process.

6.2.3 Numerical solution of the evolution problem

The discretization in space and time based on the finite element method is developed. Rate dependency (viscosity) allows the spatial differential operator in the governing equations to retain its ellipticity and the initial value problem (the Cauchy problem) is well-posed. Viscosity introduces implicitly a length-scale parameter into the dynamical initial-boundary value problem.

The viscoplastic regularization procedure assures the stable integration algorithm by using the finite element method. Particular attention is focused on the well-posedness of the evolution problem (the initial-boundary value problem) as well as on its numerical solutions. We take advantage of the Lax–Richtmyer equivalence theorem.

6.3 Identification procedure

6.3.1 Assumption of the material functions for an adiabatic process

Let us assume the plastic potential function for a material in the form

$$f = J_2, \qquad \text{where} \qquad J_2 = \frac{1}{2}\tau'^{ab}\tau'^{cd} g_{ac} g_{bd}. \tag{6.6}$$

The isotropic hardening softening material function κ is postulated in the form as follows

$$\kappa = \kappa_0^2 \left\{ \frac{\kappa_1}{\kappa_0} + \left(1 - \frac{\kappa_1}{\kappa_0}\right) \exp\left[-h(\vartheta)\, \in^p\right] \right\}^2 (1 - b\vartheta), \tag{6.7}$$

where κ_0 and κ_1 denote the yield and saturation stress of the matrix material, respectively, $h = h(\vartheta)$ is the temperature dependent strain hardening function and b is a material coefficient.

The overstress function Φ is assumed in the form

$$\Phi\left(\frac{f}{\kappa} - 1\right) = \left(\frac{f}{\kappa} - 1\right)^m, \qquad \text{where} \qquad m = 1, 3, 5, \ldots \tag{6.8}$$

6.3.2 Determination of the material constants

To determine the material constants assumed we take advantage of the experimental observations of adiabatic shear bands in a HY-100 steel presented by Cho, Chi and Duffy (1988). In this experiment the specimens used were machined in the shape of thin-walled tubes with integral hexagonal flanges for gripping, Fig. 6.9. Torsional loading at high strain rates was applied in a torsional Kolsky bar (split-Hopkinson bar). Kolsky bar provides a relatively simple and uniform state of shear stress during dynamic deformation, because the instrumentation affords an easy means of measuring average strain as a function of time, and because supplementary instrumentation can be added easily to measure local strain and the temperature distribution.

From the observations we have the results as follows:

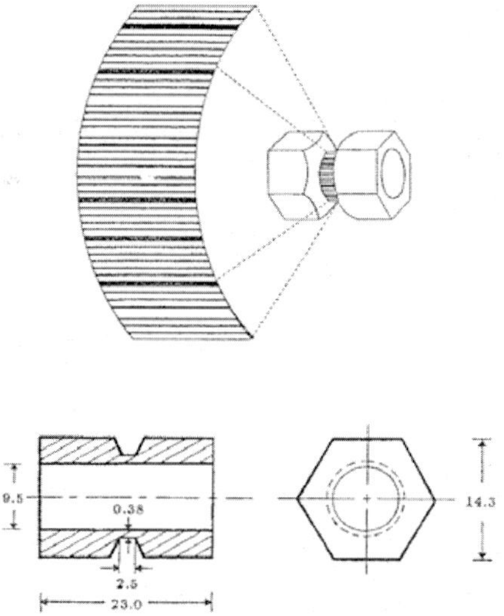

Figure 6.9: Details of the thin-walled specimen with a blow-up of the fine grid deposited photographically on the outside surface of the specimen. The dimension are in millimeters (After Cho, Chi and Duffy (1988)).

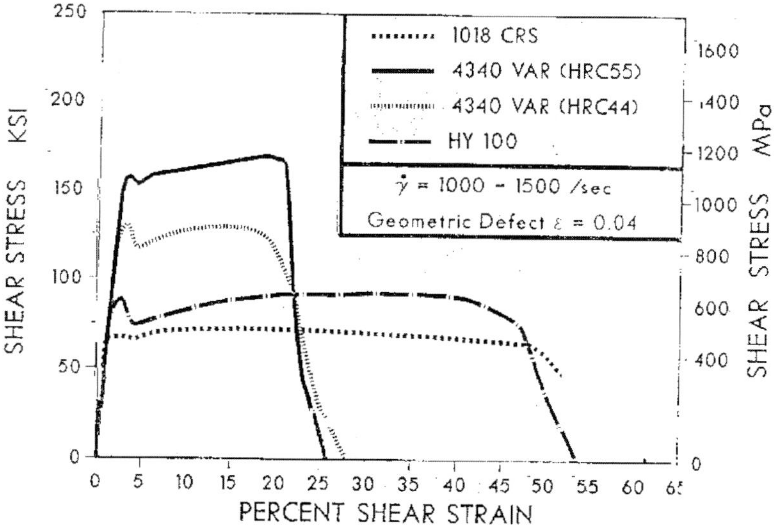

Figure 6.10: Typical stress-strain curves at a strain rate 10^3/s for each of the four steels tested with torsional Kolsky bar (After Cho, Chi and Duffy (1988)).

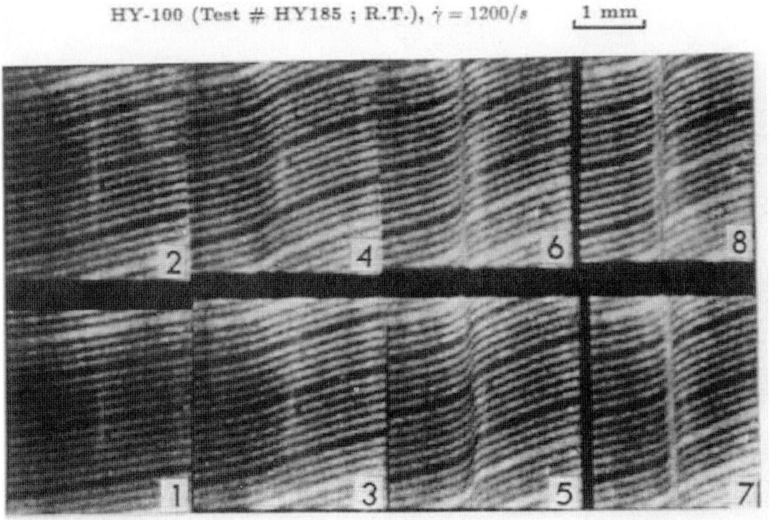

Figure 6.11: Photographs of the grid pattern during the formation of a shear band in HY-100 steel taken by a high speed image converter camera. The time interval between frames is 40 μs (After Cho, Chi and Duffy (1988)).

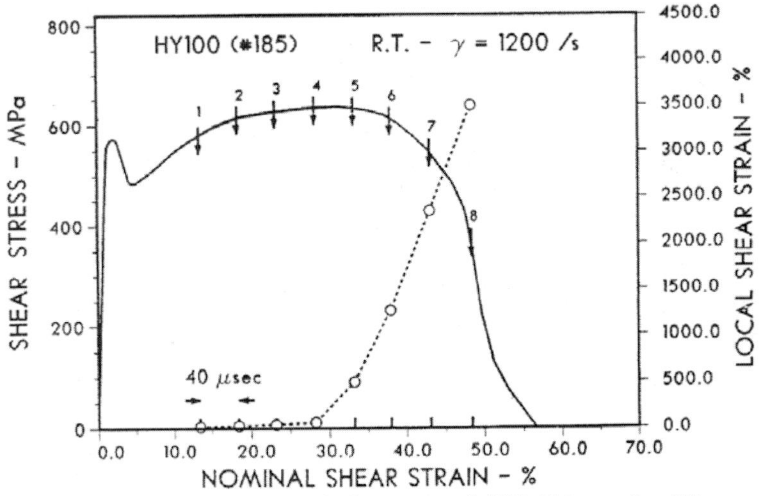

Figure 6.12: The stress-strain behaviour of HY-100 steel. The number arrows indicate nominal strain values at which the photographs shown in Fig. 11 are taken. The corresponding values of the maximum local strain are shown by the dotted line (After Cho, Chi and Duffy (1988)).

hbt

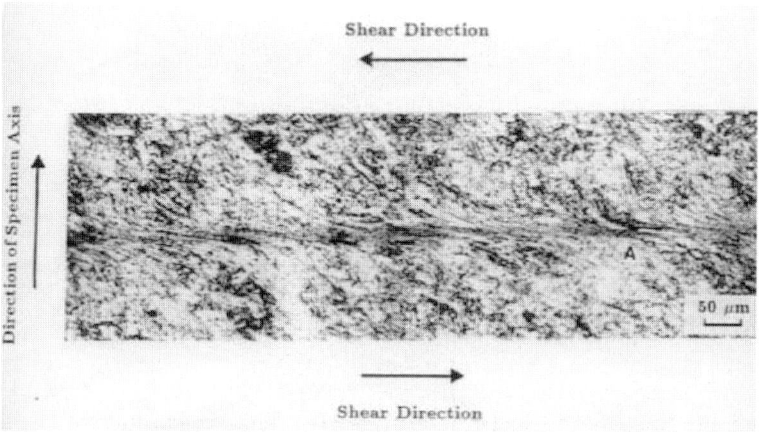

Figure 6.13: Optical micrograph of a shear band formed in HY-100 steel, after polishing and etching (After Cho, Chi and Duffy (1988)).

(i) The room temperature dynamic shear stress-strain curve, as obtained with the torsional Kolsky bar, is shown in Fig. 6.10.

(ii) The corresponding nominal strain rate was assumed constant and was about 1200 s^{-1}.

(iii) The maximum shear strain at fracture in dynamic deformation is about 1000%.

(iv) The formation of a shear band during the test of a HY-100 steel specimen has been investigated by a high speed image converter camera and is clearly shown in Figs 6.11 and 6.12. Shear bands were examined using optical microscopy, Fig. 6.13, polarized light microscopy, Fig. 6.14a, and SEM, Fig. 6.14b. Shear band area is white after atching and its edges are relatively well-defined with different surface features apparent within the shear band area and the matrix material. The average width of the shear band is approximately 20 μm, the maximum local shear strain is about 1000%, and the maximum temperature is about 590°C.

(v) The average velocity of the propagation shear band is estimated to be around 250÷500 m/s.

Let us consider the adiabatic dynamic process for a thin-walled steel tube twisted at nominal rate 1000 s^{-1}. In fact we idealize the initial–boundary

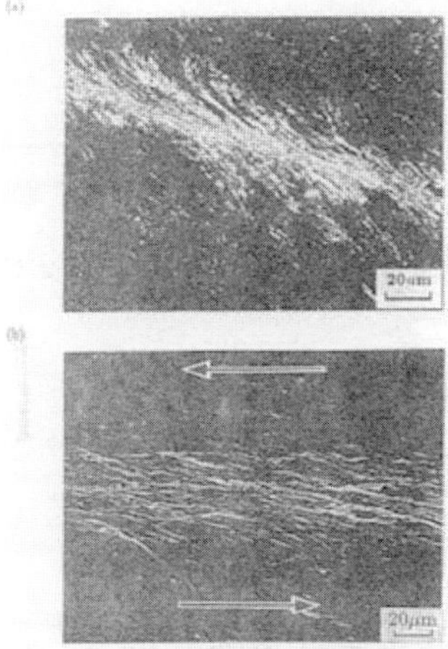

Figure 6.14: Shear band in HY-100 steel. (a) The appearance of the shear band in polarized light microscopy after polishing and etching, showing a white etched region in the center of the shear band and deformed features near its edges. (b) The surface structure of the shear band before polishing and etching, as seen with SEM. Flow lines can be seen within and outside the shear band (After Cho, Chi and Duffy (1988)).

value problem investigated by Cho, Chi and Duffy (1988). We take here advantage of our previous numerical investigation presented by Łodygowski and Perzyna (1997).

We assume an elasto–visco–perfectly plastic material and a specimen in the shape of thin-walled tube, cf. Fig. 6.15. It has been postulated that the specimen is twisted with $\omega^* = 253$ s^{-1} and the duration of the process is 100 μs. The following values for various material parameters are assumed (HY-100 steel): density $\rho = 7830$ kg/m^3, Young modulus $E = 208$ GPa, the specific heat $c_p = 460$ J kg^{-1}K^{-1}, the irreversible coefficient $\chi^* = 0.9$, the relaxation time $T_m = 2 \times 10^{-1}$ μs, $m = 1$, the yield stress $\kappa_0 = 598$ MPa for temperature 20°C, $\kappa_0 = 540$ MPa for temperature 100°C, $\kappa_0 = 483$ MPa for temperature 200°C, $\kappa_0 = 426$ MPa for temperature 300°C. The imperfection of weaker material has been introduced in one place of center plane to observe better the initiation of shear band.

It has been considered the solution in the ring form, cf. Fig. 6.16. The shear band advance and the propagation velocity of the shear band as a function of time are shown in Fig. 6.17 (at the tip of shear band it has been assumed $\in^p = 0.25$). The average width of the shear band is obtained as $80 \div 100$ μm.

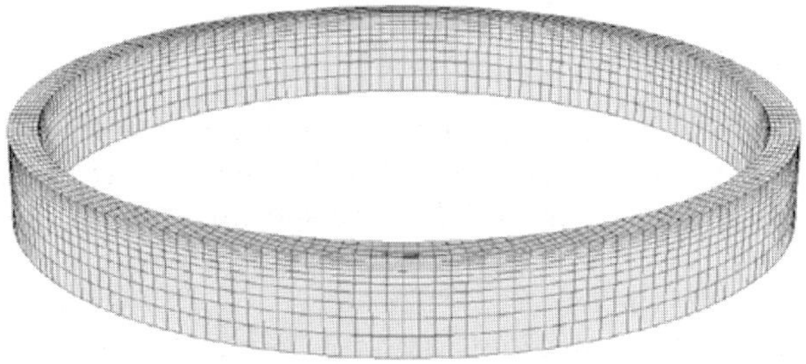

Figure 6.15: Assumed shape of the specimen.

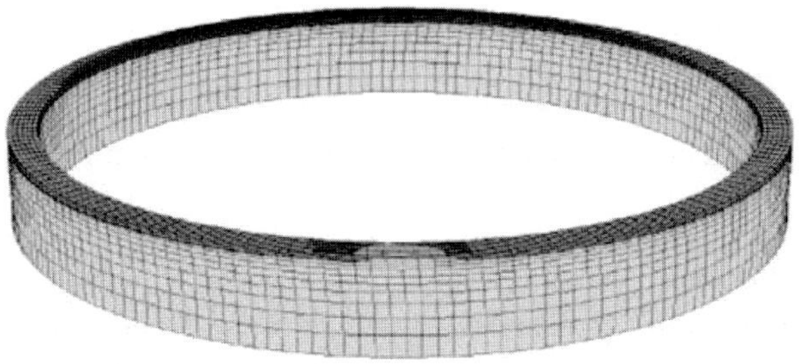

Figure 6.16: Solution in the ring form.

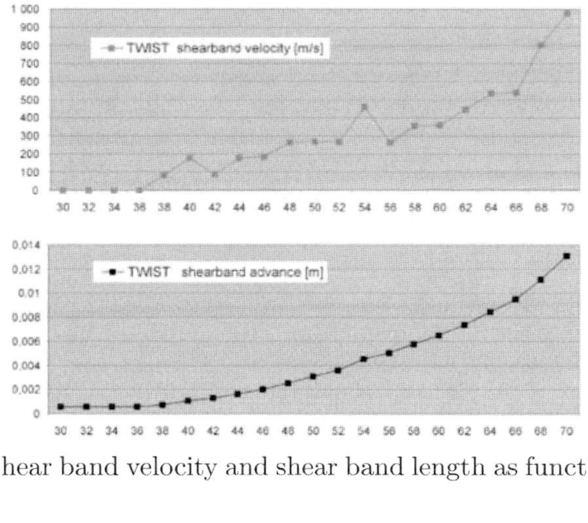

Figure 6.17: Shear band velocity and shear band length as functions of time.

6.4 Numerical examples

6.4.1 Propagation of shear bands

We idealize the initial boundary value problem observed experimentally in Guduru, Rosakis and Ravichandran (2001) by assuming the velocity boundary condition and different material of the specimen (HY-100 steel).

A plane specimen of length 254 mm and width 95 mm has the initial notch at the center of the left edge. The thickness of the specimen is 12 mm. The notch length is 27.4 mm and the initial crack 2 mm long is present in the horizontal direction (see Fig. 6.18). The dimensions of the plate, notch and the initial crack are taken from experimental data. The supports and impact projectile position are also presented in Fig. 6.18. The projectile impacts the specimen with velocity equal to 38 m/s. There are two different positions of projectile giving the unsymmetric impact and symmetric one, when central impact is considered. For both positions of projectile the boundary conditions satisfy the symmetry by taking the same distance for upper and lower edge, but not the same at left (prenotched) and right side of the specimen.

The FE discretization is defined in such the way that finite element dimensions are of 20 μm in the region of interest. The whole model consists of 49500 plain stress finite elements, 49750 nodes, 99500 degree of freedom.

The contour plots of plastic equivalent strain and temperature obtained for time instances 40, 50, 60 μm are visualized in Fig. 6.19 for unsymmetric impact. Similar results for central impact are shown in Fig. 6.20.

To show better that we obtained non-uniform temperature distribution along the shear band for both considered cases we present the magnification of the results for the field of temperature at 60 μs for unsymmetric and central impacts, respectively, cf. Figs 6.21 and 6.22. The contour value of 25% for plastic equivalent strain is taken to observe the shear band initiation and propagation. Putting down the coordinates of position of shear band advance and controlling the deformation during the process the functions of shear band length and propagation velocity of its front are calculated. That results are displayed in Fig. 6.23.

6.4.2 Discussion of the numerical results

The results of numerical simulation of shear band allow to observe the dynamic process of advance of shear band tip in material specimen under impact loading. The calculated values of equivalent inelastic deformation make possible to estimate shear band length advance and the velocity of shear band propagation. The period of time 40 μs is common for computed

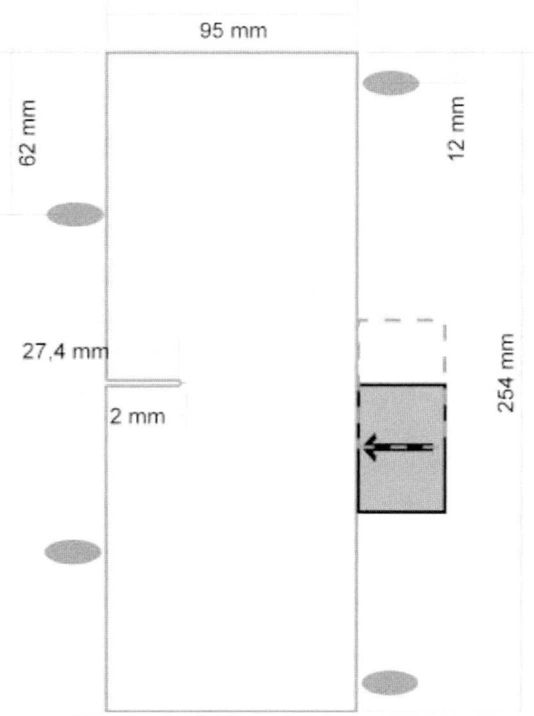

Figure 6.18: Specimen geometry.

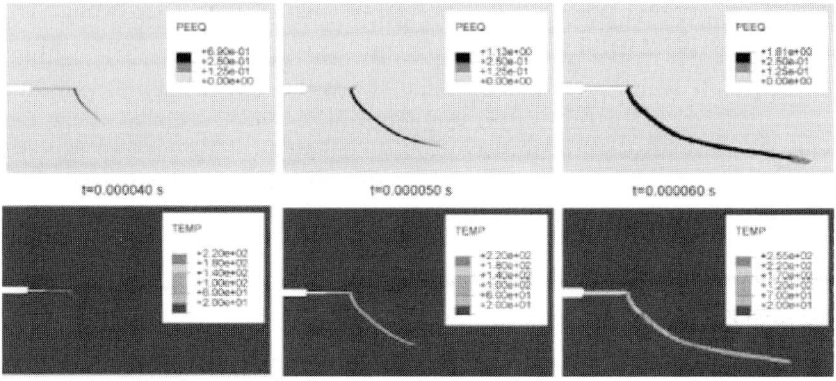

Figure 6.19: Evolution of plastic equivalent deformation and temperature along the shear band for unsymmetric impact (for 40, 50, 60 μs).

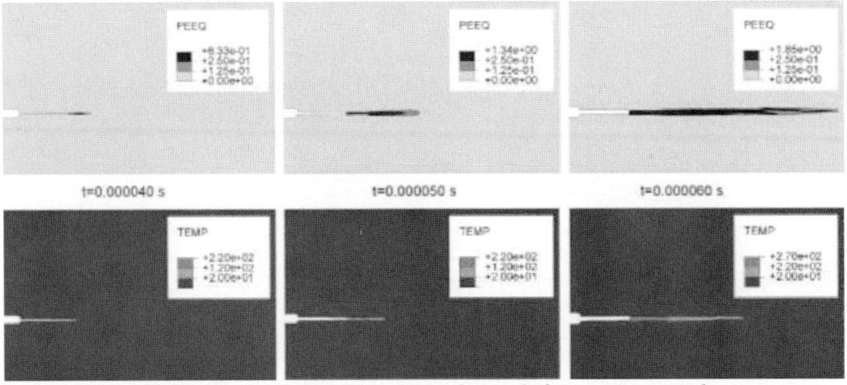

Figure 6.20: Evolution of plastic equivalent deformation and temperature along the shear band for central impact (for 40, 50, 60 μs).

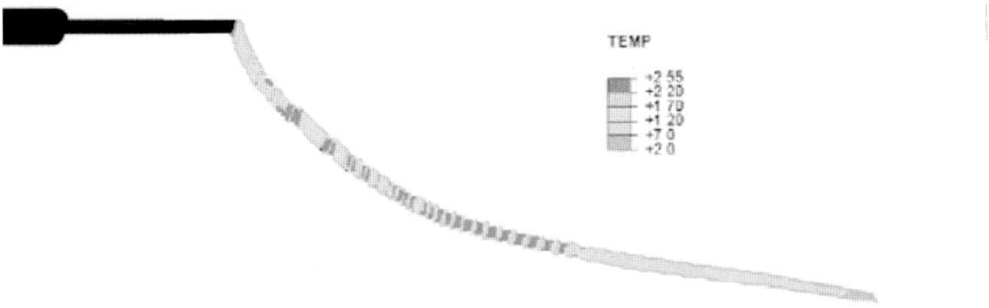

Figure 6.21: Magnification of the results for the field of temperature at 60 μs for unsymmetric impact.

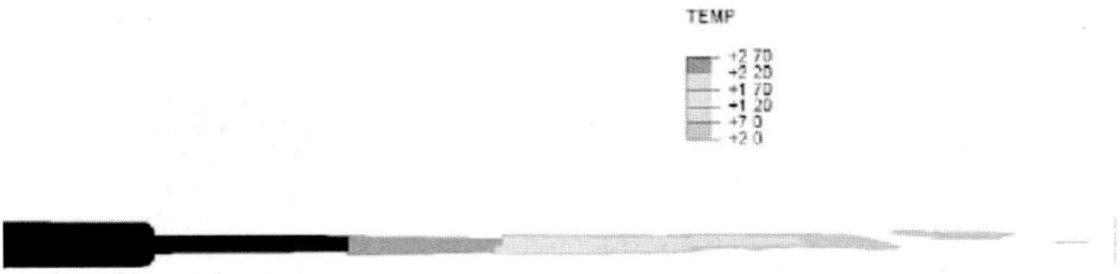

Figure 6.22: Magnification of the results for the field of temperature at 60 μs for central impact.

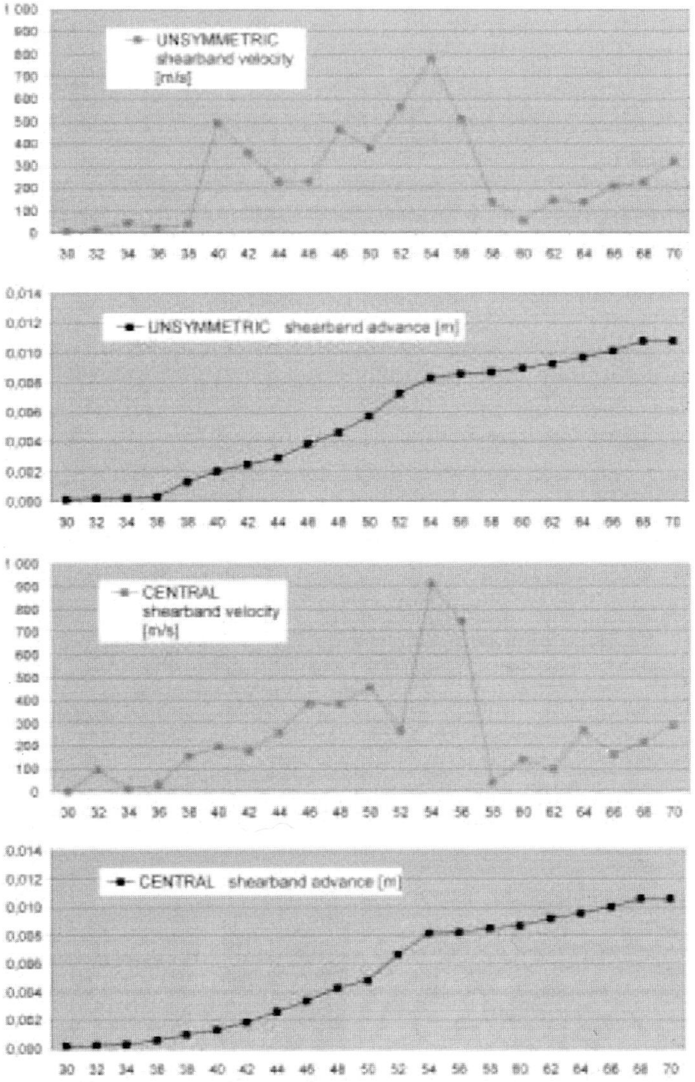

Figure 6.23: Propagation velocity of shear band front and shear band length as functions of time, for unsymmetric and central impacts.

examples to report the shear band propagation, from its initiation to considered advance. Within the same time period formation of shear band is also observed in experimental measurement. The advances of shear band length in experiments and in computations have similar final values and functions presenting its evolution are comparable. The comparison of the shear band propagation velocity reaches the same conclusion. The calculated values of $200-500$ m/s for tube twisting and from 100 m/s up to value of 900 m/s for pre-notched plate example are of the range the experimental results. Rather the qualitative agreement of experimental simulation results is verified, than strict comparison of values.

The functions of the simulated shear band propagation velocity for pre-notched plate, like the functions obtained from laboratory test, show quite variant, but increase of its value at the beginning of shear band advance, later the sudden drop of its value after reaching the maximum and one more increasing path at the end of time period. The common sudden velocity drop, observed in experiments and simulations, is specially worth to point it out. The calculation of velocity values using results from simulation is generally sensitive to the exactness of shear band tip evidence. Relatively small length of the shear band, together with the fact that material and boundary conditions do not correspond directly drive to accent the qualitative agrement of results. The remark concerning the temperature results goes in the same direction. There is valuable to expose that numerical results show the particular character of experimental ones, with occurrence of discrete regions of high temperature, that are described as "hot spots". At the beginning of shear band advance the temperature uniformly decreases around the crack tip, but next the temperature field changes the distribution and there is no more rule that temperature continuously is lower when the distance from maximum is greater. Appearance of several local extremes take place after the mentioned above drop of propagation velocity. The evolution of described variables suggests the step-wise nature of shear band propagation phenomena.

7. Numerical Investigation of Localized Fracture Phenomena in Inelastic Solids

7.1 Thermo–elasto–viscoplastic model of a material

The main objective is to develop the rate type constitutive structure for an elastic–viscoplastic material in which the effects of the plastic non-normality, micro–damaged mechanism and thermomechanical coupling are taken into consideration, cf. Dornowski and Perzyna (2006). To do this it is sufficient to assume a finite set of the internal state variables. For our practical purposes it is sufficient to assume that the internal state vector $\boldsymbol{\mu}$ has the form

$$\boldsymbol{\mu} = (\in^p, \xi), \tag{7.1}$$

where $\in^p$ is the equivalent viscoplastic deformation, and ξ is volume fraction porosity and takes account for micro–damaged effects.

Let us introduce the plastic potential function $f = f(J_1, J_2, \vartheta, \boldsymbol{\mu})$, where J_1, J_2 denote the first two invariants of the Kirchhoff stress tensor $\boldsymbol{\tau}$.

A set of the constitutive equations of the rate type has the form as presented in Section 5.2

7.2 Formulation of the evolution problem

7.2.1 Initial–boundary value problem (evolution problem)

Find φ as function of t and $\mathbf{x}$ satisfying

$$\left.\begin{array}{ll} \text{(i)} & \dot{\varphi} = \mathcal{A}(t, \varphi)\varphi + \mathbf{f}(t, \varphi); \\ \text{(ii)} & \varphi(0) = \varphi^0(\mathbf{x}); \\ \text{(iii)} & \text{The boundary conditions;} \end{array}\right\} \tag{7.2}$$

where the unknown φ takes values in a Banach space, $\mathcal{A}(t, \varphi)$ is a spatial linear differential operator (in general unbounded) depending on t and φ, $\mathbf{f}$ is a nonlinear function, and the dot denotes the material derivative.

The evolution problem (7.2) describes an adiabatic inelastic flow process provided

$$\varphi = \begin{bmatrix} \boldsymbol{\upsilon} \\ \rho \\ \boldsymbol{\tau} \\ \xi \\ \vartheta \end{bmatrix}, \tag{7.3}$$

$$
\mathbf{f}=\begin{bmatrix} 0 \\ 0 \\ -\frac{\langle\Phi(\frac{f}{\kappa}-1)\rangle}{T_m}\left[\left(\mathcal{L}^e+\frac{\chi^*\boldsymbol{\tau}}{\rho_{Ref}}\mathcal{L}^{th}+\mathbf{g}\boldsymbol{\tau}+\boldsymbol{\tau}\mathbf{g}+\mathcal{W}\right):\mathbf{P}\right]-\frac{\chi^{**}\Xi}{\rho_{Ref}}\mathcal{L}^{th} \\ \Xi \\ \frac{1}{T_m}\langle\Phi(\frac{f}{\kappa}-1)\rangle\frac{\chi^*}{\rho_{Ref}}\mathcal{L}^{th}\boldsymbol{\tau}:\mathbf{P}+\frac{\chi^{**}}{\rho_{Ref}}\Xi \end{bmatrix},
$$

$$
\mathcal{A}=\begin{bmatrix} 0 & 0 & \frac{\boldsymbol{\tau}}{\rho_{Ref}\rho}\mathrm{grad} & \frac{1}{\rho_{Ref}}\mathrm{div} & 0 \\ 0 & -\rho\mathrm{div} & 0 & 0 & 0 \\ 0 & I\!\!E:\mathrm{sym}\frac{\partial}{\partial\mathbf{x}}+2\mathrm{sym}(\boldsymbol{\tau}:\frac{\partial}{\partial\mathbf{x}}) & 0 & 0 & 0 \\ 0 & 0 & 0 & 0 & 0 \\ 0 & \frac{\vartheta}{c_p\rho_{Ref}}\frac{\partial\boldsymbol{\tau}}{\partial\vartheta}:\mathrm{sym}\frac{\partial}{\partial\mathbf{x}} & 0 & 0 & 0 \end{bmatrix}.
$$

where

$$
I\!\!E - \mathcal{L}^e - \frac{\vartheta}{c_p\rho_{Ref}}\mathcal{L}^{th}\frac{\partial\boldsymbol{\tau}}{\partial\vartheta} \tag{7.4}
$$

denotes the thermo–elastodynamic matrix for adiabatic process.

7.2.2 Numerical solution of the evolution problem

The discretization in space and time based on the finite difference method is developed. Rate dependency (viscosity) allows the spatial differential operator in the governing equations to retain its ellipticity and the initial value problem (the Cauchy problem) is well–posed. Viscosity introduces implicitly a length–scale parameter into the dynamical initial–boundary value problem.

The viscoplastic regularization procedure assures the stable integration algorithm by using the finite difference method. Particular attention is focused on the well–posedness of the evolution problem (the initial–boundary value problem) as well as on its numerical solutions. We take advantage of the Lax–Richtmyer equivalence theorem.

7.3 Identification procedure

7.3.1 Assumption of the material functions for an adiabatic process

To do the proper identification procedure we first make assumption of the material functions (cf. Dornowski and Perzyna (2000)).

The plastic potential function f is assumed in the form (cf. Perzyna (1984) and Shima and Oyane (1976))

$$
f=\left\{J_2^{'}+\left[n_1(\vartheta)+n_2(\vartheta)\xi\right]J_1^2\right\}^{\frac{1}{2}} \tag{7.5}
$$

where

$$n_1(\vartheta) = 0, \qquad n_2(\vartheta) = n = \text{const.} \tag{7.6}$$

The isotropic work–hardening–softening function κ is postulated as (cf. Perzyna (1986a) and Nemes and Eftis (1993))

$$\kappa = \hat{\kappa}(\in^p, \vartheta, \xi) = \{\kappa_s(\vartheta) - [\kappa_s(\vartheta) - \kappa_0(\vartheta)] \exp[-\delta(\vartheta) \in^p]\} \left[1 - \left(\frac{\xi}{\xi_F}\right)^{\beta(\vartheta)}\right], \tag{7.7}$$

where

$$\kappa_s(\vartheta) = \kappa_s^*(1 - \iota\overline{\vartheta}), \qquad \kappa_0(\vartheta) = \kappa_0^*(1 - \iota\overline{\vartheta}), \tag{7.8}$$
$$\delta(\vartheta) = \delta^*(1 - \iota\overline{\vartheta}), \qquad \beta(\vartheta) = \beta^*(1 - \iota\overline{\vartheta}), \qquad \overline{\vartheta} = \frac{\vartheta - \vartheta_0}{\vartheta_0}.$$

The overstress function $\Phi\left(\frac{f}{\kappa} - 1\right)$ is assumed in the form

$$\Phi\left(\frac{f}{\kappa} - 1\right) = \left(\frac{f}{\kappa} - 1\right)^m. \tag{7.9}$$

The evolution equation for the porosity ξ is postulated as

$$\dot{\xi} = \dot{\xi}_{grow} = \frac{g^*(\xi, \vartheta)}{T_m \kappa_0(\vartheta)} [I_g - \tau_{eq}(\xi, \vartheta, \in^p)] \tag{7.10}$$

where (cf. Dornowski (1999))

$$\begin{aligned}
g^*(\xi, \vartheta) &= c_1(\vartheta) \frac{\xi}{1 - \xi}, \\
I_g &= b_1 J_1 + b_2 \sqrt{J_2}, \\
\tau_{eq}(\xi, \vartheta, \in^p) &= c_2(\vartheta)(1 - \xi) \ln \frac{1}{\xi} \{2\kappa_s(\vartheta) - [\kappa_s(\vartheta) - \kappa_0(\vartheta)] \\
&\quad F(\xi_0, \xi, \vartheta)\}, \\
c_1(\vartheta) &= \text{const}, \qquad c_2(\vartheta) = \text{const}, \\
F(\xi_0, \xi, \vartheta) &= \left(\frac{\xi_0}{1 - \xi_0} \frac{1 - \xi}{\xi}\right)^{\frac{2}{3}\delta} + \left(\frac{1 - \xi}{1 - \xi_0}\right)^{\frac{2}{3}\delta}.
\end{aligned} \tag{7.11}$$

As in the infinitesimal theory of elasticity we assume linear properties of the material, cf. Section 4.1.1.

Table 7.1: An AISI 4340 steel (plate).

$\kappa_s^* = 1155$ MPa	$\vartheta = 293$ K	$\chi = 0.9$
$\kappa_0^* = 808$ MPa	$\iota = 0.1$	$c_2 = 0.067$
$\delta^* = 14$	$T_m = 1\ \mu$s	$b_1 = 0.58$
$\beta^* = 9$	$c_p = 455$ J/kg K	$b_2 = 1.73$
$v = 0.3$	$E = 208$ GPa	$\rho_{Ref} = 7850$ kg/m^3
$c_1 = 0.1$	$\xi_0 = 6 \cdot 10^{-4}$	$\xi_F = 0.25$
$n = 0.25$	$m = 1$	

7.3.2 Determination of the material constants

To determine the material constants assumed we take advantage of the experimental observations presented by Chakrabarti and Spretnak (1975). They investigated the localized fracture mode for tensile steel sheet specimens simulating both plane stress and plane strain processes. The material used in their study was AISI 4340 steel. The principal variable in this flat specimen test was the width – to – thickness ratio. Variation in specimen geometry produces significant changes in stress state, directions of shear bands, and ductility. They found that fracture propagated consistently along the shear band localized region.

Let us now consider an adiabatic dynamic process for a thin steel plate under condition of plane stress state. Additionally we simplify the considered adiabatic process by assumption that the evolution equation for temperature $(5.6)_2$ has the form

$$\dot{\vartheta} = \frac{\chi}{c_p} \boldsymbol{\tau} : \mathbf{d}^p, \tag{7.12}$$

i.e. we postulate $\chi^{**} = 0$, $\chi^* = \chi$ and we neglect the nondissipative term. In fact we idealize the initial–boundary value problem investigated by Chakrabarti and Spretnak (1975) by assuming the velocity driven adiabatic process for a thin steel plate. The problem has been solved by using the finite difference method.

Based on these assumptions we proceed similarly as in our previous paper, cf. Dornowski and Perzyna (2000). Then the material constants for an adiabatic process has been determined, cf. Table 7.1.

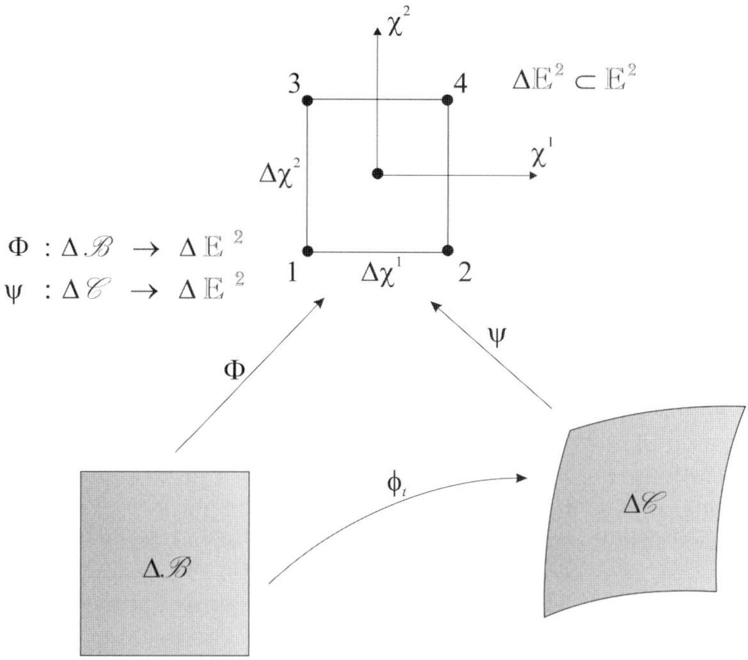

Figure 7.1: Convective finite difference mesh of nodes.

7.4 Numerical example. Investigation of localized fracture phenomena

7.4.1 Application of a finite difference method

Let us introduce in the two–dimensional Euclidean space E^2 a regular difference net of nodes (i, j) with convective coordinates $\chi_i^1 = i\Delta\chi^1$ and $\chi_i^2 = j\Delta\chi^2$, $i, j \in N$, where N is a set of natural numbers. Some of the nodes belong to the edge of the body and are used to approximate the boundary conditions. Time is approximated by a discrete sequence of moments$t_n = n\Delta t$, where Δt is the time step, $n \in N$.

In the convective description the metric tensor is a measure of deformation. This tensor depends on the displacement gradient. For the displacement field we postulate the following approximation in the domain $\Delta E^2 = \Delta\chi^1 \times \Delta\chi^2$ of a convective difference mesh (cf. Fig. 7.1):

$$u_h(\chi^1, \chi^2, t) = a_1(t) + a_2(t)\chi^1 + a_3(t)\chi^2 + a_4(t)\chi^1\chi^2. \tag{7.13}$$

The functions $a_1(t), \ldots, a_4(t)$ depend only on time and are determined by the value of the function $\mathbf{u}_w^T(t) = [u_1(t), \ldots, u_4(t)]$ in the node points of the difference mesh (cf. Fig. 7.1). Hence the approximation function (7.9) can be written in the form:

$$u_h(\chi^1, \chi^2, t) = \mathbf{n}^T(\chi^1, \chi^2)\mathbf{u}_w(t), \tag{7.14}$$

where

$$\mathbf{n} = \frac{1}{4\Delta\chi^1\chi^2} \begin{bmatrix} (-\Delta\chi^1 + 2\chi^1)(-\Delta\chi^2 + 2\chi^2) \\ (\Delta\chi^1 + 2\chi^1)(\Delta\chi^2 - 2\chi^2) \\ (\Delta\chi^1 - 2\chi^1)(\Delta\chi^2 + 2\chi^2) \\ (\Delta\chi^1 + 2\chi^1)(\Delta\chi^2 + 2\chi^2) \end{bmatrix}. \tag{7.15}$$

The equation (7.14) allows to determine values of the displacement in any point of the domain $\Delta\mathrm{E}^2 = \Delta\chi^1 \times \Delta\chi^2$. For the central point $\chi^1 = \chi^2 = 0$, $n_1 = \ldots = n_4 = 1/4$ and $n_h = (u_1 + \ldots + u_4)/4$. For the central point $\chi^1 = \chi^2 = 0$, $n_1 = \ldots = n_4 = 1/4$ and $n_h = (u_1 + \ldots + u_4)/4$.

By using (7.14) we can determine the matrix of the difference operators which approximate the first partial derivatives of the displacement field,

$$\frac{\partial u_h(\chi^1, \chi^2, t)}{\partial \chi^i} = \frac{\partial \mathbf{n}^T(\chi^1, \chi^2)}{\partial \chi^i}\mathbf{u}_w(t). \tag{7.16}$$

7.4.2 Numerical solution of the initial boundary–value problem

We analyse the impact loaded plate with a pre–notched crack (Fig. 7.2). The material of the plate is assumed as a AISI steel and is modelled as elasto–viscoplastic with isotropic hardening–softening effects. We assume that the material softening is caused by intrinsic microdamage mechanism. The height of the specimen is equal to 200 mm, width is 100 mm and length of the initial crack (notch) is equal to 50 mm. The plane stress state in an adiabatic process is considered. As it is shown in Fig. 7.2, this specimen is loaded asymmetrically to the notch axis. Similar boundary value problems for plate were analysed numerically by Li et al. (2002).

To obtain the solution of the initial–boundary value problem formulated the finite difference method for regularized elasto–viscoplastic model is used[8]. The loading condition is modelled by the velocity of nodes lying on the edge section with length equals to 50 mm, according to the relation

$$V(t) = V_0 t/t_0 \quad \text{for} \quad t \leq t_0 \quad \text{and} \quad V(t) = V_0 \quad \text{for} \quad t > t_0. \tag{7.17}$$

[8]Numerical modelling of localized fracture phenomena in inelastic solids in dynamic loading processes by means of finite element method has been presented by Łodygowski and Perzyna (1997).

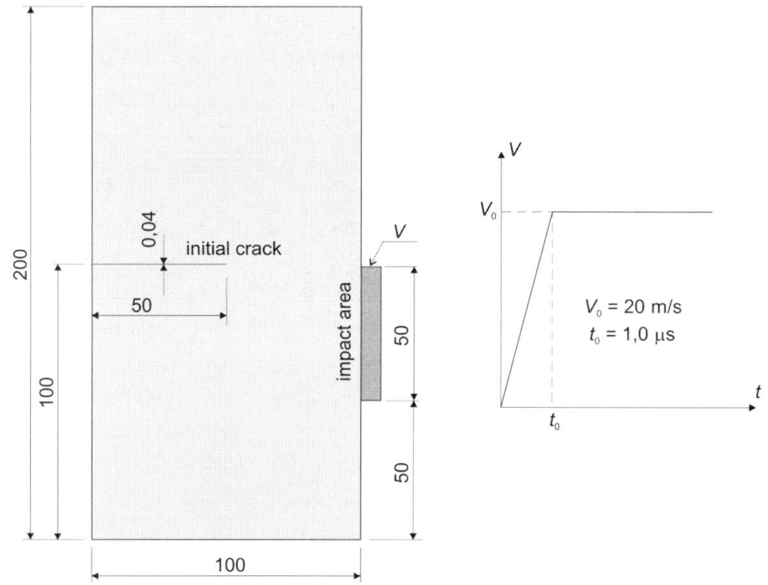

Figure 7.2: The impact loaded plate with a prenotched crack.

The rise time t_0 is fixed at 1.0 μs and the speed impact $V_0 = 20$ m/s. Initial conditions of the problem are homogeneous. In the discussion of the numerical results attention is focused mostly on the phenomenon of crack propagation.

The finite difference method with the explicit time integration scheme (conditionally stable) is used. The stress state in a nodal environment is determined by the iterative procedure of solving the dynamical yield condition with respect to the norm of the plastic deformation rate tensor. The elaborated algorithm satisfies the material objectivity principle with respect to diffeomorphism (any motion). We assume a nonuniform mesh of nodes which contain 180 000 nodes. The smallest mesh has the dimensions $\Delta\chi^1 = \Delta\chi^2 = 40$ μm, the time increment $\Delta t = 0.00356$ μs.

Figure 7.3 shows distributions of normal stresses τ^{xx}, τ^{yy} and shear stress τ^{xy} at time 10 μs, a moment later when the stress wave reaches the crack tip. The results illustrate the formation of the greatest stress zones, especially the shear stress in the vicinity of the notch. The characteristic unsymmetrical distribution of these zones is a result of the assumed way of loading.

In Figure 7.4, the evolution of the Mises stress $\tau_M = \sqrt{(2/3)\boldsymbol{\tau}' : \boldsymbol{\tau}'}$ in

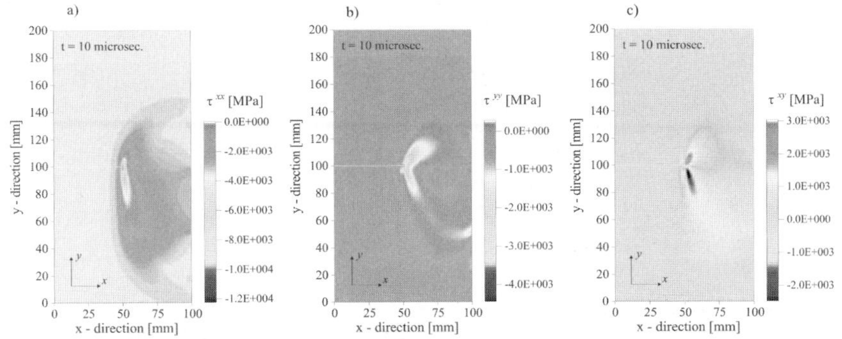

Figure 7.3: The contour of normal stresses τ^{xx}, τ^{yy} (a,b), and (c) shear stress τ^{xy} in the plate at $t = 10$ μs.

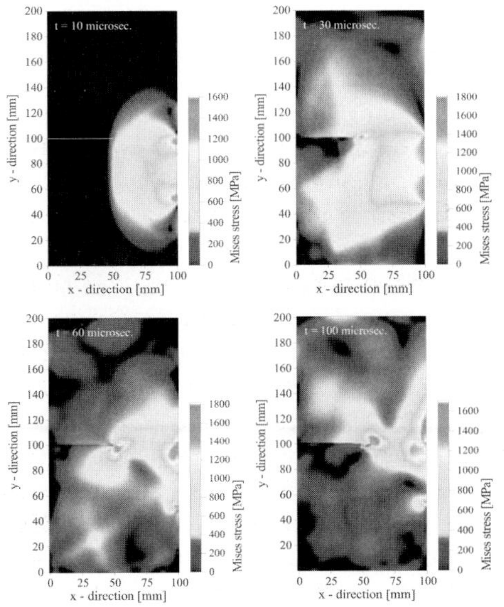

Figure 7.4: Evolution of the Mises stress in the plate.

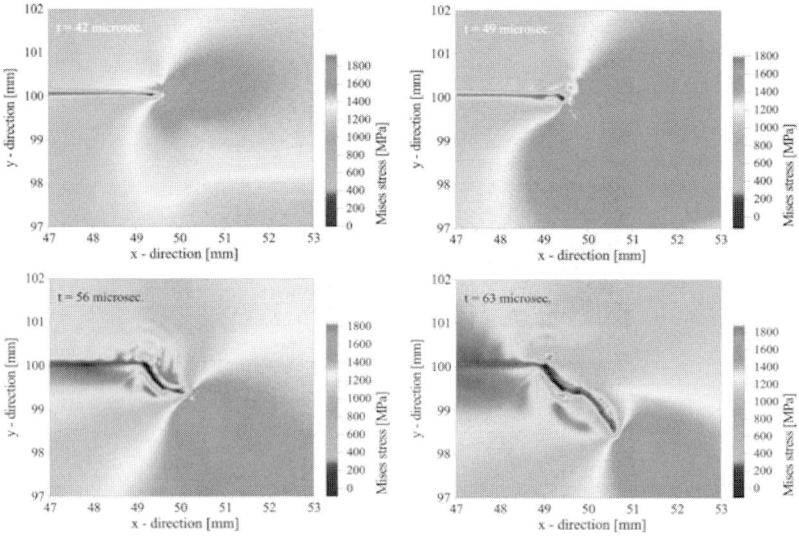

Figure 7.5: Mises stress contours in the failure region.

the plate is presented. The intense stress concentration is visible in the vicinity of the notch. The crack initiates from the tip of the pre–notch at $t = 47.64$ μs, and grows steadily. At a certain point, it suddenly changes its direction, and grows upwards. This phenomenon can be observed more clearly in the next figure.

Figure 7.5 shows the Mises stress contours in the failure region following the impact. It can be clearly observed that the crack follow the path of the greatest Mises stress (the white band). The crack path changes its direction and has irregular (ragged) edges. It results from the wave character of the considered process and from the fact that the shear band transforms into an opening crack. In this pictures the used mesh of nodes is also shown.

In Figure 7.6 the equivalent plastic strain distribution is displayed. It shows that there is a strain concentration region right in the front the pre–notch tip. Along with the shear band progress the equivalent plastic strain intensity decreases. It confirms that the shear band transforms into an opening crack.

The evolution of temperature is shown in Fig. 7.7. Zones of increased temperature correspond to the plastic zones. The maximum value of temperature is $\vartheta_{max} = 750$ K. The effect of such a strong heating of the material results from its mechanical properties, i.e. the high strength steel, $R_m = 2000$ MPa. From Fig. 7.7 one may see that the computed tempera-

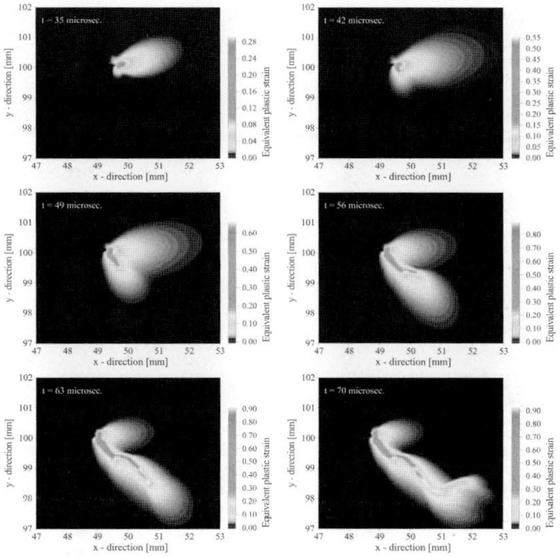

Figure 7.6: Evolution of the equivalent plastic deformation in the fracture region.

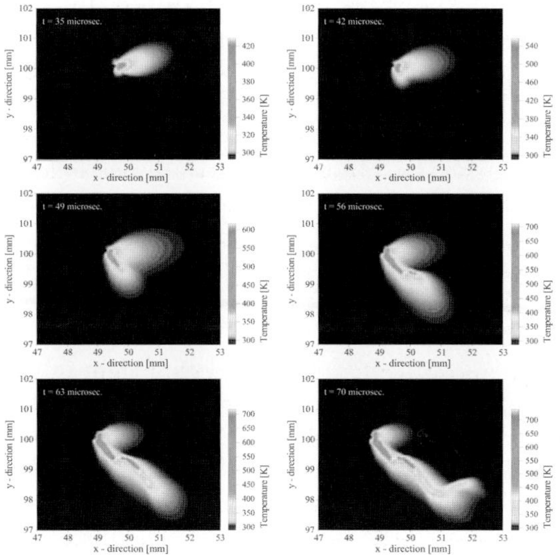

Figure 7.7: Evolution of temperature in the fracture region.

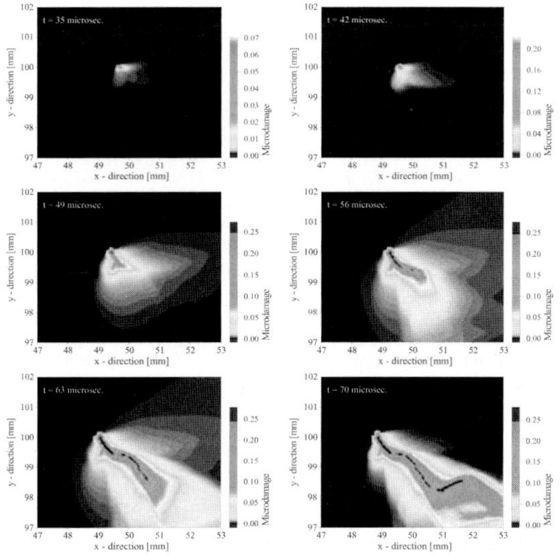

Figure 7.8: Evolution of the microdamage and the crack path in the fracture region.

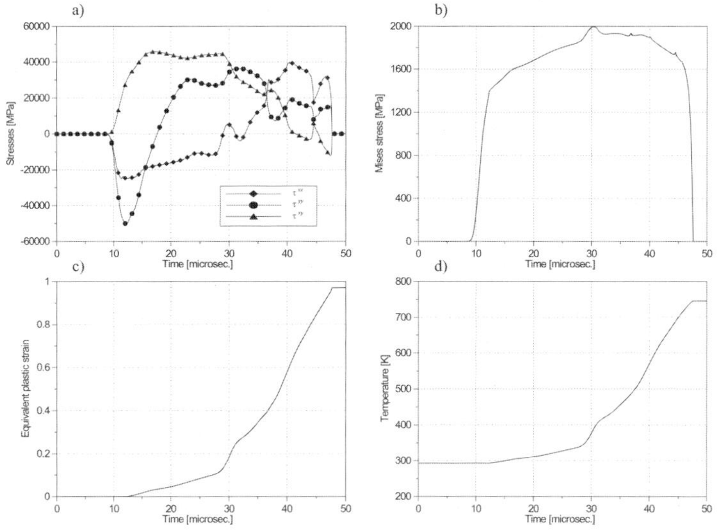

Figure 7.9: (a) Stress components, (b) Mises stress, (c) equivalent plastic strain and (d) temperature at the initial crack tip as a function of time.

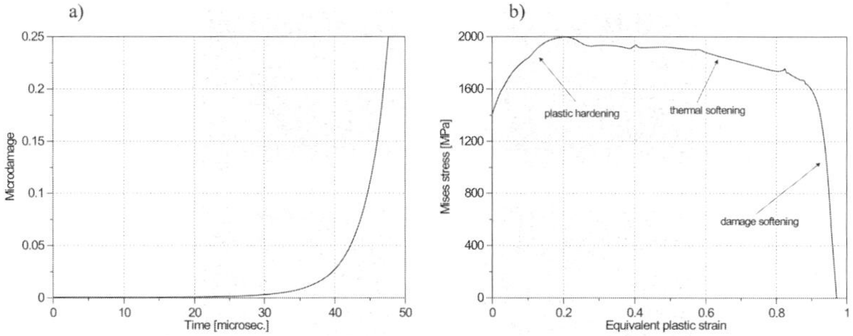

Figure 7.10: (a) Microdamage at the initial crack tip as a function of time, (b) Mises stress at the initial crack tip as a function of the equivalent plastic strain.

ture distribution is very heterogeneous. The similar effect has been noticed in experimental observations by Guduru et al. (2001).

In Figure 7.8 the evolution of microdamage is presented. It can be clearly observed that the crack path (a black line) is very irregular and it widens steadily.

A dependence on time for various process variables at the initial crack tip is plotted in Figs 7.9 and 7.10. A dependence of the Mises stress on the equivalent plastic deformation at the considered node is presented in Fig. 7.10. Figure 7.10 illustrates mainly the nonelastic stress–strain characteristic of the plate material. At the beginning of the deformation process the material undergoes isotropic hardening, and next after some time duration the material softening takes place. The evolutional softening effects are very well visible here.

7.4.3 Discussion of the numerical results

From the results presented in Fig. 7.6 we see that the localized fracture is preceded by the propagation of shear band[9]. This is also very visible in Fig. 7.8, where the propagated localized region of critical microdamage (it means the fracture front for $\xi = \xi^F$) is preceded by the shear band region along which the microdamage is smaller than $\xi = \xi^F$. Very important result we have in Fig. 7.7, where the evolution of temperature is obtained as nonuniform. This result is in agreement with the experimental observations presented by Guduru, Rosakis and Ravichandran (2001). They observed

[9]For a thorough discussion of numerical investigation of propagation of shear bands in inelatic solids please consult Glema, Łodygowski and Perzyna (2004).

that the temperature distribution along the shear band is highly nonuniform, with discrete region of high temperature, that looks like "hot spots". We suggest that the reason of this phenomenon is caused by interaction of dispersive stress waves during observed adiabatic process.

The evolution of the microdamage and the crack path in the fracture region shown in Fig. 7.8 indicates that the crack path is very irregular and it widens steadily. It looks that in some places of the crack we can expect the branching effect as it has been observed by Guduru, Rosakis and Ravichandran (2001), Of course, we have to take into account that they investigated real material (C 300 steel) which have some impurities, while we assumed fully homogeneous material of the plate for our numerical simulations.

We can conclude that the shear band branching can be generated by interaction of dispersive stress waves during adiabatic process and by existing some impurities of the investigated material.

8. Epilogue

The experimental works have brought deep understanding of the evolution of the dislocation substructure and the intrinsic microdamage mechanism during dynamic loading processes and clearly have shown that microshear banding influence a substructure of a polycrystalline material and fracture mechanism of metals does very much depend on the strain rate and wave shape effects.

The crucial idea in the theory of elasto–viscoplasticity developed is the very efficient interpretation of a finite set of the internal state variables as the equivalent plastic deformation, the microdamage tensor and the residual stress (the back stress). To describe suitably the time and temperature dependent effects observed experimentally and the accumulation of the plastic deformation and anisotropic damage during dynamic loading processes the thermomechanical coupling has been taken into account and the kinetics of microdamage and kinematic hardening law have been modified and generalized.

Since the rate independent plastic response is obtained as the limit case when the relaxation time is equal to zero hence the theory of elasto–viscoplasticity offers the regularization procedure for the solution of the dynamical initial–boundary value problems. The viscoplastic regularization procedure assures the stable integration algorithm by using the finite difference or finite element methods.

There is our hope that this thermodynamical theory of elasto–visco plasticity may be used as a basis for the description of the behaviour of modern materials, particularly nanometals (cf. Meyers, Mishra and Benson (2006), Jia, Ramesh and Ma (2003) and Korbel, Nowak, Perzyna and Pęcherski (2006)) and the investagation of plastic strain localization and localized fracture phenomena in meso–, micro–, and nano–mechanical processes.

We hope that a new constitutive model proposed in Chapter 3 is sufficiently simple in its nature that it can be applicable to the numerical solution of initial–boundary value problems under cyclic loadings.

To show how the modification of the evolution equation for the porosity parameter ξ helps to describe the accumulation of damage during dynamic cyclic loading process the evolution of the microdamage ξ_{grow} in a node near the tip of the hole (black quadrangle) for different forms of the stress intensity invariant for adiabatic process and various wave shape is presented.

The performed numerical simulations of the dynamic, cyclic loading process have proven the usefulness of the thermo–elasto–viscoplastic theory. The accumulation of damage and equivalent plastic deformation on each considered cycle has been obtained. It has been numerically found that accumulation of microdamage distinctly depends on the wave shape of the assumed loading cycle.

In a particular numerical example considered (a dynamic, adiabatic, cyclic loading process for a thin plate with sharp notch) small localized region, distributed asymmetrically near the tip of the notch, which undergoes significant deformation and temperature rise has been determined. Its evolution until occurrence of fatigue fracture in the form of a macrocrack has been simulated. The investigation of the propagation macroscopic fatigue damage crack within the material of the plate has shown that the length of the macrocrack distinctly depends also on the assumed loading cycle. This conclusion is approved by the comparison of the speed of macroscopic crack propagation for various loading processes.

Recent experimental results obtained by Rosakis, Samudrala and Coker (1999) (cf. also Lambros and Rosakis (1995a,b,c)) showed that the speed of a crack tip propagation along a bimaterial interface can exceed the shear wave speed of the more compliant constituent in the bimaterial. These experimental observations have motivated analytical and numerical investigation on fast crack growth, cf. Wang, Huang, Rosakis and Liu (1998) and Needleman and Rosakis (1999). However these analytical and numerical investigations have been based on very simple constitutive assumptions for both constituents in the bimaterial.

In this study, the edge–cracked bimaterial thin specimen performed from two identical elements joined by the cohesive band. The material of joint element (material A) and the cohesive band (material B) are modelled as elastic–viscoplastic with isotropic hardening–softening effects. It is assumed that the material softening is caused by intrinsic microdamage mechanism and temperature rise. The numerical results have proven the usefulness of the thermo–elasto–viscoplastic theory and have shown, at least qualitatively, the type of crack speed histories and crack tip fields seen in the experiments.

In Chapter 6 the numerical investigation of dynamic shear bands in an inelastic–viscoplastic solids has been presented. The elaborated numerical algorithm satisfies the material objectivity principle with respect to diffeomorphism (any motion). The discretization parameters are assumed in such a way that the problem of mesomechanics is solved properly. A thin shear band region of finite width which undergoes significant deformation and temperature rise has been determined. Its evolution until occurrence of final fracture has been simulated. Shear band advance, shear band velocity and the development of the temperature field as a function of time have been determined. Qualitative comparison of numerical results with experimental observation data has been presented. The numerical results obtained have been proven the usefulness of the thermo–elasto–viscoplastic theory in the numerical investigation of dynamic shear band propagation.

In Chapter 7 the numerical investigation of localized fracture phenomena has been presented. A thin shear band region of finite width which undergoes significant deformation and temperature rise has been determined. Its evolution until occurrence of final fracture has been simulated. Shear band advance, as a function of time and the evolution of the Mises stress, equivalent plastic deformation, temperature and the crack path in the fracture region have been determined. Qualitative comparison of numerical results with experimental observation data has been presented. Based on this comparison we can conclude that our numerical results are in accord with the experimental observations performed by Guduru, Rosakis and Ravichandran (2001). Particularly, in our numerical simulation the temperature evolution along the shear band region is very nonuniform and the obtained crack path is very irregular and is showing a tendency to branching phenomenon similarly as have been suggested by experimental observations.

References

Abraham, R., Marsden, J.E. (1978) *Foundations of Mechanics*, Second Edition, Addison–Wesley, Reading Mass., 1978.

Abraham, R., Marsden, J.E., Ratiu, T. (1988) *Manifolds, Tensor Analysis and Applications*, Springer, Berlin 1988.

Agah–Tehrani, A., Lee, E.H., Malett, R.L., Onat, E.T. (1987) The theory of elastic–plastic deformation at finite strain with induced anisotropy modelled isotropic–kinematic hardening, *J. Mech. Phys. Solids* **35**, 43–60.

Armstrong, P.J., Frederick, C.O. (1966) A mathematical representation of the multiaxial Baushinger effect, *CEGB Report RD/B/ N731*, Central Electricity Generating Board.

Auricchio, F., Taylor, R.L. (1995) Two material models for cyclic plasticity models: Nonlinear kinematic hardening and generalized plasticity, *Int. J. Plasticity* **11**, 65-98.

Auricchio, F., Taylor, R.L., Lubliner, J. (1992) Application of a return map algorithm to plasticity models, in D.R.J.Owen and E.Onate (eds.), *COMPLAS Computational Plasticity: Fundamentals and Applications*, Barcelona, pp. 2229–2248.

Barbee, T.W., Seaman, L., Crewdson, R., Curran, D. (1972) Dynamic fracture criteria for ductile and brittle metals, *J. Mater.* **7**, 393–401.

Bassani, J.L. (1994) Plastic flow of crystals, *Adv. Appl. Mech.* **30**, 191–258.

Boehler, J.P. (Ed.)(1990) *Yielding, Damage, and Failure of Anisotropic Solids*; Proc. IUTAM/ICM Symposium, Villard–de Lans, 24–28 August 1987, Mech. Eng. Public. Limited, London 1990.

Campbell, J.D., Ferguson, W.G. (1970) The temperature and strain–rate dependence of the shear strength of mild steel, *Phil. Mag.* **81**, 63–82.

Chaboche, J.L. (1986) Time – independent constitutive theories for cyclic plasticity, *Int. J. Plasticity* **2**, 149–188.

Chakrabarti A.K., Spretnak J.W. (1975) Instability of plastic flow in the direction of pure shear. *Metallurgical Transactions* **6A**, 733–747.

Chang Y.W., Asaro, R.J. (1980) Lattice rotations and shearing in crystals, *Arch. Mech.* **32**, 369–388.

Chi, Y.C., Lee, S.H., Cho, K., Duffy, J. (1988) The effects of tempering and test temperatures on the dynamic fracture initiation behaviour of on AISI 4340 VAR steel. Brown University Technical Report, August.

Cho, K., Chi, Y.C., Duffy, J. (1988) Microscopic observations of adiabatic shear bands in three different steels. Brown University Report No DAAL03-88-K-0015/3, September.

Coleman, B.D., Noll, W. (1963) The thermodynamics of elastic materials with heat conduction and viscosity, *Arch. Rational Mech. Anal.* **13**, 167–178.

Courant, R., Friedrichs, K.O., Lewy, H. (1928) Uber die partiellen differenzgleichungen der mathematischen physik, *Math. Ann.* **100**, 32–74.

Curran, D.R., Seaman, L., Shockey, D.A. (1977) Dynamic failure in solids, *Physics Today*, January, 46–55.

Curran, D.R., Seaman, L., Shockey, D.A. (1981) Linking dynamic fracture to microstructural processes, in: *Shock Waves and High–Strain Rate Phenomena in Metals: Concepts and Applications*, (Eds. M.A. Meyers and L.E. Murr), Plenum Press, New York 1981, 129–167.

Curran, D.R., Seaman, L., Shockey, D.A. (1987) Dynamic failure of solids, *Physics Reports* **147**, 253–388.

Dafalias, Y.F. (1983) Corotational rates for kinematic hardening at large plastic deformations. *J. Appl. Mech.* **50**, 561–565.

Dafalias, Y.F., Popov, E.P. (1975) A model of nonlinearly hardening materials for complex loading, *Acta Mech.* **21**, 173–192.

Dautray, R., Lions, J.–L. (1993) *Mathematical Analysis and Numerical Methods for Science and Technology*, Vol. 6. *Evolution Problems II*, Springer, Berlin.

Dornowski, W. (1999) Influence of finite deformation on the growth mechanism of microvoids contained in structural metals. *Arch. Mechanics* **51**, 71–86.

Dornowski, W., Perzyna, P. (1999) Constitutive modelling of inelastic solids for plastic flow processes under cyclic dynamic loadings, *Transaction of the ASME, J. Eng. Materials and Technology* **121**, 210–220.

Dornowski, W., Perzyna, P. (2000) Localization phenomena in thermo–viscoplastic flow processes under cyclic dynamic loadings, *Computer Assisted Mechanics and Engineering Sciences* **7**, 117–160.

Dornowski, W., Perzyna. P. (2002a) Localized fracture phenomena in thermo–viscoplastic flow processes under cyclic dynamic loadings. *Acta Mechanica* **155**, 233–255.

Dornowski, W., Perzyna, P. (2002b) Numerical analysis of macrocrack propagation along a bimaterial interface under dynamic loading processes, *Int. J. Solids and Structures* **39**, 4949–4977.

Dornowski, W., Perzyna, P. (2006) Numerical analysis of localized fracture phenomena in inelastic solids, *Foundations of Civil and Environmental Engineering* **7**, 79–116.

Dowling, A.R., Harding, J., Campbell, D.J. (1970) The dynamic punching of metals, *J. Inst. of Metals* **98**, 215–224, 1970.

Durran, D.R. (1999) Numerical Methods for Wave Equations in Geophysical Fluid Dynamics, Springer, New York, 1999.

Duszek, M.K., Perzyna, P. (1991a) On combined isotropic and kinematic hardening effects in plastic flow processes, *Int. J. Plasticity* **7**, 351–363.

Duszek, M.K., Perzyna, P. (1991b) The localization of plastic deformation in thermoplastic solids, *Int. J. Solids Structures* **27**, 1419–1443.

Duszek–Perzyna, M.K., Korbel, K., Perzyna, P. (1997) Adiabatic shear band localization in single crystals under dynamic loading processes, *Arch. Mechanics* **49**, 1069–1090.

Duszek–Perzyna, M.K., Perzyna, P. (1993) Adiabatic shear band localization in elastic–plastic single crystals. *Int. J. Solids Structures* **30(1)**, 61–89.

Duszek–Perzyna, M.K., Perzyna, P. (1994) Analysis of the influence of different effects on criteria for adiabatic shear band localization in inelastic solids, in: *Material Instabilities: Theory and Applications*, ASME Congress, Chicago, 9–11 November 1994 (Eds. R.C.Batra and H.M.Zbib), AMD–Vol. 183/MD–Vol.50, ASME, New York, 1994, pp. 59–85.

Duszek–Perzyna, M.K., Perzyna, P. (1995) Acceleration waves in analysis of adiabatic shear band localization, in: *Nonlinear Waves in Solids*, Proc. IUTAM Symposium, August 15–20, 1993, Victoria, Canada; J.L.Wegner and F.R.Norwood [Eds.], pp. 128–135, ASME Book No AMR 137, 1995.

Duszek–Perzyna, M.K., Perzyna, P. (1996) Adiabatic shear band localization of inelastic single crystals in symmetric double slip process, *Archive of Applied Mechanics* **66**, 369–384.

Duszek–Perzyna, M.K., Perzyna, P. (1998) Analysis of anisotropy and plastic spin effects on localization phenomena. *Arch. Appl. Mechanics* **68**, 352–374.

Follansbee, P.S. (1986) *Metallurgical Applications of Shock – Wave and High–Strain–Rate Phenomena*, (Murr LE, Staudhammer KP, Meyeres MA, eds.), pp.451–480, Marcel Dekker, New York, 1986.

Glema, A., Łodygowski, T., Nowak, Z., Perzyna P., Voyiadjis, G.Z. (2005) Thermo–elasto–viscoplastic model of a material with non–local and anisotropic intrinsic microdamage. McMat 2005, Mechanics and Materials Conference, June 1–3, 2005, Baton Rouge, LA, USA.

Glema, A., Łodygowski, T., Perzyna, P. (2000) Interaction of deformation waves and localization phenomena in inelastic solids, *Computer Methods in Applied Mechanics and Engineering* **183**, 123–140.

Glema, A., Łodygowski, T., Perzyna, P. (2001) The role of dispersion for the description of strain localization in materials under impact loading, European Conference on Computational Mechanics, June 26–29, Cracow, Poland, 2001.

Glema, A., Łodygowski, T., Perzyna, P. (2003) Localization of plastic deformations as a result of wave interaction, *CAM&ES* **3**, 81–91.

Glema, A., Łodygowski, T., Perzyna, P. (2004) Numerical investigation of dynamic shear bands in inelastic solids as a problem of mesomechanisc. International Congress of Theoretical and Applied Mechanics, 15–21 August, 2004, Warsaw, Poland.

Glema, A., Łodygowski, T., Perzyna, P. (2008) Numerical investigation of dynamic shear bands in inelastic solids as a problem of mesomechanics, *Comput. Mech.* **41**, 219–229.

Glema, A., Łodygowski, T., Perzyna, P., Sumelka, W. (2006) Constitutive anizotropy induced by plastic strain localization, 35th Solid Mechanics Conference, Kraków, Poland, Sempember 4–8, 2006.

Grebe, H.A., Pak, H.R., Meyers, M.A. (1985) Adiabatic shear band localization in titanium and Ti-6PctAl-4PctV alloy, *Met. Trans.* **16A**, 761-775.

Guduru, P.R., Rosakis, A.J., Ravichandran, G. (2001) Dynamic shear bands: an investigation using high speed optical and infrared diagnostic. *Mechanics of Materials* **33**, 371–402.

Gustafsson, B., Kreiss, H.O., Oliger, J. (1995) *Time Dependent Problems and Difference Methods*, John Wiley, New York.

Hughes T.J.R., Kato T. and Marsden J.E. (1977) Well-posed quasilinear second order hyperbolic system with application to nonlinear elastodynamics and general relativity, *Arch. Rat. Mech. Anal.* **63**, 273–294.

Hutchinson, J.W. (2000) Plasticity at the micron scale. *Int. J. Solids and Structures* **37**, 225–238.

Ionescu, I.R., Sofonea, M. (1993) *Functional and Numerical Methods in Viscoplasticity*, Oxford.

Jaumann, G. (1911) Geschlossenes System physikalischer und chemischer Differentialgesetze. *Sitzgsber. Akad. Wiss. Wien (IIa)* **120**, 385–530.

Jia, D., Ramesh, K., Ma, E. (2003) Effects on nanocrystalline and ultrafine grain sizes on constitutive behaviour and shear bands in iron, *Acta Materialia* **51**, 3495–3509.

Johnson, J.N. (1981) Dynamic fracture and spallation in ductile solids, *J. Appl. Phys.* **52**, 2812–2825.

Korbel, K., Nowak, Z., Perzyna, P., Pęcherski, R.B. (2006) Viscoplasticity of nanometals based on Burzyński yield condition, 35th Solid Mechanics Conference, Kraków, Poland, Sempember 4–8, 2006.

Kumar A., Kumble, R.Gg. (1969) Viscous drag on dislocations at high strain rates in copper, *J. Appl. Physics* **40**, 3475–3480.

Lambros, J., Rosakis, A. (1995a) Dynamic decohesion of bimaterials: Experimental observations and failure criteria, *Int. J. Solids Structures* **32**, 2677–2702.

Lambros, J., Rosakis, A. (1995b) Shear dominated transonic interfacial crack growth in a binaterial – I. Experimental observations, *J. Mech. Phys. Solids* **43**, 169–188.

Lambros, J., Rosakis, A. (1995c) Shear dominated transonic interfacial crack growth in a binaterial – II. Asymptotic fields and favorable velocity regimes, *J. Mech. Phys. Solids* **43**, 189–206.

Li, S., Liu, W.–K., Qian, D., Guduru, P.R., Rosakis, A.J. (2001) Dynamic shear band propagation and micro–structure of adiabatic shear band. *Comput. Methods Appl. Mech. Engng.* **191**, 73–92.

Lisiecki, L.L., Nelson, D.R., Asaro, R.J. (1982) Lattice rotations, necking and localized deformation in f.c.c. single crystals, *Scripta Met.* **16**, 441–449.

Loret, B. (1983) On the effects of plastic rotation in the finite deformation of anisotropic elastoplastic materials. *Mech. Mater.* **2**, 287–304.

Łodygowski, T., Perzyna, P. (1997) Numerical modelling of localized fracture of inelastic solids in dynamic loading processes, *Int. J. Num. Meth. Engng.* **40**, 4137–4158.

Łodygowski, T. Perzyna, P. (1997) Localized fracture of inelastic polycrystalline solids under dynamic loading processes, *Int. J. Damage Mechanics* **6**, 364–407.

Marsden, J.E., Hughes, T.J.R. (1983) *Mathematical Foundations of Elasticity*, Prentice–Hall, Englewood Cliffs, New York 1983.

Mason, J.J., Rosakis, J.A., Ravichandran, R. (1994) On the strain and strain rate dependence of the fraction of plastic work converted to heat: an experimental study using high speed infrared detectors and the Kolsky bar, *Mechanics of Materials* **17**, 135–145.

Meyers, H.C. (1994) *Dynamic Behaviour of Materials*, John Wiley, New York 1994.

Meyers, M.A., Aimone, C.T. (1983) Dynamic fracture (spalling) of metals, *Prog. Mater. Sci.* **28**, 1–96.

Meyers, M.A. Mishra, A., Benson, D.J. (2006) Mechanical properties of nanocrystalline materials, *Prog. Mater. Sci.* **51**, 427–556.

Mróz, Z. (1967) On the description of anisotropic workhardening, *J. Mech. Phys. Solids* **15**, 163–175.

Needleman, A. (2000) Computational mechanics at the mesoscale. *Acta Materialia* **48**, 105–124.

Needleman, A., Rosakis, A.J. (1999) The effect of bound strength and loading rate on the conditions governing the attainment of intersonic crack growth along interfaces, *J. Mech. Phys. Solids* **47**, 2411–2449.

Nemat–Nasser, S. (1992) Phenomenological theories of elastoplasticity and strain localization at high strain rates, *Appl. Mech. Rev.* **45**, S19–S45.

Nemes, J.A., Eftis, J. (1993) Constitutive modelling of the dynamic fracture of smooth tensile bars, *Int. J. Plasticity* **9**, 243–270.

Nowak, Z., Perzyna, P., Pęcherski, R.B. (2007) Description of viscoplastic flow accounting for shear banding, *Arch. of Metallurgy and Materials* **52**, 217–222.

Oldroyd, J. (1950) On the formulation of rheological equations of state, *Proc. R. Soc. Lond.* **A200**, 523–541.

Perzyna, P. (1963) The constitutive equations for rate sensitive plastic materials, *Quart. Appl. Math.* **20**, 321–332.

Perzyna, P. (1966) Fundamental problems in viscoplasticity, *Advances in Applied Mechanics* **9**, 343–377.

Perzyna, P. (1971) Thermodynamic theory of viscoplasticity, *Advances in Applied Mechanics* **11**, 313–354.

Perzyna, P. (1977) Coupling of dissipative mechanisms of viscoplastic flow, *Arch. Mechanics* **29**, 607–624.

Perzyna, P. (1980) Modified theory of viscoplasticity. Application to advanced flow and instability phenomena, *Arch. Mechanics* **32**, 403–420.

Perzyna, P. (1984) Constitutive modelling of dissipative solids for postcritical behaviour and fracture. *ASME J. Eng. Materials and Technology* **106**, 410-419.

Perzyna, P. (1986a) Internal state variable description of dynamic fracture of ductile solids, *Int. J. Solids Structures* **22**, 797-818.

Perzyna, P. (1986b) Constitutive modelling for brittle dynamic fracture in dissipative solids, *Arch. Mechanics* **38**, 725–738.

Perzyna, P. (1988) Temperature and rate dependent theory of plasticity of crystalline solids, *Revue Phys. Appl.* **23**, 445–459.

Perzyna, P. (1994) Instability phenomena and adiabatic shear band localization in thermoplastic flow processes, *Acta Mechanica* **106**, 173–205.

Perzyna, P. (1995) Interactions of elastic–viscoplastic waves and localization phenomena in solids, IUTAM Symposium on Nonlinear Waves in Solids, August 15–20, 1993, Victoria, Canada; (Eds. J.L. Wegner and F.R. Norwood), ASME 1995, 114–121.

Perzyna, P. (1998) Constitutive modelling of dissipative solids for localization and fracture. In: *Localization and Fracture Phenomena in Inelastic Solids* (Ed. P. Perzyna), Springer, Wien, New York, pp. 99–242, 1998.

Perzyna, P. (2001) Thermo–elasto–viscoplasticity and damage. In: *Handbook of Materials Behaviour Models* (Ed. J. Lemaitre), Academic Press, New York, pp. 821–834, 2001.

Perzyna, P. (2002) Thermodynamical theory of inelastic single crystals. *Engineering Transactions* **50**, 107–164.

Perzyna, P. (2005) The thermodynamical theory of elasto–viscoplasticity, *Engineering Transactions* **53**, 235–316.

Perzyna, P. (2008) The thermodynamical theory of elasto–viscoplasticity accounting for microshear banding and induced anisotropy effects, *Mechanics* **27**, (in print).

Perzyna, P., Drabik, A. (1989) Description of micro–damage process by porosity parameter for nonlinear viscoplasticity, *Arch. Mechanics* **41**, 895–908.

Perzyna, P., Drabik, A. (2008) Micro–damage mechanism in adiabatic processes, Engineering Transactions (in print).

Perzyna, P., Korbel, K. (1996) Analysis of the influence of substructure of crystal on the localization phenomena of plastic deformation, *Mechanics of Materials* **24**, 141–158.

Perzyna, P., Korbel, K. (1998) Analysis of the influence of variuos effects on criteria for adiabatic shear band localization in single crystals, *Acta Mechanica* **129**, 31–62.

Perzyna P., Voyiadjis, G.Z. (2005) Thermodynamic theory of elasto–viscoplasticity for induced anisotropy effects. McMat 2005, Mechanics and Materials Conference, June 1–3, 2005, Baton Rouge, LA, USA.

Pęcherski, R.B. (1998) Macroscopic effects of microshear banding in plasticity of metals, *Acta Mechanica* **131**, 203–224.

Prager, W. (1955) The theory of plasticity: A survey of recent achievements, (J. Clayton Lecture), *Proc. Inst. Mech. Eng.* **169**, 41–57.

Rashid, M.M., Gray, G.T., Nemat–Nasser, S. (1992) Heterogeneous deformations in copper single crystals at high and low strain rates, *Philosophical Magazine A* **65**, 707–735, 1992.

Richtmyer, R.D. (1978) *Principles of Advance Mathematical Physics*, Vol. I, Springer, New York.

Richtmyer, R.D., Morton, K.W. (1967) *Difference Methods for Initial–Value Problems*, John Wiley, New York.

Ristinmaa, M. (1995) Cyclic plasticity model using one yield surface only, *Int. J. Plasticity* **11**, 163–181.

Rosakis, A.J., Ravichandran, G. (2000) Dynamic failure mechanics, *Int. J. Solids Structures* **37**, 331–348.

Rosakis, A.J., Samudrala, O., Coker, D. (1998) Cracks faster than the shear wave speed, California Institute of Technology, SM Report 98–17, December 1998.

Rosakis, A.J., Samudrala, O., Coker, D. (1999) Cracks faster than the shear wave speed, *Science* **284**, 1337–1340.

Rosenfield, A.R., Hahn, G.T. (1966) Numerical description of the ambient low–temperature, and high–strain rate flow and fracture behaviour of plain carbon steel, *Trans. Am. Soc. Metals* **59**, 962–980.

Seaman, L., Curran, D.R., Shockey, D.A. (1976) *J. Appl. Phys.* **47**, 4814–4820.

Shima, S., Oyane, M. (1976) Plasticity for porous solids, *Int. J. Mech. Sci.* **18**, 285–291.

Shockey, D.A., Seaman, L., Curran, D.R. (1973) In: *Metallurgical Effects at High Strain Rates*, (Eds. R.W. Rohde, B.M. Butcher, J.R. Holland, and C.H. Karnes), Plenum Press, New York 1973, 473.

Shockey, D.A., Seaman, L., Curran, D.R. (1985) The microstatistical fracture mechanics approach to dynamic fracture problem, *Int. J. Fracture* **27**, 145–157.

Sidey, D., Coffin, L.F. (1979) Low–cycle fatigue damage mechanisms at high temperature, in: *Fatigue Mechanisms*, Proc. ASTM STP 675 Symposium, Kansas City, Mo., May 1978, (Ed. J.T. Fong), Baltimore, 1979, pp. 528–568.

Sluys, L.J. (1992) Wave propagation, localization and dispersion in softening solids. Doctoral thesis, Delft Uniwersity Press, Delft.

Spitzig, W.A. (1981) *Deformation behaviour of nitrogenated Fe–Ti–Mn and Fe–Ti single crystals, Acta Metall.* **29**, 1359 1377.

Strang, G., Fix, G.J. (1973) *An Analysis of the Finite Element Method*, Prentice–Hall, Englewood Cliffs.

Taylor, G.I., Quinney, H. (1934) The latent energy remaining in a metal after cold working, *Proc. R. Soc. Lond.* **A143**, 307–326.

Teodosiu, C., Sidoroff, F. (1976) A theory of finite elastoplasticity of single crystals, *Int. J. Engng. Sci.* **14**, 165–176.

Truesdell, C., Noll, W. (1965) The Non–Linear Field Theories of Mechanics, in: *Handbuch der Physik III/3*, (Ed. S. Flűgge), Springer–Verlag, Berlin 1965.

Van der Giessen, E. (1989) Continuum models of large deformation plasticity, Part I: Large deformation plasticity and the concept of a natural reference state, *Eur. J. Mech., A/Solids* **8**, 15.

Van der Giessen, E. (1989) Continuum models of large deformation plasticity, Part II: A kinematic hardening model and the concept of a plastically induced orientational structure, *Eur. J. Mech., A/Solids* **8**, 89.

Van der Giessen, E. (1991) Micromechanical and thermodynamic aspects of the plastic spin, *Int. J. Plasticity* **7**, 365–386.

Voyiadjis, G.Z., Ju, J.–W., Chaboche, J.–L. (1998) (Eds.) *Damage Mechanics in Engineering Materials*, Elsevier, Amsterdam 1998.

Wang, J.–D., Ohno, N. (1991) Two equivalent forms of nonlinear kinematic hardening: application to nonisothermal plasticity, *Int. J. Plasticity* **7**, 637–650.

Wang, W., Huang, Y., Rosakis, A.J., Liu, C. (1998) Effect of elastic mismatch in intersonic crack propagation along a bimaterial interface, *Engineering Fracture Mechanics* **61**, 471–485.

Zaremba, S. (1903) Sur une forme perfectionnée de la théorie de la relaxation, *Bull. Int. Acad. Sci. Cracovie*, 594–614.

Zaremba, S. (1903) Le principe des mouvements relatifs et les équations de la mécanique physique, *Bull. Int. Acad. Sci. Cracovie*, 614–621.

Zhou, M., Rosakis, A.J., Ravichandran G. (1996) Dynamic propagating shear band in impact–loaded prenotched plates. I. Experimental investigations of temperature signatures and propagation speed. *J. Mech. Phys. Solids* **44**, 981–1006.

Zhou, M., Ravichandran G., Rosakis, A.J. (1996) Dynamic propagating shear band in impact–loaded prenotched plates. II. Numerical simulations. *J. Mech. Phys. Solids* **44**, 1007–1032.

Ziegler, H. (1959) A modification of Prager's hardening rule, *Quart. Appl. Math.* **17**, 55–65.

PART IV

Safety of Concrete and Masonry Structures under Unusual Loadings

by

Tomasz Lodygowski, Tomasz Jankowiak and Piotr W. Sielicki

Safety of Concrete and Masonry Structures under Unusual Loadings

Tomasz Łodygowski*, Tomasz Jankowiak and Piotr W. Sielicki
Institute of Structural Engineering (ISE),
Poznan University of Technology (PUT), Poznan, Poland,
* tomasz.lodygowski@put.poznan.pl

Abstract In the paper the behavior of selected brittle materials and structures (concrete and masonry) subjected to explosive loadings is discussed. For concrete the accepted Cumulative Fracture Criterion (CFC) is exposed. It describes the degradation of the material under fast dynamic processes accompanied by the strong waves propagation phenomenon and large strain rates of deformation. To overcome the computational difficulty in the analyses of such complex problems, the sub-modeling technique as well as splitting of the calculations into two separate parts: analysis of acoustic wave in the air and the propagation of stresses in structures, were used. Some instructive numerical examples of concrete and masonry walls are in focus of the presentation. The numerical tools and computer simulations allow for proper estimation of the structures safety and for taking the design decisions on how to ensure their expected strength.
The support of Ministry of Science and Higher Education under the grant N 519 419435 is kindly acknowledged.

1 Some remarks on damage and localized fracture

Many materials under specific rate of loadings exhibit softening properties which usually are precursors of failure and fracture. This process is vivid for both quasi-static as well as for dynamic loadings. When dealing with softening materials one has to be very careful in mathematical formulation and then also in computer solutions of any applications. Particularly important is the consequence of well posedness lack of the system of governing equations. For example, for static cases usually governed by elliptic equations, the softening results in not positively definite constitutive matrices and suddenly change the type of equations into hyperbolic one. The further consequences for the numerical solution are crucial. The lack of mathematical knowledge on posedness of the initial boundary value problems drives in computations to so called pathological mesh dependency. It simply means that the achieved results are meaningless and should not be shown and discussed any more. There are several ways of avoiding the changing of equation types during the computations. All of them are called regularization. Sometimes the introduced regularization has a physical background (viscoplasticity or higher order media for soil mechanics) and sometimes it comes from artificial manipulations (also on the level of numerical discretization) that simply keep the type of governing equation not changed during the whole incremental process (for static problems the system remains elliptic while for dynamic hyperbolic). When talking about the type of loading (if it is quasistatic or dynamic) it is good to compare own feeling with obtained rate of deformations for typical processes, see Table 1. In the future work, our attention is focused on fast dynamic processes like

Table 1. Velocities of deformation for specific physical phenomena

Type of phenomenon	Velocity of deformation $[s^{-1}]$
Creep	from 10^{-10} to 10^{-5}
Creep beyond the yield criteria	from 10^{-5} to 10^{-1}
Hot drawing	from 10^{-1} to 10^{1}
High speed drawing	from 10^{1} to 10^{3}
Machining	from 10^{3} to 10^{5}
Drawing with use of explosion	$> 10^{5}$
Type of loading	Velocity of deformation $[s^{-1}]$
Earthquake	from 10^{-3} to 10^{-1}
Car crash	from 10^{-2} to 10^{0}
Plane crash	from $5{\cdot}10^{-2}$ to $2{\cdot}10^{0}$
Hard hit	from 10^{0} to $5{\cdot}10^{1}$
Hitting rigid body without deformation	from 10^{2} to 10^{6}
Type of deformation	Velocity of deformation $[s^{-1}]$
Geological movements	$\approx 10^{-10}$
Creeping	$\approx 10^{-6}$
One axial tension test	$\approx 10^{-4}$
Drilling, rolling, drawing	$\approx 10^{0}$
Test with Hopkinson's rod	$\approx 10^{3}$
Explosion	$\approx 10^{6}$

rigid hitting or blast which cause the rate of deformations in structures of order between 104 up to 106 1/s. Usually, in static cases the computations require any kind of imperfections, like missing element or slightly changed constitutive properties in an element, to start the localized deformations. Contrary, in fast dynamic processes any kind of imperfections are necessary. They appear in a structure or specimen as a result of wave propagation and interinfluences. That is a natural only source of imperfections which automatically chooses the places of localization without any artificial manipulations. For dynamic cases some computational results are compared with laboratory test obtained in Hopkinson bar. Plastic strain localization is usually the first sign of fracture, so using the specific failure criteria is of fundamental importance. In this part we decided to focus the attention on a description and analysis of very specific loadings (blasts and rigid hitting) and the further behavior of the structure made of brittle material.

2 Introduction

The structures like concrete or masonry walls are often subjected to unique loadings. For example, the blast wave as well as impact belong to these types of loadings. The paper presents the results for both classes of structural external forces which come as an effect of blast of explosive material and rigid hitting by projectile. To describe properly the pressure wave propagation in the air, produced by detonation, which acts on the structure, it is necessary to simulate the explosion and the waves' motion. There are many recent papers in the literature today [Alia and Souli; Henrych] which reflect the state of the art in the analysis and simulations of blast itself and the propagation of acoustic waves in the air. Some of them restrict the attention only to 2-D cases which we treat only as preliminary studies. For the whole process of blast simulation and its influence on any kind of obstacle (e.g. walls, concrete or masonry) one can propose to split it into two phases. The first is the propagation of the acoustic wave in the air which ends with the crucial information on the distribution of the pressure on the wall surface, which varies in time and space. In the second step, the analysis of the stress waves in the structure is elaborated, and finally the failure of the structure under dynamic loadings appears. For numerical simulation of the detonation process, we accept the Jones-Wilkins-Lee (JWL) equation of state for explosive material and typical equation of ideal gas for the air [Henrych; Smith and Hetherington]. Using the convenient description of the air deformations (ALE) [Jankowiak et al.] guarantees the proper solution of explosion simulation and finally the distribution of pressure loading on the structure. There are many analytical functions in literature [Smith and Hetherington], which describe the influence of time and distance from the ignition on the pressure distribution. In this study the results of numerical simulations [Jankowiak et al.] are compared with known analytical functions and particularly with few published experiments to validate the accepted models. The results how the pressure distribution in the air depends on time after explosion and the distance from the ignition point are presented. The interaction of the fluid (air) with the walls is performed using sub-modeling technique. There are two different models for both cases, the global and the local one. Once the global model consists of cubic explosive material, surrounded by the air and with the structural wall. The local models include only structural parts (concrete or masonry walls) of analyzed structure. Sub-modeling technique is accepted if the coupling exists in one way between the global to local models, but not in opposite. It is the sufficient assumption for blast simulation. When modeling 1/4 part of the space two planes of symmetry are assumed. The global model is extended up to 4 meters from the model center. It is possible to obtain the positive and negative overpressure phases acting on the structure surface like in experiment [Alia and Souli]. The first kind of the structure under consideration is the concrete wall and the second is a periodic composite masonry wall created of bricks and mortar. Both are the local models. Cumulative Fracture Criterion (CFC) introduced by Campbell for metals [Campbell] and discussed also later for concrete [Klepaczko] is used and has been added to Abaqus/Explicit environment by VUMAT procedure. The criterion describes generally the time up to failure under the stress impulse. In both, concrete and masonry structures this criterion is used. In the second one it describes only the behavior of mortar while the bricks are treated as elastic. It means that we assumed elastic-visco-brittle material. This criterion results in strain rate sensitivity of the material behavior. For one dimensional case (uniaxial dynamic tension) the fracture criterion together with constitutive parameters are verified through computer simulation of known experiments done with the use of Hopkinson bar. The criterion was also confirmed for triaxial state of stress and simulation of spalling analysis in circular concrete plate. The plate is subjected to explosion, the charge is placed in the center. The CFC is used to describe the behavior of the concrete wall, loaded with a blast wave and also for the description of perforation process. A masonry wall is the second kind of structure under consideration. The periodic model contains elastic bricks connected with the mortar.

3 Cumulative Fracture Criterion for brittle materials

3.1 Dynamic concrete behavior in codes

Using the strain rates sensitivity according to the proposition of codes CEB [CEB] is an advantage and acceptable reasonable simplification when modeling the fast dynamic structural response. CEB recommends the different values of the Dynamic Increase Factors (DIF) for dynamic tension and compression. The factor defines how many times the dynamic strength is higher than quasi-static for increasing strain rates [Bischoff and Perry; M. Klosak et al.]. The most important for dynamic failure modeling of brittle material as concrete is TDIF (Tensile Dynamic Increase Factor) because the tension is most often the source of failure during varying explosive loadings. The empirical equations which describe DIF in compression (CDIF) are as follows:

$$CDIF = \frac{f_{cd}}{f_{cs}} \begin{cases} \left(\frac{\dot{\varepsilon}_d}{\dot{\varepsilon}_{cs}}\right)^{1.026\alpha} & \text{for } \dot{\varepsilon}_d \leq 30s^{-1} \\ \gamma(\dot{\varepsilon}_d) & \text{for } \dot{\varepsilon}_d > 30s^{-1} \end{cases} \tag{1}$$

where f_{cd} is the dynamic compressive strength for strain rate $\dot{\varepsilon}_d$. Additionally, $\dot{\varepsilon}_{cs}$ is equal to $3{\cdot}10^{-5}s^{-1}$ and corresponds to static compressive strength f_{cs} and γ, α are defined by CEB particularly as $log(\gamma) = 6.156\alpha - 0.49$ and $\alpha = (5 + 3f_{cu}/4)^{-1}$, where f_{cu} is the static cube compressive strength. The most important for our consideration is TDIF that is Tensile Dynamic Increase Factor recommended by CEB. The literature and many experimental results [Klepaczko and Brara; Malvar and Ross; Schuler et al.] show that TDIF is underestimated. According to CEB, TDIF is expressed as:

$$TDIF = \frac{f_{td}}{f_{ts}} \begin{cases} 1.0 & \text{for } \dot{\varepsilon}_d \leq 10^{-4}s^{-1} \\ 2.06 + 0.26log(\dot{\varepsilon}_d) & \text{for } 10^{-4}s^{-1} < \dot{\varepsilon}_d \leq 1s^{-1} \\ 2.06 + 2log(\dot{\varepsilon}_d) & \text{for } \dot{\varepsilon}_d > 1s^{-1}, \end{cases} \tag{2}$$

where f_{td} is dynamic tensile strength for strain rate $\dot{\varepsilon}_d$. Experimental results show that DIF's curves can be extended to 10^3s^{-1}. Over this limit, the constant values of DIF for 10^3s^{-1} should be used to avoid substantial overestimation. In Fig.1 the DIFs for both compression and tension rates are plotted according to Eqs.1 and 2.

3.2 Cumulative Fracture Criterion (CFC)

The tensile strength and behavior of concrete is highly rate dependent and it is the main factor, which influences the fracture process. The Cumulative Fracture Criterion originally proposed for metals is adapted for the concrete [Jankowiak and Łodygowski; Jankowiak et al.; Klepaczko] and extended to 3D case and for numerical efficiency included into Abaqus/Explicit code. This criterion in integral form is the following:

$$t_{c0} = \int_0^{t_c} \left(\frac{\sigma_F^{eq}(t)}{\sigma_{F0}^{eq}}\right)^{\alpha(T)} dt \qquad if \qquad \sigma_F^{eq}(t) > \sigma_{F0}^{eq}, \tag{3}$$

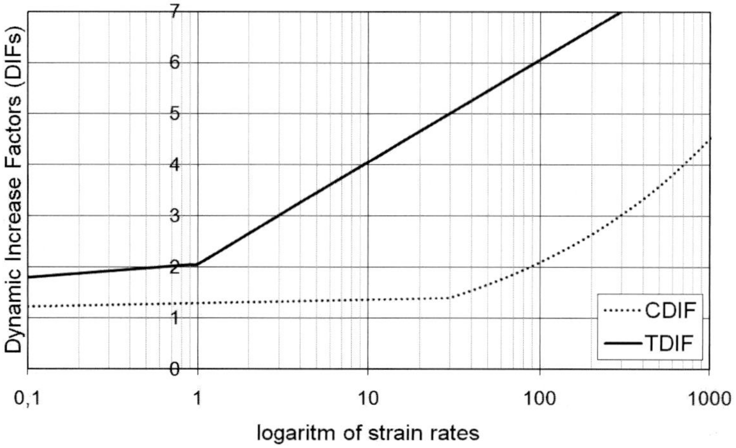

Figure 1. Dynamic Increase Factor (DIF) versus logarithm of strain rates for the tensile and compressive cases

where t_{c0} is the longest critical time, $\alpha(T)$ is the parameter connected with activated energy during the separation process and σ_{F0}^{eq} is quasi-static equivalent tensile strength of concrete. The measure of equivalent stresses σ_F^{eq} is introduced to describe advanced tri-axial state of stresses [Geers et al.]:

$$\sigma_F^{eq} = \frac{k-1}{2k(1-2\nu)} I_1 + \frac{1}{2k} \sqrt{\left(\frac{k-1}{1-2\nu} I_1\right)^2 + \frac{6k}{(1-\nu)^2} J_2}, \tag{4}$$

where

$$\begin{aligned} I_1 &= \sigma_1 + \sigma_2 + \sigma_3 \\ J_2 &= \tfrac{1}{3}\left[(\sigma_1-\sigma_2)^2 + (\sigma_1-\sigma_3)^2 + (\sigma_2-\sigma_3)^2\right] \end{aligned} \tag{5}$$

σ_F^{eq} is the generalization of the Huber-Mises equivalent stresses [Geers et al.]. $I1$ and $J2$ are the first and the second invariants of the stress tensor and deviatoric part of the stress tensor, respectively, $\sigma_i (i = 1..3)$ are the principal stresses; k is the parameter responsible for shifting of the CFC in the space of principal stresses (will be discussed later) and ν is the Poisson ratio. The physical meaning of the criterion (3) is simply the estimation of the longest time in which the material can carry the dynamic stresses that overcome the quasi-static tensile strength.

3.3 Identification and the influence of used parameters

The influence of k parameter for the shape of the criteria obtained in the space of principal stresses is shown in Figs.2 and 3. If the parameter k equals to 1.0 the symmetric (according to tension and compression) failure surface is achieved, but for k equals to 10.0 the failure surface shifts significantly into the direction of compression zone (Fig.3). Of course, the equivalent stress equation (4) which is adopted above works for the concrete properties very different for both compressive and tensile zones. For better understanding the shape of the failure surface, it is convenient to present it in meridian plane (meridian cross section). The failure surface is then represented by the curve in plane $r - \xi$, where, the important is to define the name of the axes; it means what is ordinate and what is the abscissa of the points. The ordinates of the points are computed based on following relationship $r = \pm\sqrt{2J_2}$ and $\xi = \pm\frac{1}{\sqrt{3}}I_1$. In the above ξ is the length of the interval along the hydrostatic axis and r is the distance in the plane perpendicular to the hydrostatic axis length. After taking into account the previous relationships for r and $/xi$, the equation (4) can be transformed to the following form:

$$\sigma_F^{eq} = \frac{(k-1)\sqrt{3}}{2k(1-2\nu)}\xi + \frac{1}{2k}\sqrt{\left(\frac{(k-1)\sqrt{3}}{1-2\nu}\xi\right)^2 + \frac{6k}{(1-\nu)^2}\frac{r^2}{2}}. \tag{6}$$

The substitutions and conversions, give the following form:

$$r = \sqrt{(2k\sigma_F^{eq})^2\frac{2(1-\nu)^2}{6k} - \left(8\sqrt{3}\sigma_F^{eq}\frac{(k-1)(1-\nu)^2}{6k}\xi\right)}, \tag{7}$$

for

$$\xi < \xi_{max} = \frac{k\sigma_F^{eq}(1-2\nu)}{(k-1)\sqrt{3}}. \tag{8}$$

Other shapes in the meridian plane could be also accepted and are discussed in [Stolarski]. The right end of the meridian curve is limited by ξ_{max}. It is the value for which r is equal to zero. If k equals to 1.0, the critical surface gives Huber-Mises equivalent stress. It means that in meridian plane $r - \xi$, critical state is represented by a straight line parallel to ξ axis. Cumulative Fracture Criterion predicts different Equivalent Fracture Stresses (EFS) for different loading functions. The first considered loading function is the Heaviside step function, (see Fig.5). During the loading time the level of equivalent stresses remains constant. To plot the curve, it is necessary to obtain the time up to failure and the critical stresses. The time to failure means the measure of time when the equivalent stresses obtained in dynamic

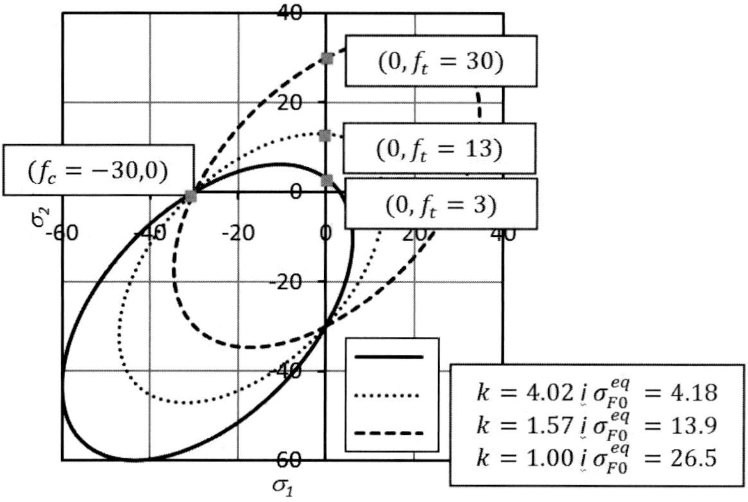

Figure 2. Iso-lines of equivalent stresses for $k = 10$ in $(\sigma_1 - \sigma_2 - \sigma_3)$ space for $\sigma_3 = 0$

process exceed the quasi-static ones. In the case of Heaviside step function, the time to failure starts at the beginning. For five different magnitudes of loadings (levels of stresses), five points, which create the critical curve (CC) are obtained (see Fig.5). The line parallel to the time axis shows the exact loading history. For higher level of stresses, the shorter critical time to failure is computed. This statement also is true for other loading histories (more complex and different from simple step function). This curve can be interpreted that dynamic strength of material is always higher then quasi-static failure strength (QFS). The material can carry higher stresses (larger forces) then quasi-static strength, however they act during shorter time. In the case of linear loading function, the constant rate of loading influences the time, for which the quasi-static failure criterion is reached. For five different loading paths (levels of stress rates), five points, which create the curve are obtained (see Fig.6). There are lines which symbolize the loading histories the values of the critical stresses as a function of the logarithm of strain rates (see Fig.6). The next case of loading history is more accurate to the real loading history (acting on structure). The magnitude is similar to $sin^2()$ function, the period of this function is about 55s. The important are the amplitudes of equivalent stresses, see Fig.7. The interesting result is, that for amplitude equals 10MPa, the failure appears during decreasing of

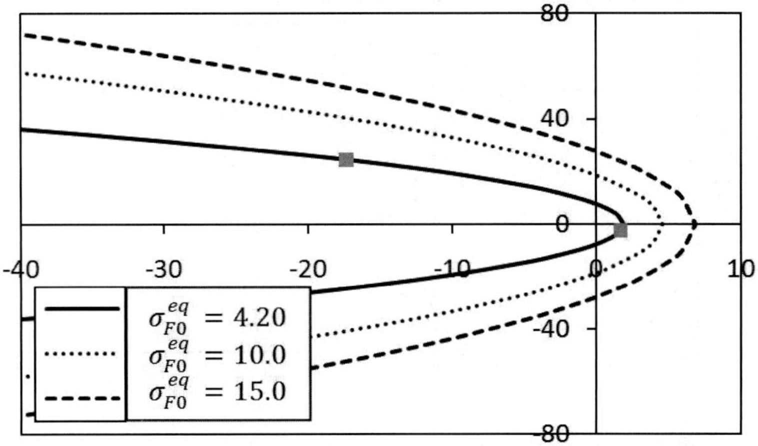

Figure 3. Iso-lines of equivalent stresses for $k = 10$ in space $r - \xi$

equivalent stresses. The time to failure is measured from time of appearing the stresses equal or higher then quasi-static strength. It is the reason that the lines do not start from the origin (0,0). The influence of magnitude history type is presented in Fig.8. All the critical curves are obtained for the same constitutive parameters (see Fig.8). The interesting is as kind of summary plot in Fig.12. Also the experimental results confirm the quality of CFC idea (compare the results in Fig.12). It is clearly seen that strain rates significantly influence the critical value and this relationship is not linear. It means that all previous plots have taken into account only the history of stresses produced by different types of loadings (magnitude and amplitude). The influence of constitutive parameters is also considered. The comparison of different values of the quasi-static equivalent stresses is shown in Fig.9. It is clearly seen that for higher quasi-static strength, the shapes of critical curves are changed. Generally, they are shifted up and to the right along the time axis. This is the comparison of results only for third type of loading history. Fig.10 presents the influence of α parameter, which in general can be a function of temperature. For our parametric study α is a constant value between 0.45 and 1.45. This value has a crucial meaning on predicted critical stresses. In Fig.11 the three curves are presented for three different values of longest critical time. These values are measured in s and vary between 49 and 149μs. Material properties for concrete specimen which we have used in the analyses are collected in Tab.2.

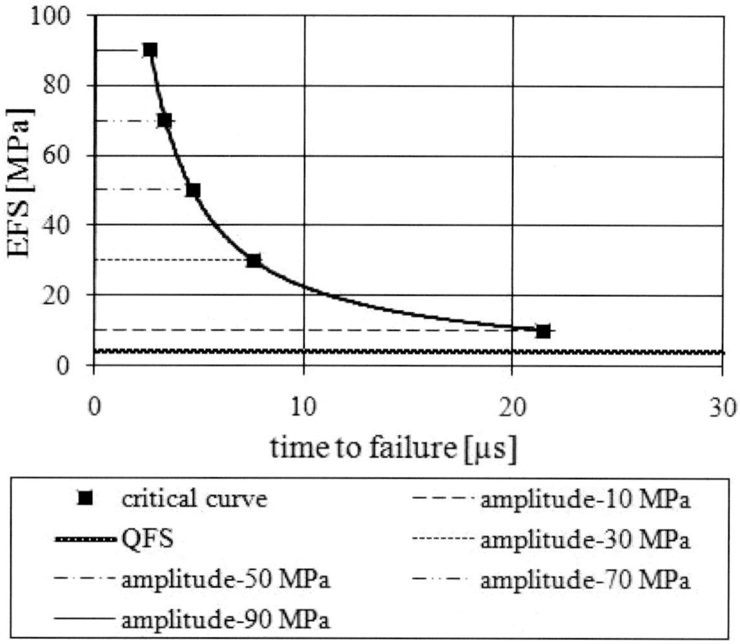

Figure 4. Critical failure curve for Heaviside step loading history for different amplitudes

4 Numerical simulation

4.1 One dimensional verification

A Hopkinson bar plays an important role in many dynamic experimental setups. The first part of the numerical analysis is limited to only one dimensional case just to verify the CFC assumptions. The well known experimental setup (Spalling Hopkinson Bar) developed in LPMM (Metz University) is applied to test a concrete specimen under fast tension. The concrete subjected to very fast loadings exhibits important strain rate dependence. Particularly, the tensile and compressive dynamic strength of concrete is much greater than quasi-static one. In the recent years the methods of dynamic testing of concrete, particularly based on wave propagation, have been developed. Such dynamic tests allow to obtain the properties of the concrete as dynamic Young's modulus and tensile strength. Description of the concrete compression strength is crucial for any analysis of the im-

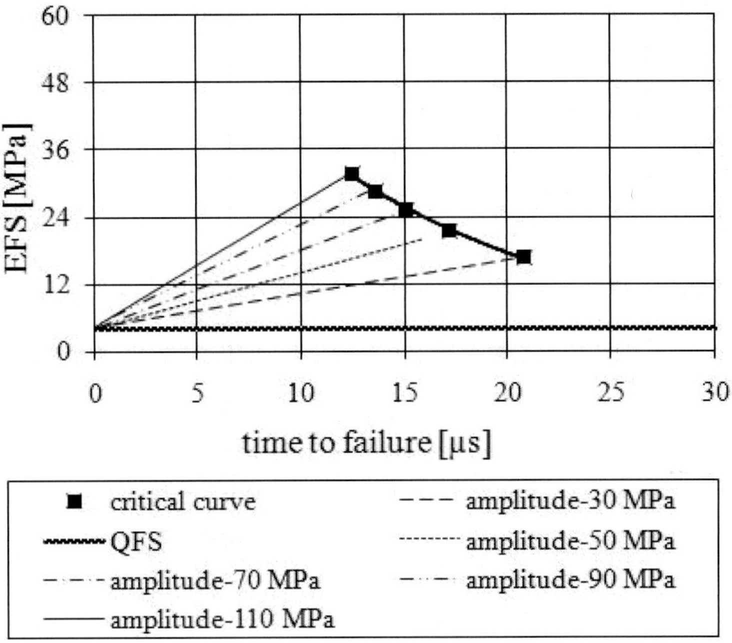

Figure 5. Critical failure curve for linear loading history for different amplitudes

pact in the concrete structure for example penetration of concrete armor [M. Klosak et al.]. Usually, as a result of the wave interaction, the tensile stresses appear. Spalling pattern due to reflection of compressive waves can be recognized as the main pattern of dynamic failure of concrete, see the bottom part of Fig.14. The laboratory tensile tests of concrete are very difficult to perform even for quasi-static loading. In the case of dynamic loading, the specification of the tensile strength is not possible to obtain without analyzing the wave propagation and interaction [Schuler et al.]. The important solution for the wave propagation analysis and fracture of semi-brittle materials is to use the arrangement with the Hopkinson pressure bar, Fig.13. Different dynamic tensile tests of concrete and other quasi-brittle materials like ceramics were reported in the literature by application of such arrangement [Gatuingt and Pijaudier-Cabot; Klepaczko and Brara]. Such laboratory experiment can be used to determine the dynamic strength of concrete under short time loading. The measurement

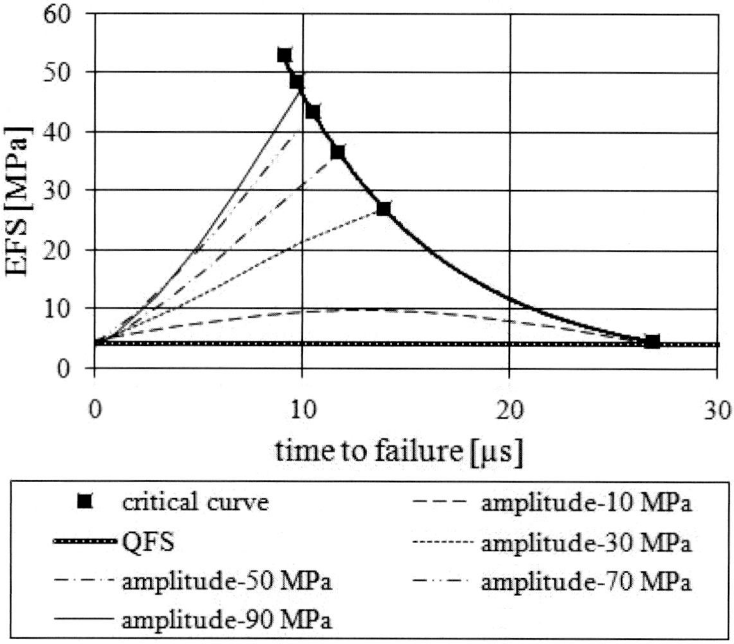

Figure 6. Critical failure curve for $sin^2()$ loading history for different amplitudes

Table 2. Material Properties of concrete specimen

E	$35{\cdot}10^9$[Pa]
ν	0.2 [-]
ρ	2395 [kg/m^3]
k	10 [-]
t_{c0}	0.000049 [s]
$\alpha(T)$	0.95 [-]
σ_{F0}^{eq}	$4.2{\cdot}10^6$ [Pa]

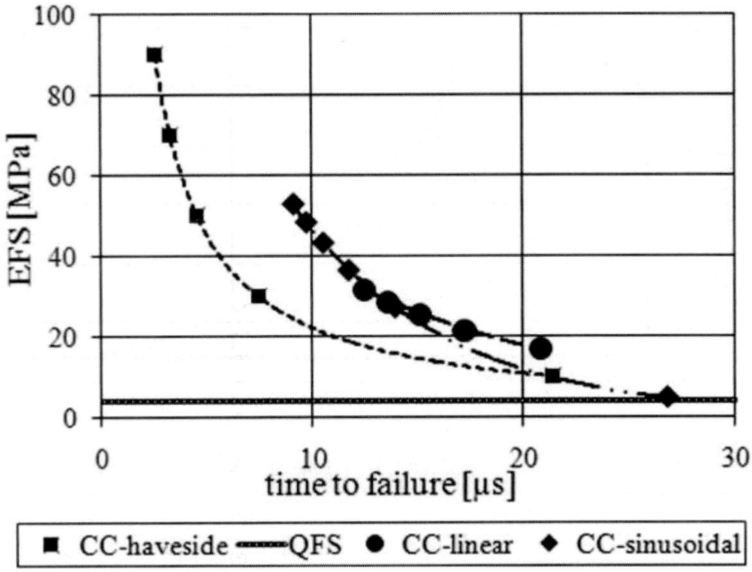

t_{c0}	0.000049 [s]
$\alpha(T)$	0.95
σ_{F0}^{eq}	4.2e6 [Pa]

Figure 7. Comparison of critical failure curves described by cumulative fracture criterion for different loading histories

part of the arrangement is the aluminum alloy bar of the length 1.0m and the diameter of 0.04m. The Hopkinson bar is hit by the striker, which is of the same diameter and the same material as the bar. The lengths of the projectiles are 0.080m, 0.12m and 0.160m, respectively. In the numerical simulation of the experiment only the 0.08m length of striker is assumed. The concrete specimen, which is investigated, has the length of 0.12m. All contact surfaces are ideally matched, this is a reason of the wave interaction between the setup components and the maximum transmission of the wave to the concrete specimen. To verify the basic pattern of dynamic fracture of concrete by spalling, the sizes of the finite elements were assumed according to the optimized study cases [Jankowiak et al.]. The dimensions of all parts are the same as in experiment reported by Klepaczko and Brara [Klepaczko and Brara]. Numerical model is axisymmetric and the contact

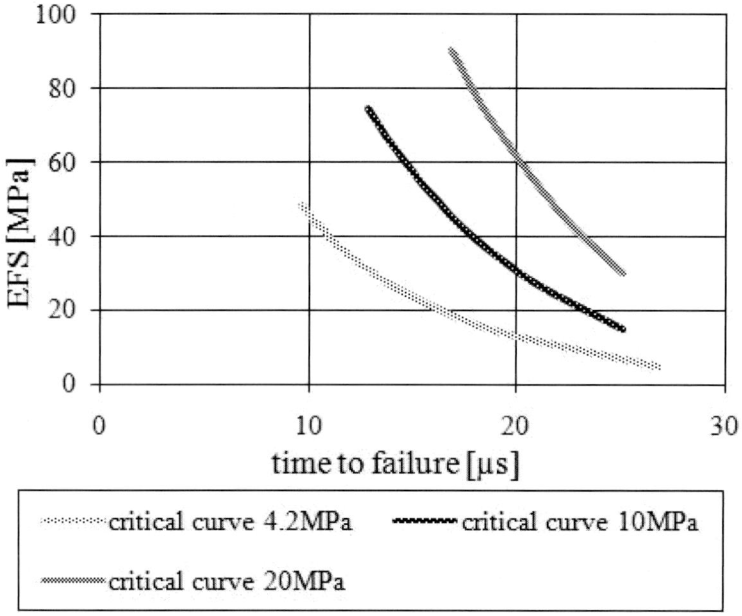

Figure 8. The influence of quasi-static strength on the critical failure curve for $sin^2()$ loading history

between the parts ensure the correct transmission of the longitudinal waves between the projectile and Hopkinson bar and further to the concrete specimen by the compression pressure. The aluminum parts, i.e. striker and Hopkinson bar are modeled as linear elastic with parameters E=70e9Pa , ν=0.28 and the density=2850kg/m^3. This is assumed because these parts during experiment always remain elastic. In the case of concrete, the elastic brittle strain rate fracture material was assumed. The material parameters for the concrete are assumed according to the data collected in Tab.1. The results for two cases of the concrete specimens tested in Hopkinson pressure bar with the velocity of projectile 7m/s and 12m/s are presented in Fig.14. Both analyses are continued up to final failure of the specimens. In the plots on the right hand sides by dots we have indicated the states of stresses in midpoint that accompany the situations which are followed. The fracture of concrete specimens is a results of tension (one dominant crack appears), although at first the whole specimen is compressed (Fig.13). The compressive

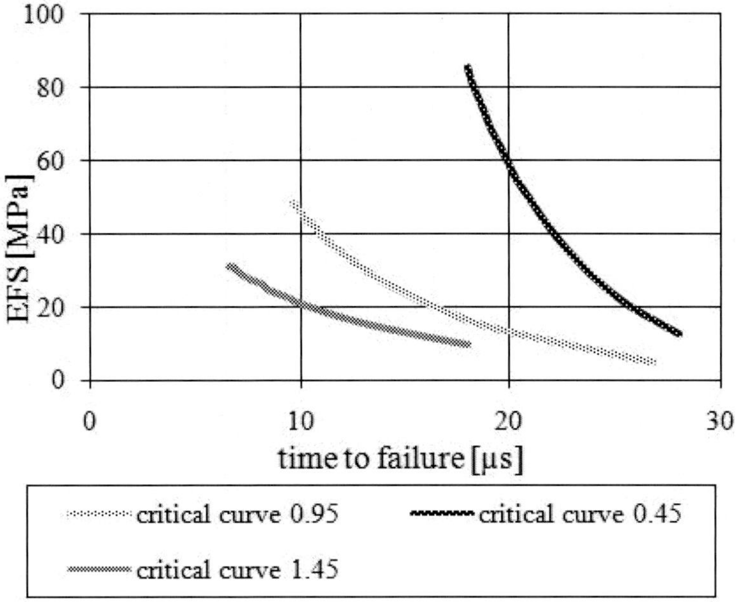

Figure 9. The influence of α parameter on the critical failure curve for $sin^2()$ loading history

wave is caused by the early wave interaction in the striker and Hopkinson bar. In Fig.14 the sequences of the distribution of longitudinal stresses are presented. The figures represent the results of computations obtained in the environment of ABAQUS/Explicit. For initial velocity 7m/s only one dominant crack appeared. It is placed 55mm from the surface hit by the projectile. In the second considered case (12m/s) the first crack appears about 80mm from the hit surface while the second about 55mm. Qualitatively the results of computations and experiments observed in laboratory are in good agreement what confirms the correctness of the constitutive assumptions and FE mesh quality.

4.2 Explosive loading

The term explosion is commonly used to describe a rapid release of energy. The physics of the of the stress waves propagation is the origin of describing the material behavior under explosion. Stress wave is in focus of

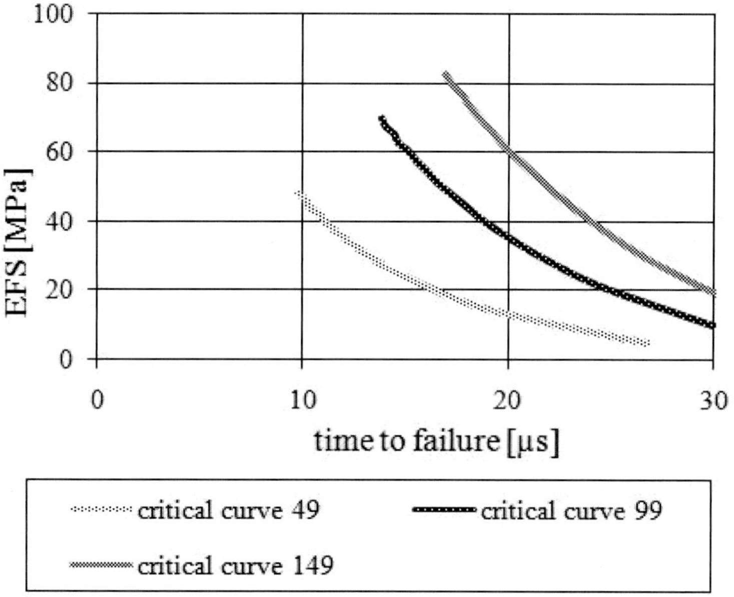

Figure 10. The influence of the longest critical time on critical failure curve for $sin^2()$ loading history

the process occurring in explosive and surrounding air [Smith and Hetherington]. An ignition of condensed charge is followed by many other events which we are not going to discuss in details. Let us only mention that the air pressure values fall with the growing distance from the ignition point. The contact of the pressure front with the obstacle is the key to the structure failure in first milliseconds. Moreover, the shape of the wave and strong increase of the pressure are significantly depended on its reflection from the ground. The discontinuity wave propagation induces a region of negative pressure, where the values are below the atmospheric (ambient). The blast pressure consists of two significantly different phases. There are: positive- and negative-phase. The positive pressure is above the barometric pressure and the negative is below [Jankowiak et al.; Kinney and Graham]. The most important shock wave parameters were published in 1870 by Rankine and Hugoniot [Henrych]. There are: blast front velocity, air density and the maximum overpressure. The values of parameters used in computations

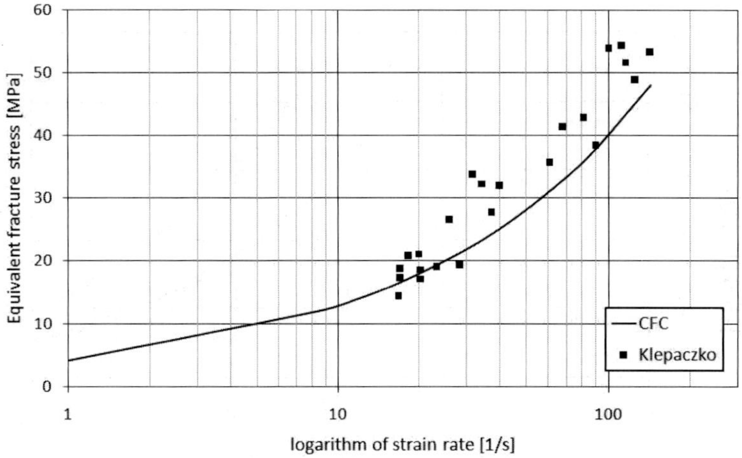

Figure 11. The critical failure curve for linear loading history for different amplitudes as a function of strain rate logarithm

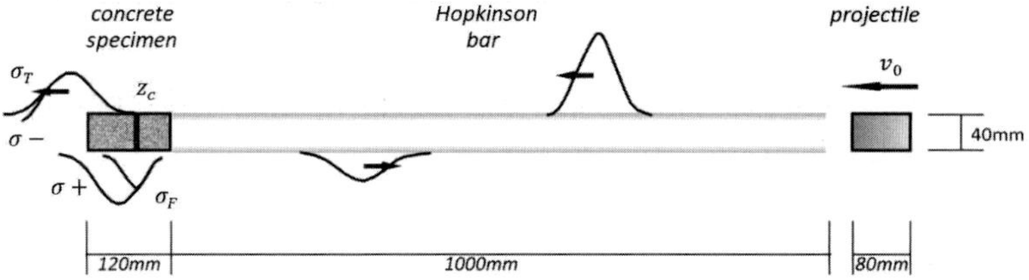

Figure 12. The scheme of experimental arrangement with Hopkinson bar

can be obtained from empirical equations and in consequence used for the analyses. The times of phases separation are also useful and can be evaluated analytically [Henrych]. The history of pressure change in time is often presented by an exponential function such as the well known Freidlander equation [Smith and Hetherington].

$$p_{FR}(t) = p_C \cdot \left(1 - \frac{t}{t_0}\right) e^{\left(-\frac{b(p_C)\cdot t}{t_0}\right)} \tag{9}$$

In this equation p_C is the peak of overpressure and t_0 the time of positive phase duration [Smith and Hetherington]. These parameters are crucial to obtain the pressure-time relation. The reliable numerical simulation of

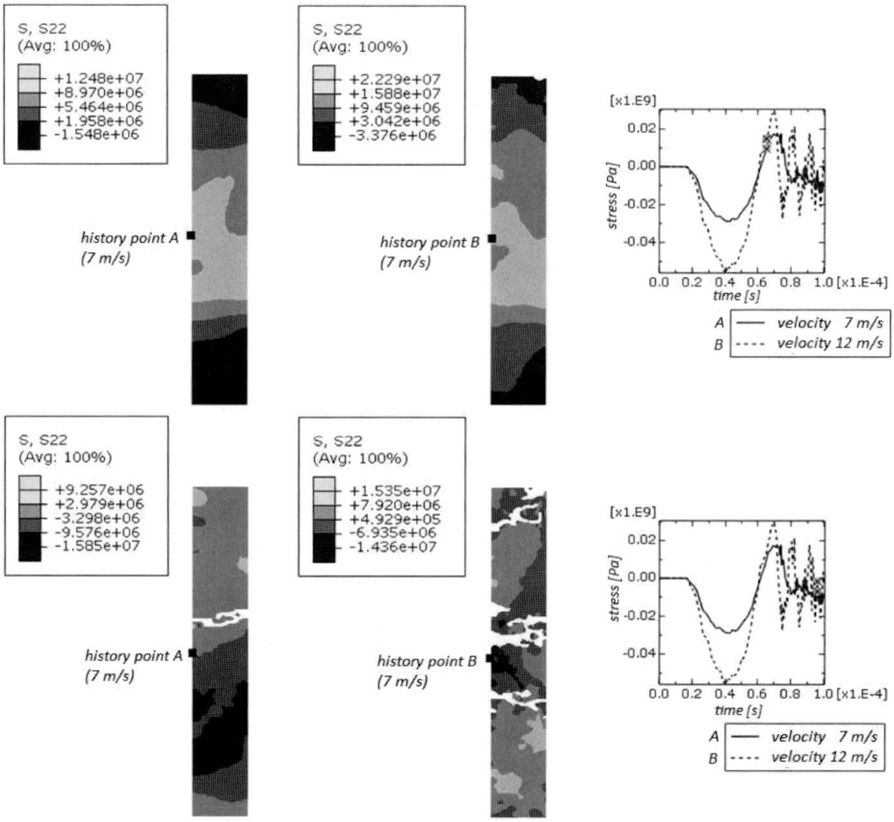

Figure 13. Sequences of spalling for two initial velocities of the projectile 7 m/s and 12 m/s

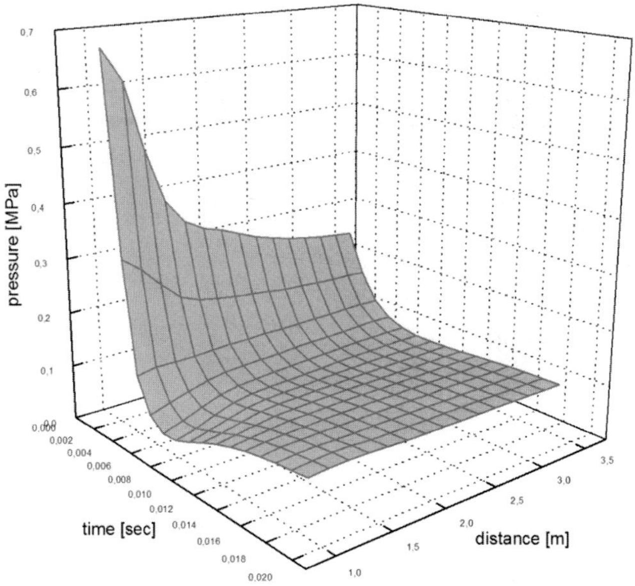

Figure 14. Typical scheme of pressure evolution in time and distance from the ignition point

any charge explosion requires the introducing the Arbitrary Lagrangian-Eulerian (ALE) formulation. ALE formulation assumes the mesh motion dependent on the material motion at free boundaries while in the other cases the material and mesh are independent. The real constitutive behavior of explosive material and the air are used in simulations. The properties of explosive material as TNT are described by the parameters collected in Tab.2. The pressure produced by chemical energy of explosion is described by so called Jones-Wilkins-Lee (JWL) formula:

$$p = A\left(1 - \frac{\omega\rho}{R_1\rho_0}\right)e^{\left(-R_1\frac{\rho_0}{\rho}\right)} + B\left(1 - \frac{\omega\rho}{R_2\rho_0}\right)e^{\left(-R_2\frac{\rho_0}{\rho}\right)} + \frac{\omega\rho^2}{\rho_0}E_{m0}, \quad (10)$$

where $A, B, R1, R2, \omega$ are material constants, E_{m0} is the internal energy per unit mass, ρ_0 is an initial density of explosive material and ρ is a current density of detonation product. The values of material constants are presented in Tab.2. The air is also modeled by the equation of state. This equation can be used as for ideal gas. The ideal gas assumption for the air is valid only for shock pressure less than 10 atmospheres,

$$\rho + p_A = \rho R(T - T^Z), \tag{11}$$

where p_A is the ambient pressure, ρ is initial density of air, R is gas constant, T^Z and T are temperatures. T^Z corresponds to -273.15 Celsius degrees. T is temperature on scale (in Kelvins degrees). An important material parameter of the air is a specific energy; which depends only on temperature and can be calculated as the following integral:

$$E_m = E_{m0} + \int_{T_0 - T^Z}^{T - T^Z} c_v(T) dT. \tag{12}$$

In the above E_{m0} denotes the initial specific energy at initial temperature T_0 and c_v is the specific heat at constant volume, which depends only on temperature (ideal gas).

4.3 Three dimensional verification for circular concrete plate

The concrete is modeled as an elastic material with Cumulative Fracture Criterion. All the constitutive parameters in this example are equal to the values from the previous 1-D case, see Tab.1. After the detonation of explosive material C-4, the direct spherical shock and reflection waves appear. Next, the compressive wave propagates to the rear of the slab and reflects as spherical, generally, tensile wave. The sphere crack appears after the equivalent stresses exceed the limit described by the Cumulative Fracture Criterion. In real situation, during the explosion of C-4 the crater on the front face of concrete slab is induced, but during this numerical simulation it is not. The high hydrostatic compressive stresses do not introduce the compressive damage, see Fig.17. The simulation results are compared with experiments presented by F.Gatuingt, G.Pijaudier-Cabot [Gatuingt and Pijaudier-Cabot]. Particularly, the axial stress histories for three points P1, P2 and P3 uniformly distributed through the thickness of the plate are shown in Fig. 18. After reflection of the wave from the free surface the localized tensile zone appears and finally drives to spectacular spalling. The evolution of spherical spall in selected time points are also presented. The prediction of behavior of concrete slab subjected to explosion is in good agreement with experimental results. Proposed failure model is accepted for this kind of computation.

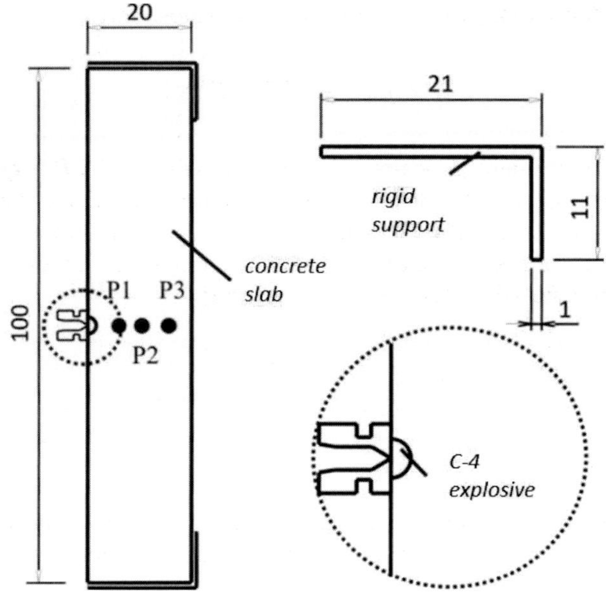

Figure 15. Geometry of the concrete slab and support

5 Engineering structure subjected to explosion

The validation of the model which describes the distribution of pressure wave in time and space in the air is necessary. As a result of this part of the analysis we obtain the loading conditions for the structure under consideration. Then, separately the failure analysis of the structure is continued.

5.1 Pressure distribution in the air

The experimental tests which allow to measure the pressure in the certain distance from the ignition point are relatively rare and very costly. Usually, they ends with the damage of the gauges and can not be repeated under the same conditions. In Fig.19, there are presented the results of field experiments [Alia and Souli] for C-4 explosive material compared with the numerical results based on empirical formulas or 2-D and 3-D FE modeling [Jankowiak et al.]. There is no doubt that empirical approximations overestimate the pressure values while 2-D modeling underestimates. Only 3-D modeling approximates the measured values with the satisfactory accuracy. This is a reason that in further computation only 3-D models are accepted.

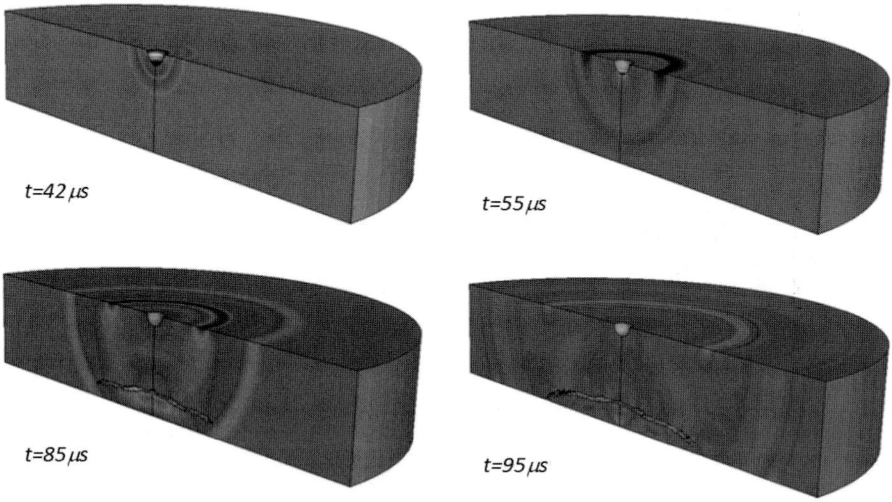

Figure 16. The process of spalling creation in rounded concrete slab. The pressure stresses as the interaction of waves for selected time of the process after explosion

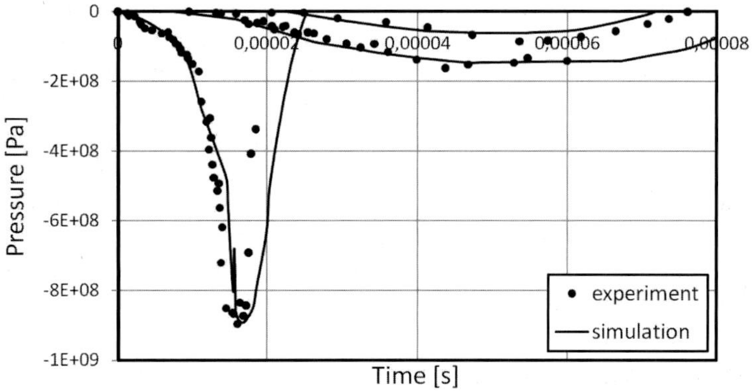

Figure 17. Axial stress during explosion in three points P1, P2 and P3 inside concrete plate

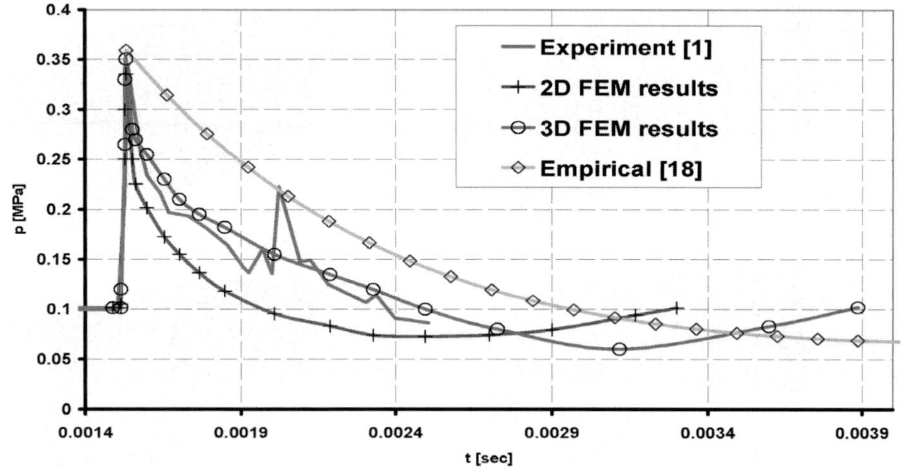

Figure 18. Pressure evolution in the point of the space

5.2 Pressure distribution on the wall front surface

There are two kinds of materials subjected to the blast loading. The loading is started by ignition of a cubic TNT charge. The first job consists of the concrete wall and the second one of the masonry wall. The computations are performed in the environment of the commercial finite element code Abaqus, using the option of sub-modeling. The possibilities of sub-modeling significantly decrease a size of the problem. This convenience is used for detailed study of a part of interest in a global finite element model. The solution process is divided into two subsequent jobs. The first outcomes are the pressure changes which make the wall loading. Further model contains only the analysis of an obstacle but with frontal loading obtained earlier. An area of interest namely the wall structure has much finer FE mesh than the global model. The scheme of analysis for both models is presented in Fig.20. The planes of symmetry are used for each numerical example, 1/4 of the air space and a half of the wall are considered. The distance between the front surface of the wall and the explosive material is constant and equals 1.0m. The simple equation (15) can be useful to obtain the velocity of the overpressure wave and then the arrival time. The time means a final moment of the global task. There is accepted p_C as peak overpressure, p_0 as ambient pressure and a_0 is sonic velocity.

$$V = \sqrt{\frac{6p_C + 7p_0}{7p_0}} \cdot a_0, \qquad t = \frac{V}{s} \tag{13}$$

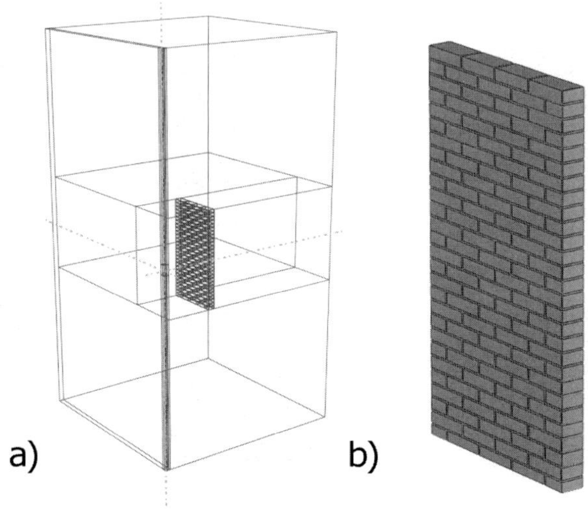

Figure 19. Analyzed model: a) 1/2 of global model, b) 1/2 of detailed part of wall

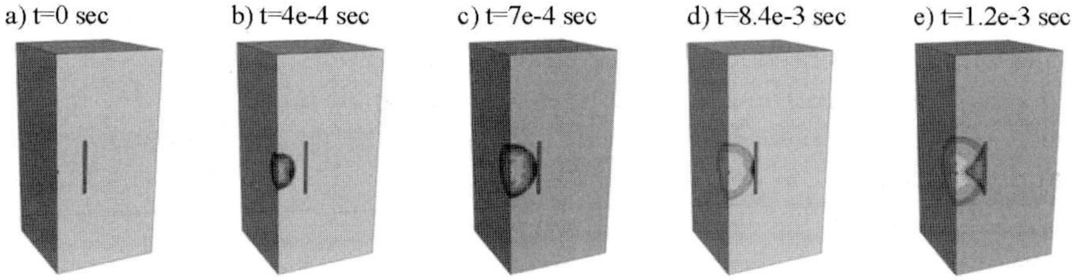

Figure 20. The propagation of blast wave

The pressure evolution is obtained for 13kg of the TNT. The propagation of blast wave in the air is presented in Fig.21. Moreover, the pressure distribution on the front of an obstacle in subsequent time maps is presented, see Fig.22. The obstacle used in the global model includes only the elastic material. The representation of pressure, just before the front surface of an obstacle, corresponds to the loading surface in a first time moment after explosion. The first map presents the ambient pressure before acting the overpressure. The explosion causes a maximal value of pressure, which

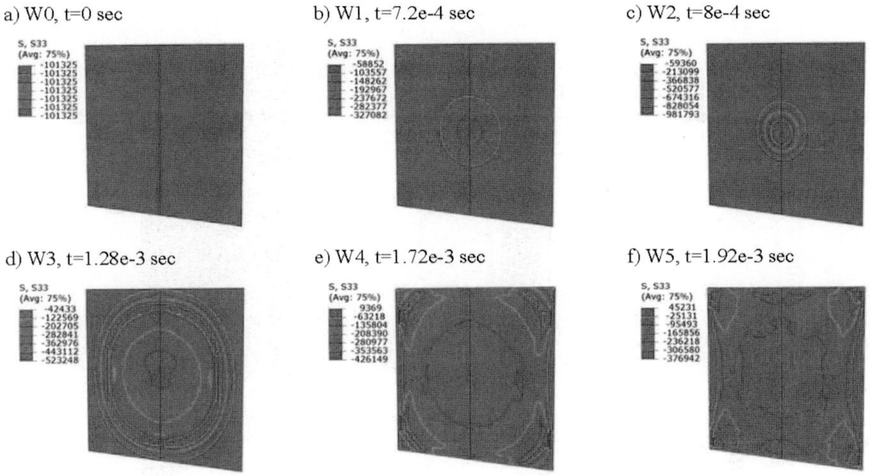

Figure 21. Loading surfaces

occurs 0.8ms after detonation. The pressure is almost 10-times higher than barometric pressure. The last map proves the existence of under-pressure phase. The presented loading surfaces are connected with Fig.22. The graph (Fig.23) describes the change of the pressure in the time, close to the front side of the obstacle. The subsequent time moments are indicated by dots and they reflect the pressure acting on this side. In the initial moment the pressure has the constant value (ambient pressure) which next is growing rapidly.

The concrete wall As the next example, let consider the concrete square 2m by 2m wall loaded by blast. The explosion of TNT material (13kg) in the distance of 1m from the wall centre is assumed. The edges of the wall are fixed. The parameters used for the analysis for both the concrete and the air are collected in Tabs.2 and 3. The sub-modelling technique is used to perform the detailed failure analysis of the concrete wall. The results of the numerical simulation are shown in Fig.24. We present the displacements of the wall in the thickness directions and the plots of principal stresses (tensile responsible for fracture in brittle materials). These plots are selected for specific time of the process T=0.00035s (when the wave in the air practically reaches the wall surface), T=0.00105s and finally for T=0.00140s when the first failure is observed. The local analysis of the concrete wall is kinemticaly

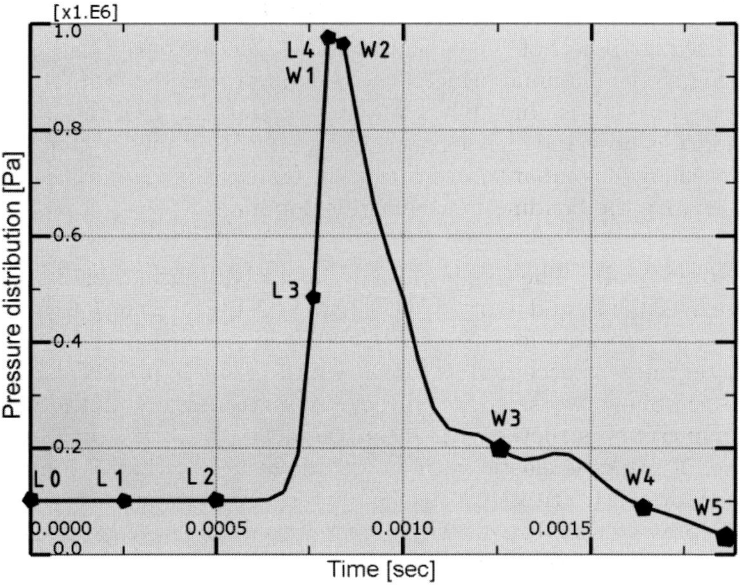

Figure 22. Pressure time history

controlled. The displacement history of all nodes is computed by global analysis and is interpolated to the nodes of the local model. In Fig.24 the displacement and maximal principal stresses for 0.00035s are shown. It is the time moment when the blast wave arrives to the wall surface. The authors present also the results for time 0.00105s and for 0.0014s. In the last frame 0.0014s first failure of the plate is observed.

Table 3. Equation of the state properties

Air		TNT	
R	287 [J/(kgK)]	A	$3.73{\cdot}10^{11}$ [Pa]
ρ	1.293 [kg/m^3]	B	$3.74{\cdot}10^{9}$ [Pa]
p_A	101325 [Pa]	$R1$	4.15 [-]
E_{m0}	$0.193{\cdot}10^{6}$ [J/kg]	$R2$	0.9 [-]
T^Z	0 [K]	E_{m0}	$5{\cdot}10^{6}$ [J/kg]
T_0	288.4 [K]	ω	0.35 [-]
c_v	1003.5 [J/(kgK)]	ν_d	6930 [m/s]
		ρ_0	1630 [kg/m^3]

The typical failure bending mechanism are observed (two perpendicular cracks). The high power of explosion causes the fracture of the wall near the edge. In Fig.25 the maximal principal stresses are presented for 0.0164s and the last one for 0.3014s. In the last Figure, the separation of the wall from the boundary is observed. For this specific kind of loading the compressive wave is very long in comparison with Hopkinson bar test and it is the reason that in this case the bending mechanism is dominant.

The masonry wall The wall is build periodically of regular bricks (0.25m by 0.12m by 0.06m) and mortar with bad and head joints 0.01m thick. The structure subjected for analysis structure is of dimensions 2m by 2m. The wall thickness is accepted as 0.12m width of single brick Fig.26. The structure combines two separate phases. In agreement with the obvious feature of masonry structures there are modelled by bricks and mortar. The model of bricks is elastic without any damage criteria. The model of mortar employs CFC criterion, specified before. Properties of the masonry phases are presented in Tab.4. The three edges of the wall are fixed while the top one remains free. This one is loaded and simulates in the analysis the loading which comes from ceiling. We assumed the ideal spherical blast (13kg) in front of the obstacle. The ground reflection is not taken into consideration, when using available in Abaqus code sub-modeling option. The local job uses the loading surfaces from the global example, see Fig.23.

The numerical model consists of 0.24e6 linear 8-node elements with 1800 finite elements per one brick and 1 on the mortar thickness. The part is composed of one solid block and is divided into two material sections. This approach allows for neglecting a contact problem on the mortar-brick bond. The mechanism of the structure destruction with the brick motion for selected steps of the analysis is presented by the sequence in Fig.28. The next Fig.29 presents the evolution of mortar failure for which CFC was accepted. The time moments are connected with the Fig.23. The level of principal tension stresses proves, for the case under consideration, that wall finally is completely destroyed. The bricks are separated and the level of principal tension in the mortar precludes the values assumed for quasi static strength (0.7 MPa).

6 Conclusions

Cumulative fracture criterion is used for concrete under fast dynamic and applied in computer simulations of the spalling process in Hopkinson pressure bar and in circular concrete slab. The crucial role plays the identification of constitutive parameters. The influence of constitutive parameters

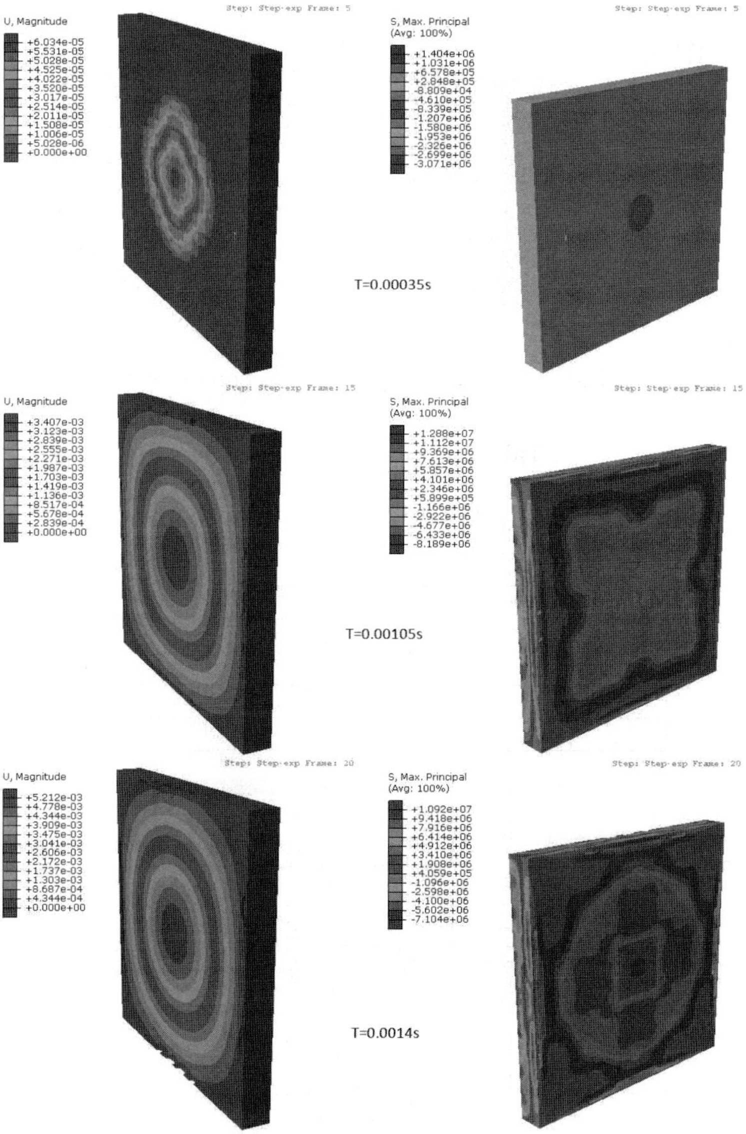

Figure 23. Displacement magnitude on the left and maximal principal stress on right for selected time points

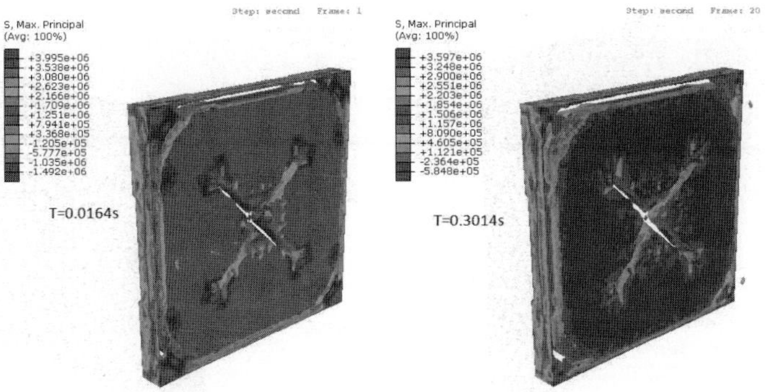

Figure 24. Failure pattern for square wall

Table 4. Masonry properties

Brick		Mortar	
E	$6{\cdot}10^9$ [Pa]	E	$3{\cdot}10^9$ [Pa]
ν	0.16 [-]	ν	0.20 [-]
ρ	1200 [kg/m^3]	ρ	800 [kg/m^3]
		σ_{F0}^{eq}	$0.7{\cdot}10^6$ [Pa]
		t_{c0}	$25{\cdot}10^{-5}$ [sec]
		$\alpha(T)$	0.95 [-]
		k	10.0 [-]

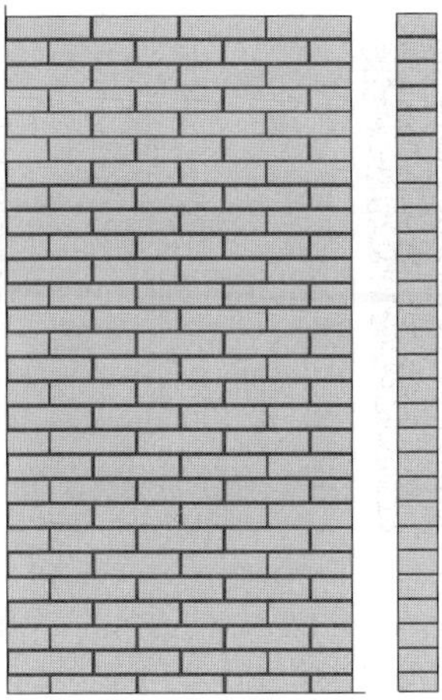

Figure 25. The half of the masonry wall

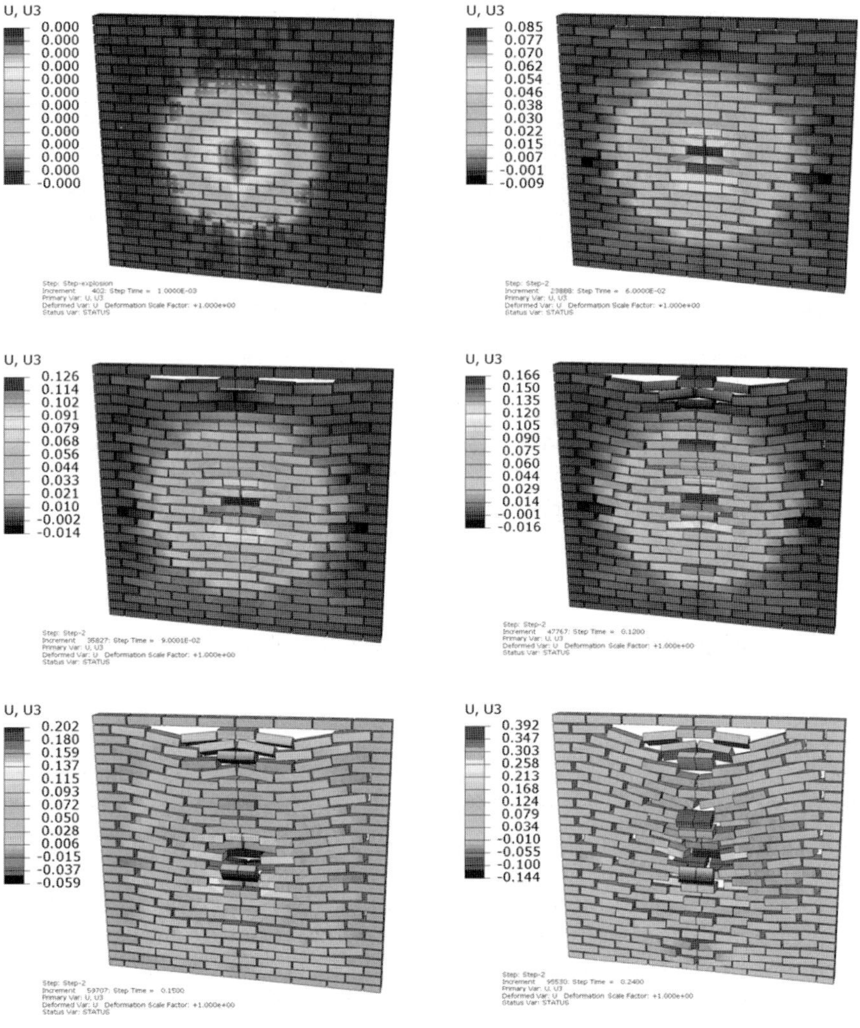

Figure 26. Mechanism of destruction of masonry wall

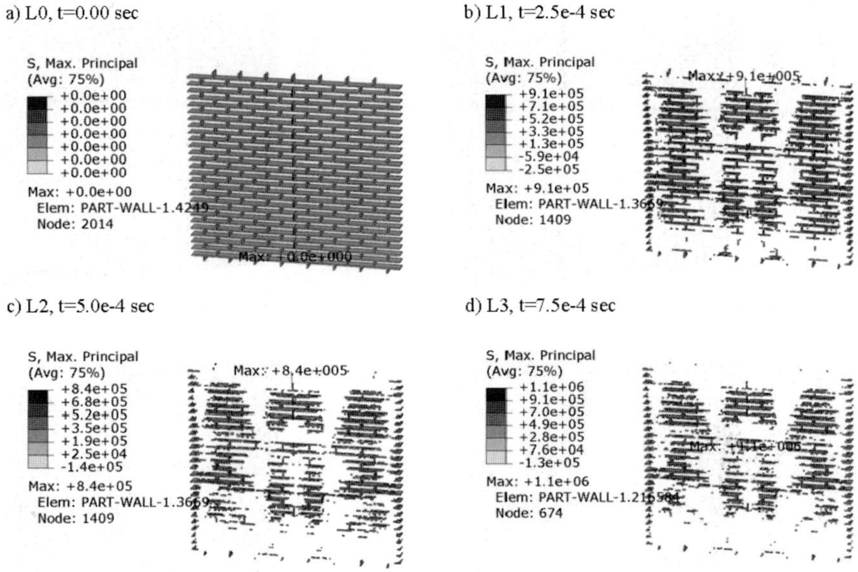

Figure 27. Mechanism of destruction of mortar phase

on critical failure curve are also presented and carefully discussed. Both the criterion and identification of parameters are fundamental for the proper description of strain rate dependency of the failure process of concrete. The evolution of spalling in circular plate state serves as the example which stresses the usefulness of CFC [Alia and Souli].

Authors discuss also the simulation of the explosion material and wave's propagation in the air using the known equation of state. The obtained results allow for studying the loading that acts at the obstacle (concrete or masonry walls) resulted from the blast. The loading is changed in space and time and in the second phase of computations acts at the structure.

Two complex boundary value problems are presented. In both cases, the sub-modeling technique is regarded. The results obtained and the methods used allow for better estimation of the safety of structures subjected to unusual loadings (blasts, rigid strikes, crashes).

Bibliography

A. Alia and M. Souli, *High explosive simulation using multi-material formulations*, In Applied Thermal Engineering, pages 1032–1042, 2006

P.H. Bischoff and S.H. Perry, *Compressive behaviour of concrete at high strain rate*, In Materials and Structures, pages 425–450, 1991

J.D. Campbell, *The dynamic yielding of mild steel*, In ACTA Metallurgica, pages 706–710, 1953

CEB, *Concrete structure under impact and impulsive loading*, 1987

F. Gatuingt and G. Pijaudier-Cabot, *Coupled damage and plasticity modeling in transient dynamic analysis of concrete*, In International Journal for Numerical Analytical Methods in Geomechanics, pages 1–24, 2002

M.G.D. Geers, R. de Borst and R.H.J. Peerlings, *Damage and crack modeling in single-edge and double-edge notched concrete beams*, In Engineering Fracture Mechanics, pages 247–261, 2000

J. Henrych, *Dynamics of Explosion and Its Use*, Academia Prague, 1979

T. Jankowiak and T. Łodygowski, *Numerical modeling of fracture in brittle material under impact loading*, In Vibrations in Physical Systems, pages 143–148, 2006

T. Jankowiak, J.R. Klepaczko and T. Łodygowski, *Numerical modeling of wave propagation and interaction in bars*, In Foundations of Civil and Environmental Engineering, pages 187–199, 2006

T. Jankowiak, T. Łodygowski, and P.W. Sielicki, *Modelling of pressure distribution after explosion*, 17th Int. Conf. on Computer Methods in Mechanics, 2007

G.F. Kinney and K.J. Graham, *Explosive shocks in air Berlin*, Berlin: Springer, 1985

J.R. Klepaczko, *Behavior of rock like materials at high strain rates in compression*, In International Journal of Plasticity, pages 415–432, 1990

J.R. Klepaczko and A. Brara, *An experimental method for dynamic tensile testing of concrete by spalling*, In International Journal of Impact Engineering, pages 387–409, 2001

M. Klosak, T. Łodygowski and J.R. Klepaczko, *Remarks on numerical estimation of the critical velocity in shear*, In Computer Assisted Mechanics and Engineering Sciences, pages 579–593, 2001

L.J. Malvar and C.A. Ross, *Review of strain rate effects for concrete in tension*, In ACI Materials Journal, pages 735–739, 1998

H. Schuler, C. Mayrhofer and K. Thoma, *Spall experiments for the measurement of the tensile strength and fracture energy of concrete at high strain rates*, In International Journal of Impact Engineering, pages 1635–1650, 2006

P.W. Sielicki, *Pressure distribution after explosion*, *1*4th Abaqus Meetings, Poznan, Poland, pages 143–148, 2007

P.D. Smith and J.G. Hetherington, *Blast and Ballistic Loading of Structures*, *O*xford: Butterworth-Heinemann, 1994

A. Stolarski, *Dynamic strength criterion for concrete*, In *J*ournal of Engineering Mechanics, pages 1428–1435, 2004